# AGRICULTURAL MARKETS BEYOND LIBERALIZATION

AF786522

# AGRICULTURAL MARKETS BEYOND LIBERALIZATION

*edited by*

**Aad van Tilburg**

**Henk A. J. Moll**

**Arie Kuyvenhoven**

*Wageningen University*
*The Netherlands*

SPRINGER SCIENCE+BUSINESS MEDIA, LLC

The Editors would like to thank their technical editor, Sandra Küçüksen-Everts

**Library of Congress Cataloging-in-Publication Data**

Agricultural markets beyond liberalization/edited by Aad van Tilburg, Henk A.J. Moll, Arie Kuyvenhoven.

p.cm.

"Selected contributions to the international seminar 'Agricultural markets beyond liberalization', held at Wageningen University in September 1998"--Introd.

Includes bibliographical references and index.

ISBN 978-1-4613-7040-6 ISBN 978-1-4615-4523-1 (eBook)

DOI 10.1007/978-1-4615-4523-1

1.Produce trade--Congresses. 2.Agriculture and state--Congresses. 3.Tarrif on farm produce--Congresses. 4.Free trade--Congresses. I. Tilburg, Aad van. II. Moll, Henk A. J. III. Kuyvenhoven, Arie.

HD9000.5+

338.1'8--dc21 00-037075

**Copyright** © 2000 by Springer Science+Business Media New York

Originally published by Kluwer Academic Publishers, New York in 2000

Softcover reprint of the hardcover 1st edition 2000

All rights reserved. No part of this publication may be reproduced, stored in a retrieval system or transmitted in any form or by any means, mechanical, photocopying, recording, or otherwise, without the prior written permission of the publisher, Springer Science+Business Media, LLC.

*Printed on acid-free paper.*

# Contents

# Part 1

## Introduction

# 1 AGRICULTURAL MARKETS BEYOND LIBERALIZATION: ISSUES, ANALYSIS AND FINDINGS

Arie Kuyvenhoven, Henk A.J. Moll and Aad van Tilburg

## 1 INTRODUCTION

Liberalization of agricultural markets has been on the agenda of policy makers and international organizations since the beginning of the 1980's. To a large extent this reflected the growing recognition that widespread government intervention in markets was much less effective than previously expected, while negative side effects resulted in misallocation of resources, reduced economic growth and an often adverse impact on equity and the environment. As a result, policies in the industrialized world became increasingly oriented towards less government interference, and were characterized by a simultaneous shift from national to supra-national regulations. Developments in Eastern Europe, the former Soviet Union and China took a dramatic change towards more freedom for individual and communal market participants. In much of the developing world, policy reforms under structural adjustment programs led to a redefinition of the role of the government in relation to agricultural markets.

Internationally, the framework within which agricultural markets functioned was affected by the results of the Uruguay Round trade reforms. For the first time in the history of the GATT (now WTO), agriculture became an integral part of multilateral trade negotiations. The proper functioning of international agricultural markets was discussed, problems of export surpluses were frankly acknowledged, and market access of developing countries was accepted as a legitimate goal.

As a result of these national and international developments, agricultural markets world-wide have entered a long-term process of liberalization, with the aim of reducing imposed market imperfections such as monopolistic public trade, entry barriers and subsidies. As markets and market channels disseminate (dis)incentives to production, trade and consumption decisions, this process affects the entire agricultural chain from primary activities to processing to final consumption. The experience of more than a decade of agriculture liberalization offers a good opportunity to review and analyze (part of) the outcome of this process and to draw lessons for the future. This experience is necessarily multi-facetted as pre-liberalization conditions, timing and sequencing of reforms, as well as the wider economic and social context within which policy adjustment took place, differ by country and by type of market.

Liberalization of markets implies changing the way markets are organized, i.e. changing the presence and combination of institutions in a market that defines its structure. It also means changing the rules and incentives governing the conduct of individual institutions in the market. Market structures have a spatial dimension, ranging from local to global, and a social, indicating those involved in the various institutions with a range from individuals engaged in a contract to world-wide parties to WTO agreements. The central topic in this volume is the relationship between market structure and how markets perform in a dynamic context during a liberalization process. The topic is studied from both a micro and macro viewpoint and refers to different types of agricultural markets. Market performance is not only measured against the classical yardslicks of effectiveness, efficiency and equity, but also against indicators such as food security, sustainability of resource use and people's live hood.

In this way, the present volume brings together the dynamics of agricultural markets in several parts of the world, with a special focus on transition economics and Africa. The different studies cover geographical areas as wide as a district as well as a group of countries, and institutions from individual contracts to multi-national organizations. The analysis of liberalization under different circumstances, and the different methods of analysis used by the authors provide a valuable foundation for the assessment of liberalization.

This book results from selected contributions to the international seminar "Agricultural Markets Beyond Liberalization", held at Wageningen University in September 1998. The seminar was organized as the 57$^{th}$ European Association of Agricultural Economists Seminar, the third of its kind dealing specifically with development issues (the first seminar took place in Montpellier, 1989, the second in Hohenheim, 1992). About 90 participants from all over the world attended the sessions and the papers selected for publication reflect this diversity. Especially encouraging has been the response of Ph.D. students affiliated with European universities and research institutes; several of their papers appear as contributions to this volume.

The remainder of this chapter is organized as follows. The next section discusses how agriculture has behaved under structural adjustment. Special attention is paid to macro and price policies, public goods and services provision, and food security. The impact of reforms on agricultural supply response, markets and institutions is briefly dealt with in section 3. Section 4 synthesizes the various contributions to this volume thematically, and attempts to characterize market structure and performance during and after liberalization. The final section brings together the main findings and identifies unresolved issues in a post-liberalization era.

## 2 AGRICULTURE UNDER STRUCTURAL ADJUSTMENT [1]

In terms of income and employment, the agricultural chain from primary producers to final consumers makes up a substantial part of any economy, especially in developing countries (LDCs). The nature of agricultural policy is therefore highly relevant because it affects major markets and institutions. Historically, agricultural policies in many (developing) countries have been complex and contradictory because government interventions have sought to achieve many, often conflicting objectives. In the industrialized world, the concept of multi-functionality of agriculture has added to this complexity.

Agricultural and rural development objectives can be achieved by a wide variety of policy instruments, which can be imposed directly at the farm level (production subsidies, food procurement), the national border (trade instruments, exchange rate), or at some other point in the market (parastatals, price instruments, public investment) (Colman and Young, 1989). Each intervention is characterized by a multitude of intended and side effects on prices, efficiency in production and consumption, fiscal and foreign exchange balances, income distribution, nature and environment, and the overall structure of the sector. Careful analysis is, therefore, a necessity to establish the quantitative link between these interventions and their effects on agricultural resource allocation and farmers' welfare.

In probably no sector of the economy has policy intervention been more pervasive than in agriculture. Sadoulet and de Janvry (1995) provide an illustrative summary of the wide variety of policy measures related to the agricultural sector:

> "…. farm subsidies in [developed countries], usually through price support programs; taxation of agriculture in the LDCs through overvalued exchange rates, industrial protectionism, and export taxes; price stabilization interventions through food stocks and variable levies; food self-sufficiency and food security objectives; minimum acreage (cotton in Egypt) and maximum acreage planted (land set-asides in the United States and Europe); consumer food subsidies through cheap food policies, fair price shops, and food stamps; input subsidies, particularly to credit and fertilizers; monopolistic control of markets through parastatal agencies; regulation of competition on agricultural markets; direct income (decoupling) and assets (land reform) transfers; and public investment in agriculture such as infrastructure, irrigation, research, and extension."

The reasons for these extensive policy interventions are related to a number of important characteristics of the sector and follow directly from the theory of public choice (Stiglitz, 1987). Agriculture not only produces strategic wage goods, depends crucially on the natural resource base and has a major spatial impact, it is also characterized by :

- a generally limited sensitivity of total demand for food to changes in real prices;
- a similarly modest short-term response of total agricultural output to changes in real prices (due to the comparative immobility of factors of production and the high proportion of fixed costs in family farms);
- a pronounced substitutability of individual agricultural products both on the demand side and in each farm's production plan; and
- the importance of physical infrastructure and institutional factors for agricultural production, processing and marketing, which is overwhelmingly characterized by public decision-making.

Theoretical reasons for intervention include such market failures as imperfect competition, public goods (infrastructure, technological innovations), externalities (pollution, erosion, common property), risk, high transaction costs and imperfect information. The last three phenomena play an important role in explaining agrarian institutions that try to cope with market failures to supply credit, insurance, management and supervision. Among the interventions to serve "non-economic" goals, improvements in the distribution of income (including poverty reduction), sustainability in resource use and food security prevail. Government's role is severely constrained, however, by information, implementation and motivation failures, especially in many LDC administrations. Complex side effects of interventions pose serious problems in predicting overall effects.[2]

**Macro-economic policies**

In many LDCs, in particular in Africa and Latin America, an implicit and explicit burden has been placed on the agricultural sector. The implicit burden is the outcome of macro-economic policies geared to the promotion of industrial activities (by means of tariff protection, quotas, subsidies, government investment, etc.), and of the resulting overvalued exchange rate. The explicit burden on agriculture consists, for example, of taxes on agricultural exports and major price disparities where marketing boards are responsible for buying and selling. These and other price interventions have resulted in an unfavorable internal terms of trade between agriculture and other sectors. Rather than sectoral policies, exchange rate, trade and tariff policies can be considered the major determinants of agricultural prices. Protection of the industrial sector has raised the prices of inputs used in agricultural production and of consumer goods. Regulation of agricultural markets, taxation and currency overvaluation have led to low prices for farm products.

The anti-agricultural bias in macro-policy in most of the developing world has left a clear mark on the sector's performance. Advances in productivity growth slowed down in much of the 1970s and 1980's, particularly in Latin America and Africa, and the gap between agricultural and other income continued to widen. The policies to counteract this phenomenon have mainly relied on instruments with a significant bearing on the budget (input subsidies, subsidized grain marketing). The down-turn in agriculture commodity prices starting in the late 1970s, the debt crisis following the two oil shocks and the reluctance to reduce public expenditures to balance lower revenues, revealed the unsustainable consequences of those policies. Fundamental macro-economic balances needed to be restored and new sectoral policy orientations to be formulated. For agriculture, this meant a change from a controlled but unprotected sector to a more free but moderately protected sector, without major claims on the budget (Norton, 1992).

Structural adjustment[3] programs sought to achieve this ambitious turnaround toward restoring equilibrium and enhancing efficient supply response. Most reforms, therefore, stressed the need for sustained growth and net export of agricultural products. To this end, efforts were made to improve the internal terms of trade for agricultural products, to adjust the real exchange rate and to increase productivity and efficiency in agriculture. While the measures taken differ among countries, a number of common features can be detected in most adjustment programs:

- Price liberalization: abolition of direct price controls of agricultural products; adjustment of guaranteed prices, commodity taxes, interest rates, and the exchange rate; reduction or better targeting of subsidies (on food consumption, fertilizers and other agricultural inputs).

- Redefining the role of government: privatization of public enterprises; increasing the effectiveness of support services; better cost-recovery.
- Stimulating the private sector: selective public expenditure measures to improve access to credit, the provision of information, transport, storage facilities.
- Removal of quantitative and administrative trade barriers and tariff structure reform: abolition of quotas and licensing, and the introduction of a more uniform system of (modest) tariffs to ensure comparable effective protection in industry and agriculture.

In many Latin American and Asian countries, reforms have worked out positively (Corbo and Fisher, 1995). The 'East Asian miracle' in agriculture has been sustained by avoiding overvalued exchange rates, targeting of subsidies, concessional credit - usually tied to prescribed input packages for high-yielding varieties, and sensible output support packages - if necessary with protective trade provisions. African countries have shown mixed results. Among the reasons are incomplete price reforms, insufficient abandoning of negative protection of agriculture due to falling world market prices and fiscal constraints, and the failure of governments to make complementary investments in infrastructure, technology, human capital, and to initiate specific support programs.

**Price policies**

Agricultural price policies have been largely insufficient to compensate for the strong price-depressing effect of macro-policies. Other compensating interventions (i.e. input subsidies) have often been regressive and discriminating against the poor. The high costs in terms of efficiency and distribution of price distorting interventions in agriculture, insufficient targeting and under investment in overhead provisions are by now well documented (World Bank, 1986a, b; Timmer, 1988, Norton, 1992; Sadoulet and de Janvry, 1995). Sectoral policies heavily relied on fiscal measures, price stabilization, market policy and the subsidization of consumers and producers. These policies mostly resulted in adverse effects on agriculture and, consequently, on the incomes of those who depend on this sector for a living.[4]

Compensating policies to support input acquisition through price subsidies and/or subsidized credit supply are commonly used in developing countries. Taking into account suppressed farm-gate prices, these instruments fulfilled an important role in stimulating market supply of agricultural products. Institutional arrangements through marketing boards, parastatal trading agencies and state-managed rural development banks, have been used to link up input or credit subsidies with compulsory delivery contracts.

Allocational consequences of input and interest rate subsidies have been widely analyzed. Adams et al. (1984) demonstrate how subsidized credit provisions tend to undermine rural development due to perverse incentives for efficient resource use, and a tendency of high default as screening of reliable clients becomes impossible (Hoff and Stiglitz, 1993). Similarly, input subsidies on fertilizers, seed and the purchase of equipment distort information on marginal costs and may lead to substitution of locally available resources by externally purchased inputs. Both instruments have a clear regressive influence on rural income distribution. Price control for industrial consumer goods is sometimes used as an additional incentive to enhance the trade ability of agricultural commodities. The effects may be limited if government price control or export taxation result in rationing peasants' demands for consumer goods and eventually lead to reduced supply response (Bevan et al., 1987).

**Public goods and services**

In many LDCs the price- and income-depressing effect of a decline in the internal terms of trade of agricultural products has not been the only factor influencing rural development. The limited importance attached to agriculture for many years – certain Asian countries aside – has also been disadvantageous for agricultural structural policy. Investment in research, services, rural infrastructure, training and extension, and the creation of institutions and other facilities that can be regarded as public goods, has often been insufficient to promote the development and adjustment of the agricultural sector.

Access to public goods and services is usually considered an important condition to improve supply responsive in agriculture. Antle (1983) and Binswanger and Kandler (1993) demonstrate the positive effects of investment in infrastructure for transport and communication on agricultural productivity and on the functioning of commodity markets. Similarly, Hayami and Ruttan (1985) relate public investment in agricultural research to agricultural growth. Whereas their approach relies on induced innovation, de Janvry *et al.* (1989) refer to the explicit state biases in budget allocation and in the delivery of public goods. Technology development is conditional on the distribution of assets and is highly influenced by collective action of farmers' groups, occasioning a fairly regressive distribution of welfare gains. Without structural policies aiming at a reduction in transaction costs for access to market and information, the allocative role of prices tends to be restricted and supply response and input-intensity will remain low.

The explanation for the decline of public outlays for agricultural development is complex. A classical argument of Lipton (1977) refers to the political power structure, which is characterized by 'urban bias' and so imposes limits on price and structural policy in agriculture. Little (1982) and Timmer (1988) stress the lack of understanding in policy-making bodies of the specific factors that characterize agriculture. Rent-seeking government behavior explains much of the limited attention for raising agricultural efficiency and addressing rural poverty.

**Food security**

Depending on pre-adjustment distortions in agriculture, the implementation of structural adjustment programs can be progressive in terms of income distribution. Producers and traders stand to gain from better price incentives, whereas urban consumers' welfare declines. Rural workers are hit by higher food prices in the short run, but face better employment prospects if supply response is forthcoming. Food security is, therefore, likely to increase on balance in the rural areas; urban areas face immediate declines. The disappearance of black markets and the elimination of rents result in an almost immediate reduction in income disparities (Azam, 1996).

Production increases – the major objective of agricultural policy – are a means to improved food and nutritional security, although neither a necessary nor a sufficient one. Efficient (implying favorable) producer prices serve both production and rural incomes.[5] Many would agree that rural welfare is best served by improved nutritional status of the rural population. Higher rural incomes are a major condition for improved nutritional conditions, provided basic staple foods are sufficiently available. With land fully utilized, higher production can only be realized through productivity increases per hectare, i.e. higher yields and an economically more attractive crop mix.

## 3 IMPACT OF REFORMS

The recognition of both market and policy failures has led to a more balanced view of the effectiveness of policy intervention in agricultural markets. At the micro level, this view is supported by a growing understanding (and relearning) of farm households' behavior and their potential supply response to relative prices. Interventions are considered to be most effective if they create an enabling (i.e. improving access and correcting for classical market failures) environment for basically private decision-making, and focus on structural and competition policy rather than direct price and market interventions. The latter can be confined to policies to stabilize prices, restricted to a few strategic crops or inputs, and to make better use of international (or large national) markets.

Has this change in policy stand, triggered by structural adjustment, liberalization and privatization worked? Given the concern of the reformers with adverse agricultural incentives, this question is basically one of supply response. A second area regarding the impact of reforms relates to the question how markets and institutions themselves have been affected. Both types of impact are discussed below.

### Supply response

Two remarkably candid studies by the World Bank (Umali-Deininger and Maguire, 1995; Meerman, 1997), both supported by a wealth of country experiences, have attempted to come up with an overall assessment of recent liberalization efforts. Meerman (1997) concludes that for a large number of LDCs the impact of agricultural adjustment on supply has been significant and positive, though small. Forceful reformers like Chile, China and New Zealand have shown a much more dramatic impact.

Three more findings are interesting. First, supply response is "symmetrical". Where there has been heavy agricultural protection, as in the former Soviet bloc, liberalization leads to output contraction and resource outflow as well as reallocation of resources within the sector towards more attractive crops. Second, supply response is synergistic. Where adequate rural infrastructure and support services are deficient, getting prices right, even in an enabling macro environment, will not suffice. Third, supply response depends on the credibility of reforms. Where reforms are not sustained over longer periods, private investment will not be forthcoming.

These findings by the Bank's Operation Evaluation Department in fact quietly close a fierceful debate during the late 1980's and early 1990's prompted by the alleged "pricist" and "state-minimalist" approach adopted in the 1986 World Development

Report on world agriculture and subsequently associated with adjustment policies (Lipton, 1987 and 1988). In a Bank reply to this criticism, Ray (1988) emphasizes that "It is not a question of whether the government should intervene in agriculture: rather, it is a question of how it should intervene, in what forms, and in what areas. Our review of the past experience suggests the need for a fairly significant qualitative change in the role of the government. This does not mean that public spending in agriculture should necessarily be reduced. In fact large increases may well be warranted in many cases. The point concerns allocation priorities, not total size."

Lipton's critical analysis points, among other things, to the political decision-making process, which makes it difficult for many governments to undertake price reforms to agriculture's advantage while increasing investment in rural services and infrastructure. In the absence of a consensus on the feasibility of simultaneous action in both spheres, Lipton chooses to place the emphasis on structural policy. Ray, on the other hand, refers to the empirical finding that rural development, particularly in Africa, is doomed to failure unless accompanied by an effective macro-economic policy and sectoral market incentives.

Supply response analyses are not unambiguous about price elasticities of agricultural production. Individual commodity studies show considerable short-term response to higher real prices, but short-term aggregate supply effects are mostly fairly small (Binswanger, 1989). The results of supply response models are strongly influenced by aggregation procedures (countries, regions or crops do not necessarily face the same binding constraints), and by the use of market prices instead of subjective reservation prices. For locally grown, non-tradeable food crops, demand depends on purchasing power, which in turn is determined by supply, giving rise to a simultaneity problems. While recorded values for short-run price elasticiticies are low, elasticities for other factors influencing supply response (rainfall, land quality, road density, rural finance, literacy and population density) are higher and their inclusion strongly improves the explanation of supply response.

Further methodological problems arise if aspects of technological change (usually captured by a time trend) and substitution between crops are addressed. Supply response is expressed in terms of adjustment of factor demand, production and cultivated area. Implications for changes within the farm household cropping systems can, however, not be directly derived. Since price changes give rise to both income and substitution effects, choices between labor and leisure, substitution of internal by external inputs, and fine-tuning of decisions related to changes in cropping frequency and cropping density deserve additional attention, the more so as these aspects have a decisive impact on the sustainability of resource use.

A lack of synchronization in input and output price adjustment has often been observed to retard a positive supply response. Even if both prices rise proportionally, higher input costs inevitably precede higher output revenues. Credit-constrained farmers may, therefore, refrain from adjusting their current production plans. This effect is reinforced if adjustments in input prices are stronger than in output prices, moderating the real increase in producer's margins. When inputs are imported and subsidies are simultaneously eliminated with exchange rate adjustment, such developments are not unlikely for non-export crops.

It is generally understood that supply response structurally increases with the level of development. However, two important observations can be made (Binswanger, 1989). The first concerns price versus structural policy. Empirical research has shown that for a given technology, production increases in response to sustained price increases are limited. Almost all growth in agricultural production is due to investment in capacity expansion and to technological advances. Price policy is important in this context, but primarily as a means of supporting favorable expectations by producers regarding the future profitability of their private (investment) activities. Other forms of price policy may also be important in the context of structural adjustment. For example, where price controls lead to the rationing of consumer goods in rural areas, their abolition may be followed by a sharp increase in agricultural production.

A second issue refers to the choice between government and on-farm investment. Both types feature a high level of complementarity. In most societies, decision-making at the level of the individual farm is a highly decentralized and private activity. At the same time, this decision-making takes place in an environment that is heavily dependent on government activities in the areas of research, infrastructure and the regulatory framework. These activities are vital to the promotion of private investment in agriculture, as are satisfactorily functioning rural financial markets, with forms of security accessible to as many farmers as possible.

**Markets and institutions**

After the discussion of supply response, we can be brief about the impact of liberalization on markets and institutions. Agricultural markets in developing countries can easily fail: inadequate price information; problems with product quality, non-standardized weights and measures; access barriers; price collusion; poorly defined contract laws and rights. Structural adjustment offers, according to Meerman (1997), good opportunities to improve market performance. Examples are better collection and dissemination of domestic prices and supplies; better analytical surveillance of patterns of (international) agricultural price and trade interventions (Valdés, 1996); analysis of transaction costs and of spatial and temporal co-integration of markets; improved standards and legal systems; removal of trading constraints; and accompanying investment in infrastructure and support services. Efforts to reform agricultural parastatals have on the whole been disappointing and have led to a strong preference (by donors) for privatization. Although not always explicitly mentioned, this raises the question of competition in those markets. Not unexpectedly, improving agricultural public expenditure and investment programs can be shown to have large pay-offs.

Reviewing extensive World Bank experience, Knudsen and Lindert (1995) confirm the widespread domestic pricing and subsidy reforms, but emphasize their often incomplete nature. Many governments have been reluctant to fully liberalize prices for fiscal reasons and out of fear to destabilize food prices. Strengthening rural institutions and modifying the regulatory framework has reportedly progressed, but encouraging competition has remained difficult to implement. Despite the substantial literature on the complementary role of non-price factors, little attention has been paid to those factors in adjustment lending. Both Knudsen and Lindert (1995) and Meerman (1997) conclude that the impact of agricultural adjustment and reforms on food security, poverty, and the natural resource base has largely been ignored in agricultural sector lending.[6]

## 4 AGRICULTURAL MARKETS BEYOND LIBERALIZATION

Having analyzed the major issues and indicated the likely impact of structural adjustment and liberalization, the obvious question arises how markets (and institutions) actually perform in liberalizing economies. What do we know about the functioning of these markets, can we establish their performance, which lessons can be learned, can we formulate clear results and identify remaining weaknesses? These are among the questions addressed in the contributions to this volume reflecting the 1998 seminar on agricultural markets beyond liberalization.

Two issues need to be emphasized from the outset. First, as the various contributions show, country and policy experiences vary widely, and these differences in outcome are confounded by differences in initial conditions, timing and sequencing of reforms, and concurrent external developments. Generalizing and clarifying post-liberalization experiences is therefore inherently difficult. Secondly, it is essential to have an operationally useful definition of agricultural markets. For this purpose, Thorbecke proposes the concept of *exchange configuration*, which is characterized by the commodity traded, the actors involved, and the institutional environment in which exchange takes place. Exchange configurations include both market and (informal) non-market transactions, and can be conceived as nodes in a commodity–cum-marketing chain.

In many countries the impact of liberalization in the 1990's can be characterized by the partial nature of the reforms, an often long transition period, and hence benefits that are far from fully realized (Badiane). Resistance from vested interests and uncertainty about which institutions are needed to enable the functioning of efficient markets (and how to realize them) play a major role here. Evidence from Africa shows that liberalization has been largely of the pricist type aiming at a rise in real prices of agricultural commodities. However, commodity taxation has in many cases remained high, so that discrimination against agriculture has been reduced but not eliminated. Depending on the type of commodity and developments on the world market, liberalization is in reality accompanied by both declines and increases in real prices (Thorbecke, Acosta).

A major problem in the various structural adjustment programs for Africa is the lack of complementary policy measures in rural infrastructure and support services, the state-minimalist problem. The disappointing supply response in much of the private sector is likely to be closely related to failing to provide the necessary public goods in agriculture. As a result, marketing and transaction costs have remained high, and farmers may have rationally decided to continue producing for subsistence, rather than attempting to enter the market. By contrast, the experience of China shows how a sensible combination of price reforms, public goods provision and institutional changes has triggered a remarkable performance in food production and a striking increase in the integration of agricultural markets. With the gradual deregulation of factor and commodity markets, market functions also gain importance in the allocation of rural land in mainland China, emphasizing the need for further institutional development of the land market (Huang *et al.*). At a more theoretical level, Nuppenau and Badiane demonstrate the essential role of public goods provision for attaining sustainable intensification in the process of increasing agricultural productivity.

Thorbecke addresses the dilemma of how to reconcile fiscal austerity with the need for public expenditures to improve infrastructure and support services. His suggestion to increase the share of agricultural sector adjustment loans at the expense of general program loans accords well with the recent World Bank findings and experiences discussed in the previous section. Periodic public expenditure and investment reviews of agricultural budgets may help to weed out (or at least identify) less cost-effective items of public outlays. At the micro level, Hoogeveen cautions against supposed but unfounded market failures in the area of rural financial markets. Zimbabwean rural households successfully apply a buffer stock strategy to stabilize consumption, thereby reducing the need to diversify consumption risk through financial markets.

As several case studies included in this volume indicate (Badiane, Kuiper *et al.*, Dedehouanou and Quarles van Ufford), many local agricultural markets have markedly improved their performance after liberalization. Price instability has decreased, price spreads across markets have become lower in both spatial and seasonal terms, and costs of intermediation have gone down. However, due to the various constraints discussed above, private sector response has been mixed and has not always been able to effectively replace previous state activities. Private traders apparently find it difficult to cover wider areas because of market segmentation and thinness. Hence, with few private market entrants, not oligopoly as often feared, but a regional breakdown in the marketing system has emerged as a new risk.

The relevance of exchange configurations is clearly brought about by Kuiper *et al.* for staple foods in West Africa. Retailers in two major towns, who can still directly buy from farmers, do not allow wholesalers to behave as vertical price leaders. However, wholesalers in two larger rural centers, involved in arbitrage among urban markets, are able to influence price formation. Elsewhere in West Africa, Dedehouanou and Quarles van Ufford show how farmers themselves provided the conditions for a successful withdrawal of the state from the draught animal market, while most agricultural input markets remained spatially fragmented and thin in demand. Only the provision of an appropriate institutional environment could improve market performance in the cases reported. Two contributions dial explicitly with the approach to such institutions. The organization of private sector provision through voluntary marketing institutions is discussed by Meulenberg. He reviews the literature, proposes a classification scheme and analyses the evolution of voluntary marketing institutions in food markets. Problems related to the institutionalized strong pro-male bias in many integrated grain marketing systems, leading to constraints upon female commercial accumulation and endangering female livelihoods, are analyzed by Harriss-White.

In his conclusion on improved market performance, Badiane lists three success facors: (1) the extent of private trading before the reforms, (2) the sustainability of reforms, and (3) the absence of partial liberalization (with parastatals operating alongside private traders). The experience of agricultural markets in transition economies adds an interesting dimension to these findings. Kopsidis argues that the growing supply of imports to the Russian grain market, following the desintegration within the Federation that halved domestic grain trade, has done little to develop domestic market structures. A return to rigorous state control of agricultural marketing is no option as it would probably decrease production even further. Similar problems are analyzed by Spoor for two newly independent Central Asian states, where minimal market-oriented and political reforms prohibit agricultural diversification based on comparative advantages. Missing markets now threaten to force the two countries into costly and inefficient self-sufficiency of food following a previous overspecialization in export cotton.

The opportunities for regional agricultural specialization are illustrated by Acosta for South-East Asia. Freeing trade in rice among ASEAN countries would reduce its real price, stimulate intra-trade at the expense of extra-ASEAN exports, and allow further specialization among member states. By contrast, freeing trade in maize would raise it price, but leave extra-ASEAN trade largely unaltered. Le Goulven shows how initial disparities in access can bring about the danger of dualism when markets are freed. Although after withdrawal of the state from hog marketing in Northern Vietnam the market appeared close to perfect, strong disparities between those having access to capital and those who lack it emerged. Lack of rural credit institutions and lack of income buffering possibilities among dissaving farmers threaten the long-run performance of the hog market.

## 5 UNFINISHED BUSINESS AND CONCLUSIONS

What can we learn from the post-liberalization experiences reported in this volume, and elsewhere? Although some of the findings are mixed and even contradictory, a number of results stand out. First, in view of the varying experiences of countries involved and policies applied, there is a clear need to define markets and liberalization in an operationally meaningful sense. The concept of exchange configurations, including both market and non-market (informal) transactions in the framework of the agricultural marketing chain, serves that purpose well. It clarifies the functioning and evolution of different agricultural markets in their institutional context, and easily accommodates other than economic considerations. Adjustment and liberalization policies need to be carefully specified in terms of instruments, rules of operation and institutions.

Second, although many domestic agricultural markets perform better after liberalization, their operation is often far from efficient because of weak or absent institutions and an improper regulatory framework. The necessary institutions are often known: better price information, uniform standards, better access, improved risk management, credit discipline in rural finance, decentralization of the delivery of support services, to mention a few. Third, to improve market structure and performance, a range of complementary public expenditure and investment measures to provide public goods is necessary. There is now ample evidence that "adjustment had an anti-agricultural effect on expenditures" (Knudsen and Lindert, 1995). As a result, private investment has often not been crowded in or revitalized, transaction cost remained high (and markets thin), and supply response limited.

Fourth, combining these findings, the policy design of liberalization has often been weak. Sequencing of measures has been particularly neglected, and state withdrawal without strengthening existing institutions or the creation of new ones (property rights, risk management) has done little to improve the functioning of markets. Finally, whereas liberalization is primarily meant to improve efficiency, its effect on equity, including poverty, and food security has been largely ignored. In almost all agricultural World Bank programs no attention has been paid to food security and poverty (Meerman, 1997), and approaches to address it (through local governments, NGOs, voluntary participation) have been neglected.

Several issues remain therefore on the agenda as unfinished business and more can be added. Competition in thin markets and public versus private provision of support services remain areas of attention. How much public activities can be decentralized to gain efficiency but without losing fiscal control? Financing of public goods provision remains problematical, not in the least when political-economy considerations are taken into account. How to approach soil degradation and water management in a post-liberalization era deserves more attention. Price risk management in freeer markets, insurance schemes and effective rural finance institutions can equally be seen as unfinished business. The post-liberalization agenda does not lack challenges.

## NOTES

[1] This section relies heavily on parts of van Keulen, Kuyvenhoven and Ruben (1998).

[2] The effectiveness of government intervention depends on the ability and willingness to act according to shared objectives. Most problematic is the rent-seeking aspect of government behavior: while paying lip-service to efficiency and distributional objectives, many governments intervene for reasons of narrow self-interest and are responsive to discriminative lobbying activities (Ellis, 1992).

[3] Corbo and Fisher (1995) define structural adjustment as "a process of market-oriented reform in policies and institutions".

[4] Knudsen and Lindert (1995) mention that "a modest estimate of the accumulated losses to the world of this unfavourable agricultural policy stand is conservatively estimated at $ 6 trillion, or twice the aggregate annual GDP of developing countries .... The tragedy of these policies is ... [the] negative impact on the poor." Krueger, Schiff and Valdés (1992) find over a period of 25 years that high taxation of agriculture in LDCs is correlated with low growth in both the sector and the economy is a whole, and *vice versa*.

[5] Efficient prices also mean that prices are in line with the world market with a protective correction for EU-type budgetary price distortions.

[6] Interestingly, food and agricultural policies under structural adjustment was the theme of the previous EAAE development seminar at the University of Hohenheim, 1992 (Heidhues and Knerr, 1994).

**REFERENCES**

ADAMS, D.W., D.H. GRAHAM, & J.D. VON PISCHKE (1984) *Undermining Rural Development with Cheap Credit.* Boulder: Westview Press.

ANTLE, J.M. (1983) 'Infrastructure and Aggregate Productivity: International Evidence.' In *Economic Development and Cultural Change* , 31, pp 609-619.

AZAM, J.P. (1996) 'The Diversity of Adjustment in Agriculture.' In S. Ellis (ed.) *Africa Now.* London : James Curry and Heineman, pp.136-154.

BEVAN, D.L., A. BIGESTER, P. COLLIER & J.W. GUNNING (1987) 'Peasant Supply Response in Rationed Economies.' In *World Development* , 15, pp.431-439.

BINSWANGER, H.P. (1989) 'The Policy Response of Agriculture.' In *Proceedings of the World Bank Annual Conference on Development 1989.* Washington D.C.: World Bank.

BINSWANGER, H.P. & S.R. KANDLER (1993) 'How Infrastructure and Fnancial Institutions Affect Agricultural Output in India.' In *Journal of Development Economics*, 41, pp.337-366.

COLMAN, D. & T. YOUNG (1989) *Principles of Agricultural Economics: Markets and Prices in Less Developed Countries.* Cambridge: Cambridge University Press.

CORBO, V. & S. FISHER (1995) 'Structural Adjustment, Stabilization and Policy Reform: Domestic and International Finance'. In J. Behrman & T.N. Srinivasan (eds.) *Handbook of Development Economics*, vol. III, Amsterdam: North-Holland, pp.2845-2923.

DE JANVRY, A., SADOULET E. & M. FAFSCHAMPS (1989) 'Agrarian Structure, Technological Innovation, and the State.' In P. Bardhan (ed.) *The Economic Theory of Agrarian Institutions*, Oxford: Clarendon Press, pp.356-382.

ELLIS, F. (1992) *Agricultural Policies in Developing Countries.* Cambridge: Cambridge University Press.

HAYAMI, Y. & V.W. RUTTAN (1985) *Agricultural Development: An Interntional Perspective.* Baltimore: The Johns Hopkins University Press.

HEIDHUES, F. & B. KNERR (eds.) (1994) *Food and Agricultural Policies under Structural Adjustment.* Frankfurt am Main: Peter Lang.

HOFF, K. & J.E. STIGLITZ (1993) 'Imperfect Information and Rural Credit Markets: Puzzles and Policy Perspectives.' In K. Hoff , A. Braverman & J.E. Stiglitz (eds.) *The Economics of Rural Organization: Theory, Practice and Policy.* Oxford: Oxford University Press for the World Bank, pp.33-52.

KEULEN, H VAN,., A. KUYVENHOVEN & R. RUBEN (1998) 'Sustainable Land Use and Food Security in Developing Countries: DLVs Approach to Policy Support.' In *Agricultural Systems*, vol. 58(3) , pp.285-307.

KNUDSEN, O. & K. LINDERT (1995) 'Agrcultural Lending in Transition: Lessons from Experience and Future Approaches.' In D. Umali-Deininger & C. Maguire (eds.) *Agriculture in Liberalizing Economies: Changing Roles for Governments.* Washington, D.C.: The World Bank, pp.379-427.

KRUEGER, A.O., M. SCHIFF & A. VALDÉS (1992) *The Political Economy of Agricultural Pricing Policy.* Baltimore: Johns Hopkins University Press.

LIPTON, M. (1977) *Why Poor People Stay Poor. Urban Bias in World Development.* London: Temple Smith.

LIPTON, M. (1987) 'Limits of Price Policy for Agriculture: Which Way for the World Bank?' In *Development Policy Review,* vol. 5, no. 2, pp.197-215.

LIPTON, M. (1988) 'A Rejoinder to Ray.' In *Development Policy Review* ,Vol. 6, no. 2, pp.215-222.

LITTLE, I.M.D. (1982) *Economic Development: Theory, Policy, and International Relations.* New York: Basic Books.

MEERMAN, J. (1997) *Reforming Agriculture: The World Bank Goes to Market.* A World Bank Operations Evaluation Study. Washington, D.C.: The World Bank.

NORTON, R.D. (1992) *Integration of Food and Agricultural Policy with Macro-economic Policy: Methodological Considerations in a Latin American Perspective.* ESD Paper III. Rome: FAO.

RAY, A. (1988) 'A Response to Lipton's (June 1987) Review.' In *Development Policy Review,* Vol. 6, no. 2, pp.211-215.

SADOULET, E. & A. DE JANVRY (1995) *Quantitative Development Policy Analysis.* Baltimore and London: The Johns Hopkins University Press.

STIGLITZ, J. (1987) 'Some Theoretical Aspects of Agricultural Policies.' In *The World Bank Research Observer,* 2, pp 43-60.

TIMMER, C.P. (1988) 'The Agricultural Transformation.' In H.B. Chenery & T.N. Srinivasan (eds.) *Handbook of Development Economics* , vol. I, Amsterdam: North-Holland, pp.275-331.

UMALI-DEININGER, D. & C. MAGUIRE (eds.) (1995) *Agriculture in Liberalizing Economies: Changing Roles for Governments.* Washington, D.C.: The World Bank.

VALDÉS, A. (1996) *Surveillance of Agricultural Price and Trade Policy in Latin America during Major Policy Reforms.* World Bank Discussion Paper, no. 349, Washington, D.C.: The World Bank.

WORLD BANK (1986a) *World Development Report 1986.* New York.: Oxford University Press.

WORLD BANK (1986b) *Poverty and Hunger.* Washington, D.C.: World Bank.

# 2 AGRICULTURAL MARKETS BEYOND LIBERALIZATION: THE ROLE OF THE STATE

Erik Thorbecke

## 1 INTRODUCTION

The purpose of this paper is to analyze the impact of liberalization and other policies and state institutions on agricultural markets.[1] The first issues that need to be elucidated are what is meant by liberalization and by agricultural markets. Liberalization tends to be a catch word that can be used to describe a whole different range of policies and institutions. In a narrow sense, it entails trade liberalization (i.e. a package consisting of a devaluation, reduction and harmonization of tariffs, and the elimination of quantitative restrictions against imports) and elimination of state commodity boards and state monopolies for export crops. In its broader sense, liberalization can be interpreted as being almost synonymous with structural adjustment – including a combination of trade liberalization measures in addition to a significant state withdrawal from intervention in the pricing of domestic food crops and imports, as well as a whole constellation of structural measures designed to improve the competitiveness and functioning of agricultural markets, such as public investment in physical infrastructure (particularly irrigation and farm to market roads), research, extension, antitrust regulations and the privatization of many marketing activities previously performed by the state. I believe that the more conventional usage is to interpret liberalization in the more minimalist sense and particularly as measures required for "getting the prices right."

An attempt is made in this paper to distinguish clearly among the different combinations of policy measures and institutions that can be defined as constituting "liberalization"-- ranging from the narrow "pricist" definition to the all-encompassing "adjustment" definition. In turn, the impact of these different combinations of measures and institutions on various performance criteria (such as efficiency, equity, adaptability, contribution to food security and sustainability) is analyzed.

Next, it is essential for analytical purposes to come up with an operationally useful definition of agricultural markets. I will use for this purpose the concept of exchange configurations developed in Thorbecke and Cornelisse (1991). An exchange configuration is defined by the characteristics of three sets of elements: the item traded, the actors trading, and the environment within which the exchange takes place. Exchange configurations include both market and non-market (informal) transactions. As argued elsewhere (Thorbecke & Cornelisse, 1991; Thorbecke, 1992), the functioning of a market must be understood in both terms of its elements and in terms of its interactions with related exchange configurations (both market and non-market configurations).

The concept of a market is meaningless unless one specifies the specific item to be exchanged. There are agricultural markets for products, inputs, services, factors of production and even knowledge. In this paper we shall concentrate on agricultural product markets. There are literally hundreds of different traded products ranging from such typical export crops as cocoa, coffee and cotton, to export-cum-domestic food crops such as rice, wheat and corn, all the way to essentially nontradable food crops such as sorghum, millet and cassava.

Each item possesses its own set of characteristics, as do the other elements of an exchange configuration, the actors and the environmental setting. Since the marketing chain between initial producers (farmers) and ultimate consumers may involve many intermediaries, it is useful to think in terms of specific commodity systems. During the marketing process, agricultural commodities gain in value as they are moved through space (e.g. transported from the farm to a retail outlet); held over time (e.g. storage between harvest season and lean season); and transformed (e.g. through processing and packaging).[2]

Each commodity system has its own particular marketing chain and network and set of transactions corresponding to the various functions performed by different actors as a commodity progresses from producer to final consumer. In fact, a given commodity system can be broken down into a hierarchical and sequential set of exchange configurations embracing the different intermediate functions occurring between production and ultimate consumption. In this way the distinct transactions taking place along the marketing chain can be determined and evaluated within the specific configurations in which they are shaped. It is the attributes and endowments of the actors, together with the characteristics of the environment, that determine the specific channels (i.e. exchange configurations) along a commodity chain. Thus, a commodity system-cum-marketing chain can be thought of as being constituted by a series of configurations consisting of their own respective actors with their attributes and underlying environmental constraints focused on a specific item that gains value as it moves along the chain.[3] For example, the successive activities and transactions occurring along a wheat marketing chain could take the form of coarse wheat exchanged at the farm gate, wheat moved from one location to another by traders, the consolidation of small batches into larger batches by commission agents, the milling of the grain into flour, and the packaging and selling of the flour to consumers in retail outlets.

In Thorbecke (1993) the Pakistan Wheat Market is decomposed into a series of specific configurations and transactions to illustrate the applicability of the exchange configuration approach to a specific and concrete case. Thus, in the 80's it was possible to identify eleven distinct exchange configurations along the marketing chain. These configurations are shown graphically in Figure 2.1.

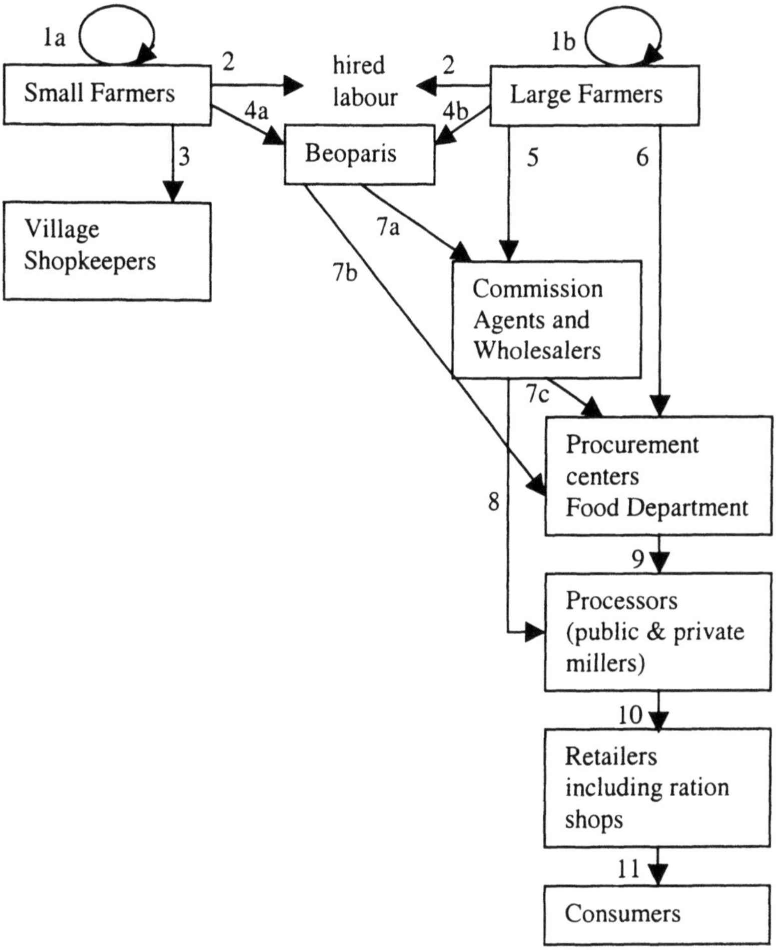


*Figure 2.1 Major configurations and transactions along the wheat market chain*

Each number corresponds to a distinct configuration with its own item (i.e. wheat at different stages of processing and in different locations), actors and environment. Each set of actors confer additional time, form and place utility to the item being exchanged. Crow (1989) for wheat and rice in Bangladesh and Harriss (1982) for the circulation of rice around a district neighboring Calcutta -- although not formalizing the concept of exchange configuration -- follow a similar approach of decomposing the system of circulation of rice into subsystems, as the following quote indicates:

> The system of circulation of rice is not simple, nor is it homogeneous. It can be divided into a number of subsystems, each affecting others and some interlocking through elements, relationships or disturbing factors, with others. One subsystem has as its elements the members of classes related through agricultural production. Rice circulates in return for cash, as rent, in return for labor services, as a medium for loans to be repaid at a future date in cash, kind or by labor services. It is a medium for barter exchange....Another subsystem

> of circulation is the sale of paddy for transfer out of the locality. A third subsystem is the "official" system of bulking by purchase agents for rice mills who yield a portion up to the state in exchange for the right to trade the residual more freely. A fourth subsystem is that of petty trade in husking machines for local retail sales or for wholesaling and long distance trade. (Harriss, 1982)

Each exchange configuration can be conceived as being a node in the marketing chain. Depending on the characteristics of the elements, a commodity (chain) system can be decomposed into a variable number of configurations--the lower limit being one which would be approximated by the American Supermarket Model, where all functions from production to consumption are vertically integrated into one giant corporation. In contrast, the chain may be very long and consist of many distinct configurations. For example the wheat marketing chain in a relatively poor developing country might include such exchange configurations as small wheat farmers selling to village traders and consumers, village traders and large farmers selling to regional traders, regional traders selling to wheat processors, the latter selling to urban wholesalers, urban wholesalers selling to urban retailers and finally the latter selling to urban consumers.

In fact, in many instances, the above marketing chain should be complemented by the two extreme exchange configurations that are likely to prevail at either end. At the beginning of the chain belongs the farm household configuration. The family farm combines a whole set of market transactions (such as selling the marketed surplus, off-own farm wage employment by family members and the purchase of manufactured inputs and consumption goods) with virtual, implicit and internal (nonmarket) transactions occurring among members, such as the exchange of labor services by a member for food grown on the farm. It is the hybrid nature of this configuration that makes it such a fascinating organization to study. At the other end of the chain--as long as the item is a tradable--one finds the world market configuration where the main actors are the exporters and importers.

Section 2 introduces an operational methodology of analysis of market and non-market exchange configurations and transactions and explores the impact of government intervention on the operation of these configurations. Section 3 is devoted to describing and analyzing the characteristics of items, actors, and the environment in agricultural product markets with emphasis on food grains.

Section 4 reviews the policy environment in the developing world prior to and at the outset of liberalization. It attempts to capture the status-quo-ante in the mid-eighties. It reviews the overall discrimination against agriculture in the Third World in terms of policies and treatment of the agricultural surplus (in 4.1); the relative size of the marketable surplus (in 4.2); the extent of market integration for agricultural products (in 4.3); and, government interventions and their impact on food and cash crop configurations (in 4.4).

Section 5 is devoted to the impact of liberalization on agricultural product markets in the 1990's. The final section concludes.

## 2 TOWARDS AN OPERATIONAL METHODOLOGY OF ANALYSIS OF MARKET AND NON-MARKET EXCHANGE CONFIGURATIONS AND TRANSACTIONS

A general approach to distinguish specific exchange configurations was developed in Thorbecke and Cornelisse (1991) and subsequently applied to agricultural markets by Thorbecke (1992, 1993). This general methodology is presented next before it is applied more specifically to agricultural product markets.

Economic activity is based and relies on transactions. Transactions can take any number of forms. They can occur within or outside markets. Parties to a transaction must enter into voluntarily and agree to its terms. Transactions occur and are endogenously determined within distinct exchange configurations. Each configuration is shaped by a unique combination of specific elements. The elements constituting the building blocks of configurations are the characteristics of a) the item transacted; b) the actors; c) the physical, spatial and locational environment; d) the technological and organizational environment; e) the cultural environment; f) the policy, institutional and legal environment; and, finally, g) the socioeconomic structure.[4]

Exchange configurations can be market or non-market configurations. An initial necessary step to understanding the process of exchange in developing countries is to identify the most important specific characteristics of elements that influence the items exchanged, the behavior of the actors and the properties of the environment and, thereby, the form of the transactions that take place. Different exchange configurations and corresponding transactions obtain depending on such characteristics of, first, the item exchanged as the degree of variability and uncertainty in the pattern of production and whether the good is perishable and storeable; second, the attributes of the actors as they relate e.g. to their endowment of land and labor, their access to information, their attitude vis-à-vis risk, their access to assets that can be used as collaterals and the nature of the interrelationships among actors; and third, the environment e.g., the structure of production and of production relationships, the prevailing conventions and norms, the system of property rights, the underlying legal and institutional framework, current government policies and the phase of development.

Exchange configurations can be thought of as channels through which specific transactions are effectuated. Actors, given their own attributes, the properties of the item transacted and the environment they face, will choose to operate within (or invent) that configuration and corresponding transaction that minimizes the sum of production and transaction costs. Surrogate (non-market) transactions are created and evolve in response to "missing markets", or existing market configurations characterized by high transaction costs. Generally, only a small number of key and discriminating elements determine the specific nature of a given configuration and corresponding transaction (Thorbecke & Cornelisse, 1991).

The difference between market and non-market configurations (and their corresponding transactions) is not entirely clearcut.[5] Some exchange configurations are clearly market ones, such as the market for a given food crop (for example, wheat or rice in a given setting) at the village level or the market for a cash crop (for example, cocoa, coffee, rubber) in the world market. In contrast, other configurations lead to transactions that occur essentially outside market configurations. The family farm's decisions regarding whether and how much of its output to sell in the food crop market configuration depend on the ruling price in that market. In turn, non-market (internal) transactions (i.e. the provision of old age and state contingent insurance within the household, barter type exchanges of labor effort for food among household members) are influenced by such variables as the price of output and the wage rate for casual labor determined in the food crop market and rural labor market configurations, respectively. Generally speaking, market transactions entail the use of money and tend to be impersonalized (taking the limiting form of arms'-length trading) while non-market transactions more often than not are of a barter type, interlinked and highly personalized. In traditional societies, multi-stranded relationships prevail, leading to multiplex types of transactions. Furthermore, non-market transactions almost always occur between members of a given (stable) organization or institution, while the actors in a market transaction can be perfect strangers. Thus, the continuum of possible configurations ranges from a competitive organized market (for example, for a staple crop) to sharecropping and interlinked institutions that can be subjected to a certain degree of monopolistic competition while being strongly influenced by norms and conventions.

## 2.1 IMPACT OF GOVERNMENT INTERVENTION ON THE OPERATION OF MARKET AND NON-MARKET EXCHANGE CONFIGURATIONS

An approach based on exchange configurations highlights the importance of the interaction among the different elements of configurations, as well as the interaction between different exchange configurations. It is the combination of an entire set of relevant elements that determines the transactions that result. The impact of changing the value of a given element (e.g. the policy environment) must be understood in conjunction with the specific characteristics of the remaining elements (items, actors, and other properties of the environment) defining a particular exchange configuration. The world operates, at best, in a second best framework. Relaxing such elements as a given policy or institutional constraint may not lead to improved market performance (as measured by efficiency, equity and adaptability considerations) in light of other constraints which exist in a given market. Thus, for example, the prescription of getting the prices right, while a necessary condition, is hardly a sufficient condition to improve the performance of markets, as is amply demonstrated subsequently.

The scope for generalizations and blanket policy recommendations to improve market functioning is limited by the context-specific nature of the particular characteristics of the elements shaping given exchange configurations. Prototype non-market configurations are likely to respond even more variably to shocks across settings than are market configurations.

Government intervention or institutional changes affecting a given market (or non-market) configuration are likely to modify at least some of the elements constituting other exchange configurations. Changing the elements affecting the functioning of a given configuration (e.g. by altering the response of actors) may result in a new equilibrium transaction (e.g. a new price and quantity) that, in turn, will feed back to influence and alter some of the key elements shaping other configurations. Thus, in addition to understanding the interaction of elements within a configuration, it is also important to understand the interaction of configurations within an economy. Market and non-market configurations do not operate independently of one another. Their linkage is most obvious in the case of the interrelationship between official and parallel markets. Official markets (transactions) breed parallel markets (transactions). This means that government or institutional actions occurring in one exchange configuration may spill over to influence and affect other configurations. Thus, for example, artificially low food prices in the official market normally encourage farmers to sell on the parallel market. Many such spillovers between and among different configurations can be thought of. In some cases the spillover is between a market and a non-market configuration. Perhaps, the prime example of the latter is that when the price of the food crop falls within the food crop market configuration, a farm household that was previously selling some of its food output may revert to pure subsistence and consume the totality of its food output. In this case food becomes a nontradeable within the household farm configuration and its internal shadow price is determined within the farm household configuration.[6]

Where the exchange configuration is a parallel one, i.e. was initially generated by government interventions that created a situation of excess supply or demand in a particular product or factor market, then the removal of interventions can unify an otherwise parallel market with its counterpart official market. In contrast, if the underlying causes of the "segmented" configurations are structural, liberalization policies will have much less impact on the behavior and performance of those market configurations.

Thus to summarize: changes in any elements, such as the implementation of structural adjustment policies and reforms by governments, are likely to lead to differential effects, first, on the various exchange configurations within the same region or country; and, secondly, on the same general type of configuration across different regional or country settings. Certain characteristics of the elements constituting such configurations may constrain the responsiveness of the market to policy changes and, at the limit, become binding and lead to market failure, or even perverse results. A few examples suffice to illustrate this phenomenon. In the case of a domestic food market configuration, a very inadequate (farm to market) road network, or a badly functioning marketing system for inputs create bottlenecks that could negate farmers' supply response to higher producer prices following liberalization. Similarly, the unavailability of manufactured consumption goods and inputs, notwithstanding higher farm gate prices, tends to induce a return to production for subsistence (and a corresponding reduction in the marketed surplus) within a farm household configuration. However (small) producers and traders at the beginning of the marketing chain are likely to be affected more negatively than are larger farmers and regional traders further along the chain. In this sense the same set of policies can affect different configurations differently.

In the second case, the possible occurrence of differential effects of changes in similar elements (e.g. adjustment policies) on the same general type of configuration across different regional and national settings, results, essentially, from the fact that some key (structural) elements (e.g. property rights, norms and conventions, physical and topographical environment, and infrastructure) may be more restrictive in one particular setting than in another, affecting thereby the degree of market response.

The performance of a given exchange configuration (be it a market or an "agrarian institution") can be evaluated according to economic criteria. Efficiency and equity are two such criteria. A third criterion is the adaptability of the structure to adjust to short-run changes (as a consequence, for example, of external shocks originating abroad, or new policies and institutional reforms) or more long-term changes in the nature of elements during the structural transformation which characterizes the development process. Still other criteria are food security and sustainability.

Clearly some elements are more flexible and maneuverable than others. Also, the time required to alter "structural" elements that impede the performance of given markets can differ markedly. The provision of crucial infrastructure projects, such as building large scale irrigation, required to relax a binding bottleneck can be a long, expensive, and arduous process. In contrast, certain administrative changes in rules, regulations and sometimes even property rights can be implemented quickly. If the goal is to improve the performance of markets on the basis of the above criteria, the key issue is not just to identify those elements which most constrain the successful behavior of markets, but even more importantly, to identify and concentrate on those elements over which some influence can be exerted.

It is essential to review and analyze the characteristics of the item exchanged, the actors and the environment across different Third World settings to understand how different initial conditions helped shape specific agricultural product configurations. This is undertaken in the next section. Section 4 explores how these configurations were affected by different combinations of initial characteristics of elements at the outset (just prior to) liberalization, while Section 5 discusses how they appear to be affected by different liberalization packages presently.

## 3 CHARACTERISTICS OF ITEMS, ACTORS, AND THE ENVIRONMENT IN AGRICULTURAL PRODUCT MARKETS WITH EMPHASIS ON FOOD GRAINS

### 3.1 CHARACTERISTICS OF AGRICULTURAL PRODUCTS AS AN ITEM

There are some key characteristics of food and cash crops that distinguish them from other items exchanged: 1) food plays a vital part (literally!) in the consumption pattern of households and is the main basic necessity and determinant of material well being and health; 2) staple production and cash crop production by smallholders provide employment and a livelihood to a large share of the population; 3) foodstuffs and cash crops are major tradable items; and, 4) these crops are subject to a high degree of risk and seasonality. For these reasons a very high degree of government intervention can be expected in these commodity markets. This means that the policy and legal environment is absolutely crucial in affecting the functioning of these markets, as is discussed subsequently.

Even though there are significant differences in the characteristics of the key elements shaping the markets for food crops relative to cash crops, the distinction between subsistence and cash crops as items can be overdrawn. Even the smallest farmer needs to generate a marketable surplus to satisfy his cash needs. Often both staple food and cash crops are produced by small and large farmers, alike, and are, in some instances, simultaneously intercropped. All farmers respond to an existing effective demand which can be for exports, to satisfy urban consumers, village level consumers, and farm households' own consumption needs. What matters then is to explain the operation of the various commodity systems, which exchange configurations different farmers participate in, and the degree of their market orientation.

## 3.2 CHARACTERISTICS OF ACTORS

The actors and their attributes constitute the next key set of elements. They include all the participants along the marketing chain. The chain includes a number of sequential activities starting with agricultural production (or even with the provision of inputs as a prior stage to production) followed by transportation, transformation and packaging, processing, wholesaling, retailing and ending with final consumption. The typical actors operating along the various links of this chain include, as we have seen, small farmers, tenants, hired workers paid in cash or in kind, large farmers, private and state owned and operated plantations and enterprises such as collective farms, village store keepers, various traders providing transportation and transformation services, processors, wholesalers, retailers, and ultimately consumers.

How do the characteristics of the private actors affect the choice of configurations, the type of exchange transactions, and the performance of the commodity systems? Starting at the production end of the chain, a first crucial determinant of the exchange behavior of farmers is their resource endowment. In many parts of the Third World land distribution is very uneven, giving rise to distinct groups in agriculture such as landless, small farmers, and large farmers. The landless households have as their main, if not only, asset their own family labor, while small farmers possess a disproportionately high ratio of family labor to land and other capital and vice versa for large farmers.

The fact that small, and near landless, farmers are poorly endowed with land and other assets, such as agricultural equipment and draft animals, signifies that they are deprived of the necessary collateral to borrow in the formal credit market. In fact, this class of actors is sealed off from that particular market configuration and is forced to enter into alternative credit configurations, contracts and transactions, such as borrowing from money lenders in the informal unorganized market, or from landlords or storekeepers, in essentially non-market configurations--giving rise to interlocked transactions. Furthermore smallholders, typically, face missing or incomplete markets for insurance futures more so than do better endowed farmers. Different size producers differ also in their attitudes vis-à-vis risk and in their capacity to bear risk. Larger farmers, possessing collateral, may be in a better position to bear risk than are smaller farmers. Finally, one can expect larger farmers to have better and cheaper access to information than their poorer counterparts. This leads to "asymetric information", where one group of participants involved in the exchange of a particular commodity has better access to information and resulting bargaining power than does another group.

The above differences in the characteristics of small vs. large farmers in terms of resource endowment; access to credit, insurance and information; and size of marketed surplus affect in a crucial way their choice of exchange configurations and the resulting nature of transactions and exchange contracts. This choice is, of course, further influenced by the underlying environmental elements (physical, technological, cultural and policy) confronted by the producers in different settings, as is discussed subsequently.

Many of the same attributes that influence the exchange behavior of producers (farmers) affect, likewise, the behavior of traders and other intermediaries. Individual traders involved in the marketing system are exposed to a variety of risks. They face the same constraints as producers do in terms of incomplete or missing credit and capital markets, and the lack of well functioning insurance or future markets. As Smith and Thompson have noted

> "The transfer of produce from producers to ultimate consumers takes time. Marketable produce is therefore an item of capital that must be financed during the marketing process. The textbook role of "middlemen" is to purchase produce from farmers and add value to it by transforming it in time, space and form before selling it to the next person in the marketing chain or to the ultimate consumer. Implicit in this scenario is the assumption that the middlemen have unlimited access to capital, their main constraints being their managerial ability, the availability of supplies at prices that allow them to make an expected profit on their transactions and their degree of risk aversion to the possibility of making a loss on their transactions. But what would happen if capital or credit for marketing transactions were limited?" (Smith & Thompson, 1991, pp.24-25)

The rate of turnover of the product is largely a function of access to credit and capital, particularly as it is embodied in storage capacity. Small traders lacking in credit and storage capacity have to operate within an exchange configuration consisting typically of only one link (function) in the marketing chain and allowing a high rate of turnover of the commodity.

Information relevant to marketing decisions is generated largely as a by-product of engaging in trading activities. Larger traders are involved in many more, larger and more continuous transactions than their smaller counterparts and thereby generate a much greater flow of information. This is further compounded by the fixed-cost nature of information gathering leading to scale economies favoring the large transactor.

Access to information brings about greater bargaining power; it can be collected and property rights attached to it. Information can, in turn, be shared with others in a mutually beneficial fashion. Since information is valuable, it pays to internalize its benefits. In this context, it has been observed that small groups linked by kinship, caste, tribe or nationality--all characteristics subsumed under our element "cultural environment"--tend to monopolize the dissemination of the available information (Smith & Thompson, 1991, p.41).

## 3.3 CHARACTERISTICS OF ENVIRONMENTAL ELEMENTS

After having analyzed the impact of attributes of private actors on exchange behavior, the next step is to explore the effects of key environmental elements (in particular, physical, technological and cultural) on the agricultural marketing process.[7]

### Access to Land

The initial resource endowment, particularly in terms of access to land, is likely to be a crucial determinant of the pattern of agricultural development a country or region will follow. Platteau and Hayami (1996) provide evidence that Asia is characterized by scarcity of land resources relative to population and labor force as compared with Africa. As they put it,

> "The high population density and the unfavorable land-labor ratio have induced more intensive land use, resulting in high percentages of land used for agricultural production, ..., by building better land infrastructure... above all, irrigation. The better land infrastructure created suitable conditions for the introduction of modern land-saving technologies such as high-yielding varieties and chemical fertilizer." (p.4)

If the amount of arable land per agricultural worker is taken as a measure of access to land, Platteau and Hayami (1996) show that the latter ranged from 0.3 (hectares of arable land to agricultural workers) in East Asia to 0.8 in South Asia (and 0.5 for the whole of Asia) and 1.2 in Africa in the early 1990's. Khan (1997) using somewhat different sources comes up with somewhat different estimates: 0.43 for Asia (with 0.20 for China and 0.73 for India) contrasting with 0.96 for SSA. Khan (1997, Table 5) demonstrates the wide diversity of land endowments among African countries, ranging from a ratio of 0.27 in Kenya to 1.88 in Nigeria, which prompts him to state that "this overall measure of land endowment hides a great deal of difference among individual countries of SSA. In many countries land scarcity is worse than the Asian average and in some it is as bad as in quintessentially land-scarce China" (p.8). Furthermore, Khan argues that "once the higher cropping intensity due to irrigation and the better land quality in Asia is taken into account, the relative advantage of SubSaharan Africa (SSA) over Asia...becomes much narrower. ... (and) SSA should perhaps be considered just as land scarce as India" (p.8).

Even if the contrast in relative land access between Africa and Asia is significantly less pronounced today than some authors would argue, the initial conditions that prevailed in the past--say at the outset of the post colonial era and before the greater population growth trends in Africa than in Asia that reduced the differential in the land/man ratio over time--would appear to be consistent with greater land scarcity in Asia as a stylized fact.

### Quantity and Quality of Infrastructure

There is a great scarcity of physical infrastructure in SSA--particularly road networks within rural areas (as well as farm to market roads), and between rural and urban areas. There is also tremendous underinvestment in irrigation projects (only 4.6% of SSA agricultural land is irrigated compared to 38.4% in Asia:Khan, 1997,

Table 5). The quantity and quality of the road network play a crucial role in facilitating trade at all levels (intraregional, interregional, and international). This network is tremendously underdeveloped in SSA and is a major cause of i) the very high transportation costs that prevail; ii) the high price spreads between initial agricultural producer prices and ultimate consumer prices; iii) segmented agricultural product markets; and iv) very limited market-orientation on the part of the small African farmers who produce largely for subsistence with low marketable surpluses.

In short, the above interrelated factors--together with technological constraints and discriminatory policies against agriculture (which are discussed subsequently)--go a long way in explaining the essentially stagnant agricultural production picture in SSA over the last three decades or so.[8]

A comparison of Africa's road network to that of Asia is especially illuminating in this regard. Very large interregional and intercountry differences in the extent of transport infrastructure can be observed.[9] Ahmed and Rustagi (1984) documented the underdeveloped stage of road infrastructure in Africa: Africa possessed only between 0.01 to 0.11 kilometers of road per square kilometer of land area compared to 0.30 to 0.45 kilometers of road per square kilometer of land area in Asian countries.[10] Furthermore, as of the early 1980's, only about ten percent of the road network in African countries consisted of paved roads, compared to about 35% of the road network in Asia. The relatively poor state of physical infrastructure in much of SSA compared to Asia is directly related to another characteristic of the physical environment, mainly a much lower population density in the former. Asian countries are likewise significantly better off in terms of railways and river transport networks. Because of greater reliance on trucks and railways in Africa, the import content of transportation marketing costs in Kenya and Tanzania, for example, is about 50%, compared to an estimated average import intensity of only 17% in Indonesia and Bangladesh (Ahmed and Rustagi, p.43). The absolute transport costs in marketing was also found to be twice as high in Africa as in selected Asian countries. To the above list, Platteau and Hayami (1996) add the low quality of the rural road network in Africa with about half the rural road network requiring "substantial rehabilitation".

## Production Technology and Form of Organization

In many parts of the developing world significant differences in agricultural *technology and form of organization* are observed that are highly correlated with farm size. The structure of food production and, to some extent, cash crops can range from strongly dualistic and bimodal, to unimodal in terms of type of technology adopted and form of enterprise.[11] In many developing countries, part of the agricultural production is supplied by the traditional sector and part by the modern sector. The traditional subsector consists typically of small farms cultivated by owners, tenants and sharecroppers, or communally-cultivated land tracks (such as the ejidos in Mexico) producing food for subsistence (self-consumption), while selling or disposing of the rest of their production on the commercial food market. The prevailing technology tends to be highly labor-intensive, relying on traditional methods and traditional varieties. The form of organization, in turn, is the family farm (as owner-cultivator) or, alternatively, a tenancy or sharecropping arrangement.

On the other hand, the modern subsector tends to be constituted by large holdings that are commercially managed, often as incorporated enterprises, using relatively capital-intensive techniques, chemical inputs, high yielding varieties and producing mainly for the market (often for exports). This bimodal structure, which is found in many Latin American (e.g. minifundia vs. latifundia) and African (subsistence farms vs. crown land) countries, is largely a heritage of the colonial period that resulted in a very uneven land distribution and institutions favoring the colonial settlers at the expense of the indigenous population.[12]

The structure of food production can also be unimodal. Countries such as Taiwan, South Korea, Japan (during the Meiji period), Sri Lanka and, to some extent, Indonesia, display a relatively very even distribution of land--with a more or less unimodal farm size. In these countries, the successful implementation of an initial land reform allowed the land to be more evenly distributed among the rural population and tended to convert tenants and landless households into owner-cultivators. In addition, these countries established institutions and followed policies designed to achieve the gradual development of traditional agriculture from "the bottom up" as it were.

**Agro-climatic Diversity and Technological Constraints**

It has been well documented that the "physical environment for agriculture (and cattle-rearing) in SSA is marked by an exceptional diversity of agro-climatic and soil characteristics, of farming systems and socioeconomic conditions" (Platteau & Hayami, 1996, p.19). This diversity is not only across SSA countries but also within countries and even regions. Another characteristic of SSA agricultural production is that it occurs almost completely on rain-fed land--less than one-twentieth of the total arable land is irrigated. Still another feature is the lack of congruence between the large number of locally produced foodstuffs (such as coarse grains) and preferred foodstuffs (such as wheat and rice) that have to be largely imported (Oyejide, 1996).

Given the above characteristics, SSA is at a great technological disadvantage compared to Asia. The Green Revolution technologies have been extremely successful in creating new and highly productive high yielding varieties of rice, wheat and maize grown on irrigated land but have had only very limited success coming up with improved new varieties applicable to rain-fed land and other crops. Thus, given the diversity of products grown in SSA on rain-fed land that is itself agronomically heterogeneous, a standard technical package comparable to single rice varieties (such as IR36) that worked so well in Asia has no chance to succeed within the context of SSA (Platteau & Hayami, 1996). Latin America would appear to be better off than Africa but worse off than Asia. The highly bimodal land endowment (large coastal farms on irrigated land producing export crops) vs. small subsistence rainfed farms producing domestic food crops along the Andes and in Northeast Brazil resulted in mainly the former benefiting from the new technologies.

What is perhaps surprising is that, notwithstanding the bleak picture described above, there is evidence that expenditures on agricultural research have had high returns in SSA. In a study of total factor productivity (TFP) in SSA agriculture--based on a data set of physical output aggregates (where different products are converted into wheat-equivalent units) and corrected for artificial price and exchange-rate effects, Block (1994) found that after fifteen years of stagnation, African agricultural TFP increased substantially during the mid-1980's, growing at about two percent per year from 1983 to 1988.[13] In turn, taking the real exchange rate depreciation as a proxy for policy reform (i.e. adjustment), his suggested finding is that policy reform and lagged research expenditures explain most of the improvement in agricultural TFP growth. One possible explanation for the very limited expenditures on agricultural research in SSA provided by Block (1994) is that cuts in domestic absorption following structural adjustment programs have come largely from public investment--a critical source of funding for agricultural research. This issue is discussed in Thorbecke and Koné (1995).

The condition described above also help explain 1) the "pitifully low level of fertilizer consumption in SSA", amounting to only 14% per hectare of the average consumption in low income LDCs in 1992-93 (Khan, 1997); and 2) the very limited scope of extension services provided in the light of the topographic and physical constraints.

**Land Tenure and Titling**

The typical land tenure pattern in SSA is collective land ownership at the village or tribe level. Village chiefs allocate land to individual members of the community who maintain their land use rights throughout their life times and often can pass it on to their descendants. This pattern is in direct contrast with the Asian model where small farmers-cultivators own their land.

There is a school of thought--perhaps best reflected by the World Bank--that subscribes to the so-called Evolutionary Theory of Land Rights (ETLR) as being applicable to SSA in largely the same way that it applies to other parts of the world. According to this thesis, growing population pressure and increasing commercialization of agriculture give rise, as it were endogenously, to changes in land tenure practices in the direction of enhanced individualization of tenure (Platteau, 1996, p.32). In turn, land titling and the security of tenure create the necessary incentives for small farmers in Africa to invest in their land through a variety of activities such as land leveling, terracing and other types of improvements that would increase yields and output. In other words, the absence of clear titling and property rights is seen as a major institutional constraint to the growth of agricultural production in SSA.

Platteau (1996) shows, in a very incisive piece, that the ETLR, based on the theory of induced innovation, does not lead to the expected institutional innovation in the form of land titling in the African context. Platteau (1996) demonstrates that in order to be valid that theory requires two crucial conditions to be fulfilled: first, new technical packages must be available so as to create attractive investment opportunities for people willing and able to invest, and second, efficiency and equity considerations must be separable. Since neither of these conditions hold in SSA, enhanced land titling will not evolve endogenously as an induced institutional innovation.

If, in the specific setting of SSA, land titling is not to evolve naturally, what about imposing it by fiat? Platteau (1996) provides strong arguments against the alteration of customary rights under the aegis of governments. In short, individual land titling and property rights within much of the context of SSA are not the panacea its supporters claim.

**Cultural and Community Norms**

The differences in cultural and community norms and customs between SSA and Asia and their impact on the divergent rural development paths followed by these two regions in recent times have been perceptively analyzed by Platteau and Hayami (1996). Their thesis can be summarized as follows: 1) cultural and social norms under land-abundant conditions in Africa compared with land-scarce Asia have constrained capital accumulation in Africa; 2) "the critical role of community as an economic organization is to guide its members to voluntary cooperation.... to insure the subsistence of all community members"; 3) in Asia, high population density has shaped community norms so as to prevent free riders from depleting scarce natural resources, in contrast with land-abundant Africa, where norms did not adapt quickly enough to prevent serious degradation; 4) rural communities in SSA are "typically tribal, lineage-based societies" relying on "production activities characterized by spatial mobility such as shifting cultivation and nomadic grazing" in contrast with "strongly immobile village communities based on settled agriculture in much of Asia." (p.24); 5) "since land commands relatively low value in SSA, private property rights on land have not become well established" resulting in little social stratification resulting from unequal land ownership; 6) as a consequence, tribal communities in SSA are characterized by strong egalitarianism; no causal link is seen between effort applied and resulting output and success is attributed to "luck" and hence it is expected that some type of balanced reciprocity norm should lead to redistribution from alleged "lucky" to "unlucky" individuals; 7) Asian rural communities are based on conjugal owner-cultivator farm households living together in villages and having to cooperate for their security and survival leading to the emergence of community norms based on cooperative and collective actions aimed at the conservation of the common-property resources; 8) under those circumstances reciprocity norms evolved in Asia and were consistent with the acceptance and full recognition of the link between effort and outcome (such as the yield-increasing effects of careful water control), in contrast with the redistributional norms in Africa that tend to deny the relationship between effort and outcome.

The bottom line of the Platteau-Hayami (1996) thesis is that the redistributive norms in SSA have growth-retarding effects in contrast with the growth-enhancing effects of reciprocity norms in the Asian rural sector.

# 4 THE POLICY ENVIRONMENT IN THE DEVELOPING WORLD PRIOR TO AND AT THE OUTSET OF LIBERALIZATION: THE STATUS-QUO-ANTE IN THE MID 1980'S

## 4.1 THE OVERALL TREATMENT OF AGRICULTURE: POLICIES AND THE AGRICULTURAL SURPLUS

The major mechanism to obtain the resources needed for industrialization at an early stage of development is through an intersectoral transfer out of agriculture. It is important to identify the major components of this transfer. A first component consists of the resources that tend to flow out of agriculture, automatically, through the market mechanism wherever the rate of return on resources is higher in agriculture than in nonagriculture (typically in the incipient industrial sector). Teranishi (1997) has called this flow a "market-based resource shift". In addition there are resource flows that are policy-induced through the direct intervention of the government. Therefore, it is useful to make a distinction as Teranishi (1997) does between 1) market based resource flow; and 2) policy based resource flow further broken down into a) net direct taxation; b) net indirect taxation; and c) infrastructure investment in agriculture.

Typically, developing countries tax their agricultural sector heavily through direct taxation (usually by turning the internal terms of trade against agriculture through such interventions as artificially low consumer prices for food and high input prices, e.g. the hidden rice tax through high fertilizer prices in Taiwan); and indirect taxation (mainly through the impact of an overvalued exchange rate on agricultural tradables).

In a careful empirical study of intersectoral resource flows, Teranishi (1997) showed that there was no significant difference in the (high) degree of direct and indirect taxation on agriculture among the four regions, East Asia, South Asia, Latin America, and SSA, but that the regional differences in infrastructure investment in agriculture were enormous. Teranishi (1997, p.289) concluded that

> "In East Asia, the adverse effects of indirect taxation (real exchange rate over-valuation and industrial protection) and direct taxation of agriculture were counterbalanced by government efforts in agricultural development, particularly in the area of infrastructure investment, resulting in the relatively low level of total policy-based resource shift from agriculture."

The explanation that is given for the radically different treatments of agriculture in Asia and in SSA and the consequent very disparate performances is that in SSA governments used "divisible benefits" in a very selective way to keep or win over agricultural actors who supported the incumbent political regimes regardless of their contribution to production. Furthermore, Teranishi (1997) provides an interesting political economy explanation of why small farmers in SSA do not react collectively against the effects of policies detrimental to agriculture, in contrast with farmers in East Asia. The answer lies in the shifting mode of cultivation of small African farmers that does not provide incentives to invest in land improvement, a situation made worse by the fact that most small farmers do not own their land in contrast with Asia. Given the very different production and tenure conditions in East Asia, incentives for small farmers to resist policies detrimental to agriculture are much larger in the former than in Africa.

In the 1980's, the Development Center of the Organization for Economic Cooperation and Development (OECD) embarked on a large-scale research project to evaluate the effects of policies and institutions on agricultural performance over time in six poor developing countries: Mali and Burkina Faso in West Africa; Kenya and Tanzania in East Africa, and Nepal and Sri Lanka in Asia. Six individual case studies following the same conceptual framework were undertaken in connection with this project.[14] For each of these countries, a careful attempt was made to measure the agricultural surplus (i.e. the net transfer out of agriculture) over time. The main lesson to be drawn from the experience of a large set of developing countries (including the above six) was summarized by Thorbecke and Morrisson (1989, p.1490):

> "The process of capturing the surplus is quite delicate. The goal should be to generate a reliable and continuous flow of net resources from agriculture into the rest of the economy throughout much of the structural transformation. A lesson learned from those countries which were most successful in achieving both growth and equity throughout their development history is that a continuing gross flow of resources should be provided to agriculture in the form of such elements as irrigation, inputs, research and credit, combined with appropriate institutions and price policies to increase this sector's productivity and potential capacity of contributing an even larger flow to the rest of the economy. It is much easier to extract a net surplus from increasing production than from stagnant or falling output."

One interesting finding of the comparative analysis is that in those countries in which foodstuff prices were most depressed as a result of the actions of the government, aggregate output either fell or stagnated. For example, in Tanzania, the sheer magnitude of the burden imposed on both the domestic food crop and cash crop export sectors in the 70's was shown to have short-circuited the development process and, more specifically, jeopardized the desired industrialization. The ridiculously low regulated food price in the official market led to a booming parallel market, where at one time prices were eleven times higher than the official food price. Likewise, in Nigeria (approximately half of SSA in terms of population) agriculture was seen as a sector to be squeezed and taxed with impunity to provide an agricultural surplus to finance the incipient industrial sector.

## 4.2 THE RELATIVE SIZE OF THE MARKETABLE SURPLUS

This is a critical issue if one wants to evaluate the impact of agricultural markets on equity and poverty alleviation. It is well known that, on the whole, the relative size of the marketable surplus (i.e. the proportion of farm household output sold out of total farm household production) is significantly higher in Asia than in SSA. As the World Bank (1997) emphasized, most farmers in SSA operate on a small scale "often producing commodities that, because of their type and small quantities, are not part of the market economy. In Côte d'Ivoire, Ghana and Malawi, the rural poor grow 60% of their food; in Tanzania the poor produce 50% of what they consume" (p.31). In other words, African small farmers tend to be much more subsistence-oriented than their Asian counterparts.

Why is the proportion of farm household output consumed within the farm household typically larger in SSA than in Asia and, conversely, why is the size of the relative marketable surplus smaller? To answer this question, we return to the exchange configuration approach elaborated above to analyze the behavior of peasant households in terms of their reliance on intra-household (nonmarket) transactions vs. market transactions. We ask how farm households decide on the extent to which they engage in intra-farm household transactions such as production for own consumption, and family farm labor applied to own farm production, as opposed to participating in transactions in existing market configurations for the same items.

De Janvry, Fafchamps and Sadoulet (1991) have provided a formal framework within which this question can be answered.[15] They start by offering an interpretation of market failure for food and labor that is specific to the household and not to the commodity. They proceed to derive within an integrated farm household model (acting as a producing and a consuming unit) the household response to changes in the price and productivity of cash crops, changes in the price of manufactured and consumption goods, the levying of a monetary tax, and availability of new technological opportunities in the production of food. They postulate that for commodities such as food and labor that can be sold and bought by peasant households, the sales price is a fraction of the purchase price. In turn, the width of this band depends on a whole set of transaction costs (such as transportation costs and marketing margins). "The poorer the infrastructure, the less competitive the marketing systems, the less information is available, and the more risky the transactions, the greater the size of this band" (de Janvry, Fafchamps & Sadoulet, 1991, p.1402). When the shadow price of a product, or of labor produced and used by a farm household, falls within this price band, no trade takes place and the household reverts to self-sufficiency (subsistence) and relies on intra-household transactions.

The key finding is that the chronic inelasticity of supply response--particularly within the context of SSA – may be explained "as a structural feature associated with missing markets and not as an inherent behavioral trait of peasants" (de Janvry, Fafchamps & Sadoulet, 1991, p.1410). This implies that a number of specific characteristics of the environmental element of existing exchange configurations, such as the previously discussed large price spread from farm gate to ultimate consumer reflecting high transportation and transaction costs and the scarcity of road infrastructure, operate as binding constraints on the behavior of actors within the farm household configuration. In turn, the more inelastic supply response in the African context, relative to Asia, can be attributed to the fact that most, if not all environmental and physical elements are less structurally rigid in the latter case.

The key policy question that flows from the above analysis is how to relax the structural constraints (to yield an upward shift and narrowing of the price band) so as to elicit greater market responsiveness on the part of peasant actors. de Janvry,

Fafchamps and Sadoulet (1991) mention a number of potentially desirable interventions, such as infrastructure investment, increased competitiveness among local merchants, better access for peasants to credit markets, technology transfer and a more elastic and low price supply of manufactured consumption goods such as textiles, footwear, processed foods, and some inputs. We shall return to these policy implications in the last section of this paper.

### 4.3 EXTENT OF MARKET INTEGRATION FOR AGRICULTURAL PRODUCTS

Market integration can take different forms, i.e. spatial integration, intertemporal integration, and intercommodity integration. In the present discussion, the emphasis is on spatial integration.[16] A market is spatially integrated when price differences between any two regions (or markets) that trade with each other just equal transfer (mainly transportation) costs. Alternatively, markets will be spatially segmented if the interregional price differences are less than their transfer costs. Integrated markets have been defined as "markets in which prices of differentiated products do not behave independently" (Monke & Petzel, 1984, p.482) – the assumption being that identical products are differentiated by location.

In evaluating the extent of market integration and efficiency along a commodity system (marketing chain) and interregionally, two types of price spread indicators suggest themselves. The price spread between the producer and consumer end of a commodity system represents the overall marketing margin. Its relative magnitude, as well as its decomposition among components, yield insights about the efficiency of the product market and the degree of integration among the various exchange configurations constituting the marketing chain. A second category of price spreads, i.e. spatial price spreads, reflects the differences in prices prevailing in various regional markets at a particular time.

These two types of price spreads in food grain markets were estimated for five African and four Asian countries, respectively, by Ahmed and Rustagi (1984).[17] Three major empirical findings emerge from an analysis of the data: 1) average producer prices expressed as a percentage of final consumer prices in the African countries ranged from 30% to 60%; whilst in Asia they ranged from 75% to 90%. Thus, African farmers received a significantly smaller proportion of final consumer prices of marketed food grains than did their Asian counterparts; 2) the regional price differences within each country were also substantially larger in Africa than in Asia; in some African countries the lowest price in one region was only one-fourth to one-third that of the highest price in another region. In contrast, the corresponding ratio in Asia ranged from 64% to 83%; and, 3) the absolute size of the regional price spread in Africa was significantly larger than the marketing margin, (i.e. the producer/consumer price spread). From the above quantitative analysis, Ahmed and Rustagi, (1984, p.109) concluded that as of the early 80's

> "Many markets may not be linked with one another in African countries because of high transport costs resulting from poor transport and communication infrastructure or government restrictions. In the Asian countries, the regional price spreads are quite close to the marketing margins, which indicates that the markets scattered over various regions are probably well integrated with one another".

The example of Zaire may be enlightening in this respect. Koné and Thorbecke, (1996, p.303) in a detailed study of sectoral investment priorities in Zaire found that

> "Owing to chronic transport and marketing problems, about 40% of total production is consumed by the farmers themselves, while urban markets are increasingly supplied by imports. Clearly, there is great potential for increased production in agriculture through exports and further increase in domestic demand once the major obstacles, both on the production and the distribution side, are removed."

However, the producer/consumer and interregional price spreads are not only determined by transportation and marketing costs, they are also influenced by government taxes, profit margins of parastatals and private traders and transaction costs. Ahmed and Rustagi (1984) concluded that almost two-thirds of the larger marketing costs in Africa compared to Asia are accounted for by transport and transaction costs. The latter reflect the greater degree of government intervention in grain marketing in Africa, through such measures as bans on the interregional movement of commodities by private traders and a variety of licensing schemes imposed on these same traders.

At this stage, we can summarize the main factors that have been identified as being responsible for Africa's relatively low levels of market infrastructure development and market integration compared to Asia, and the associated marketing inefficiencies and significantly greater price spreads before liberalization. (See Ahmed & Rustagi, 1984; and FAO, 1992, particularly p.226). These factors are 1) the much lower population density in most African countries (15 to 30 persons per square kilometer compared to 500-750 persons per square kilometer in Asia), resulting in a wider dispersion of production and consumption centers in Africa; 2) road, railway, and river transport systems are generally much less developed in Africa than in Asia, as some of the earlier statistics indicated; 3) transport modes in Africa are less diversified and more import intensive; 4) some African countries generate a small volume of marketable surplus in food grains because of the predominance of subsistence production, which reduces the scope for scale economies in transport and marketing; 5) a bimodal structure in agriculture is typical of many African countries which results in market dualism; 6) economies of scale in Asian marketing have enabled separate specialization in transport services and grain trade, whereas in Africa the more typical pattern is for truckers to combine transport services with wholesaling and retailing; 7) the more extensive spread of rural electrification in Asia allows more small scale milling and processing to occur close to the production location, with concomitant lower transportation costs.[18]

## 4.4 GOVERNMENT INTERVENTIONS AND THEIR IMPACT ON FOOD AND CASH CROP CONFIGURATIONS

This heading covers, in fact, four sets of distinct exchange configurations 1) domestic food (official); 2) domestic food (parallel); 3) cash crops (official); and 4) cash crops (parallel). The main reason for treating these exchange configurations under one heading is that they are or were often subjected to the same range of interventions--although the impact of these interventions, as will be seen, may be quite different. Furthermore, the strong interconnections that exist between the official and the parallel markets, for both food crops and cash crops, make it convenient to analyze these exchange configurations jointly.

Traditionally, marketing boards, or their equivalent, have been the principal form of government intervention in cash crop markets and, to a somewhat lesser extent, in staple food markets. Commodity boards fulfilled two main objectives 1) price stabilization; and 2) revenue raising and surplus extraction. However, in the great majority of the cases, the real objective of commodity boards was not price stabilization per se but the extraction of the agricultural surplus (Bates, 1981; Lecaillon, Morrisson, Schneider, & Thorbecke, 1987). There is much evidence that artificially low prices for cash crops and foodstuffs discouraged production which, combined with disappointing output performance, triggered the somewhat myopic emphasis on liberalizing markets and "getting the prices right". Certainly the latter became a necessary but hardly a sufficient condition for output to rise.[19] By now there exists a large body of empirical and econometric evidence showing that farmers' supply response to higher prices--although usually significant--tends to be relatively weak. It is stronger in those settings where complementary resources are present, such as roads, irrigation schemes, well functioning marketing channels and accessibility to and availability of credit, modern inputs, and manufactured consumption goods. Hence, generally speaking, the supply response has been found to be substantially more elastic in Southeast Asian settings than in sub-Saharan African countries (Bond, 1983; Lecaillon et al., 1987).

In particular, it is clear that public investment in agricultural infrastructure--when the other conditions above are met--can be very effective through its impact on productivity in attracting private investment (a phenomenon which has been referred to as "crowding-in"). In fact, the rice miracle in Indonesia between the mid-seventies and the mid-eighties, which converted the country from the largest rice importing country in the world to virtual self-sufficiency, is a prime example of the successful interaction of a whole set of complementary and mutually self-reinforcing policies and institutions to yield an almost phenomenal rise in output (paddy production increased at an annual growth rate of more than seven percent between 1979 and 1984).

Any number of sub-Saharan African countries could be taken as counter examples. For example, in the case of Senegal, the following evaluation of the relationship between the World Bank and the government of Senegal over the period 1965-85 concluded[20]

> "For the two key crops, ground nuts and millet, constituting 86% of the total area under cultivation, a completely static picture is apparent. A fundamental cause for this is the weakness of the research effort and the absence of technical packages which would have affected yields positively...It is difficult to understand the basis upon which the government and the Bank could expect a significant growth in aggregate agricultural output with neither of the above two conditions having been fulfilled. To compound this problem, our evaluation of the agricultural sector...brings out that the institutional weaknesses on the marketing and distribution sides act as an important brake

> to increased output so that even in the presence of real price incentives and the availability of viable technological packages, significant aggregate output response seemed quite unlikely." (World Bank, 1989, pp.40-41)

Two rather obvious lessons emerge from the above discussion. First, a comprehensive and joint package of policy measures addressing institutions and other initial conditions to relax the constraining effects of binding elements is a sine qua non to the successful performance (according to our three criteria: efficiency, equity and adaptability) of cash crops as well as of staple food markets. Pushing the price button in a setting where one or more complementary measures or conditions are absent is fruitless. This is a lesson which had previously been learned in the context of the diffusion of the Green Revolution technology but, strangely enough, appears to have sometimes been overlooked in the context of the formulation of structural adjustment policies applied to the agricultural sector. Thus, a devaluation, the major instrument under a stabilization and structural adjustment regime, raises the prices of tradeables (i.e. cash crops and many foodstuffs, among others) vs. non-tradeables. However, it may only have a negligible impact on output when other key elements are missing. Conversely, a devaluation may result in highly differentiated effects under different sets of initial conditions obtaining in the cash crop relative to the domestic food configurations. Furthermore, by definition, farmers producing non-tradeable foodstuffs will not be helped directly by any devaluation.

A second lesson is that certain types of infrastructure projects, such as rural roads, irrigation schemes, and research are examples of activities that have strong public goods aspects--with high potential payoffs--falling largely within the public domain. Therefore, these activities are highly unlikely to be undertaken on any significant scale by the private sector.

A natural consequence of government interventions that affect either the price or quantity of the item traded is to encourage the development of parallel markets. Official markets breed parallel markets. The greater the degree of intervention, the greater the scope for parallel markets. The differential between the parallel price and official price for staple foodstuffs can be enormous (in Tanzania the ratio between these two prices reached eleven fold in the seventies (Lecaillon et al., 1987)). The official and parallel market configurations are intrinsically linked. Nguyen and Whalley (1986) have shown that in a parallel market consumers will compete for the rationed good until transactions costs equal the difference between the official and the parallel prices (they labelled these costs "endogenous transactions costs").

These experiences further illustrate the need to understand the combinations of elements – the characteristics of items, actors and environment – that make up exchange configurations that are targeted for government intervention if the benefits of such intervention are to be maximized. The linkages among the different elements of an exchange configuration, as well as the linkages and spillover effects between different exchange configurations, will profoundly affect the outcome of market interventions.

## 5 IMPACT OF LIBERALIZATION ON AGRICULTURAL PRODUCTS MARKETS IN THE 1990'S

In what follows, an attempt is made to evaluate the impact of liberalization on the functioning of agricultural product markets in the Third World and to discuss how changes in the latter affected, in turn, market performance criteria such as efficiency, equity and food security. The evaluation is not intended to be comprehensive. Rather the emphasis and magnifying glass scrutinize SSA for two reasons 1) it is by far the most vulnerable region in terms of stagnation, if not retrogression, in food production per capita and hence potentially the most food insecure region; the magnitude of poverty appears furthermore to be rising; and limited growth of export crops and all types of structural obstacles to agricultural growth are endemic--as has been amply demonstrated in previous sections of this paper; and 2) Africa in the 1990's appears to have been more systematically studied in this context than have other developing regions.

A first robust finding is that liberalization in SSA has been largely of the "minimalist" "getting the prices right" type. This comes out clearly in a detailed assessment of the liberalization of food marketing in SSA by Seppälä (1997), as the following quote indicates:

> "Structural adjustment has addressed the problems of agricultural pricing in many ways. Changes have been geared to getting prices "right" through flexible exchange rates, competitive liberalized crop marketing, reduction of tariff and non-tariff barriers, and reduction of agricultural taxation and subsidies. At the same time, fiscally constrained governments have been ill-equipped to give support for agricultural production in terms of infrastructure maintenance, market stabilization and farmer services (input supply, extension, etc.)." (p.4)

Likewise, Sahn et al. (1997) concluded that

> "While the evidence indicates that adjustment programs during the 1980's failed to spur rapid growth, it is premature to declare Africa's experience with responsible macroeconomic management and market liberalization a failure... besides getting markets to function and macroeconomic policies in order, the role of the state in making complementary investments in human resources and infrastructure also merits closer attention." (p.254)

Finally, Barrett (1997), in a number of studies of Madagascar, finds strong empirical evidence for Lipton's (1990, 1991) "Market-relaxation-state compression hypothesis". The example Barrett (1997, p.156) gives is that dismantling state trading monopolies may foster competition and spatial market integration, but decreased government revenues from those monopolies may reduce government investment in rural infrastructure.

If liberalization consisted largely of getting the "prices right", perhaps the first question to ask relates to an important market efficiency criterion, namely, how effective was the transmission of price signals along the whole marketing chain, as well as spatially, following key adjustment measures such as a currency devaluation.

What impact did it have on producer prices and on price variability? Did these measures reduce market segmentation?

Although the evidence is somewhat sketchy and therefore one ought to be careful not to overgeneralize, the following findings are suggested. First, while one would expect the impact of a devaluation and most trade liberalization measures to increase the producer price of agricultural tradables (i.e. export crops and food crops such as wheat and rice) Sahn et al. (1996) concluded that although macroeconomic reforms affected real producer prices positively in the early 1990's, these prices did not rise proportionately to the observed depreciation. While a devaluation removes the implicit taxation, it does not remove the marketing costs of the commodity plus explicit taxation via export duties. Sahn et al. (1996) found that explicit tax rates remain high "indicating there is ample room for policy change to improve incentives to farmers" (p.739).

This suggests that the discrimination against agriculture and the traditional policy of squeezing a significant agricultural surplus from agriculture that characterized the pre-reform era may have been reduced but was certainly not eliminated in the recent past. In a comparative study of many SSA countries, Seppälä (1997) concludes that there is no clear line to be drawn between the change in food marketing policies and resulting food prices (p.47). In some cases, liberalization is accompanied by a decline in real prices, while in others a modest increase has taken place. There are many confounding variables such as whether the item in question is importable and the evolution of world prices for the affected commodities.

Thus, in this context, in several countries in Latin America, agricultural price and trade reform is taking place against a backdrop of significant declines in the profitability experienced by agriculture (Valdés, 1996). In an analysis of the evolution of producer prices between 1986 and 1995, Valdés (1997) found that in seven out of the eight Latin American countries studied, all major agricultural producer prices had declined in real terms. Unfortunately for this set of countries, the initiation of trade reforms leading to the opening up of the economy coincided unexpectedly with the fall in border prices of most agricultural commodities and an appreciation of the exchange rate. In turn, lower domestic farm prices led to strong pressure in favor of a policy reversal towards more protection for agriculture (Valdés, 1997, p.2). The same study found that four out of the eight countries taxed agriculture (i.e. Argentina, Ecuador, Paraguay and Uruguay), three protected agriculture (Chile, Colombia and the Dominican Republic) and Brazil was a mild protector. Net income transfers out of agriculture during the pre-reform period of 1985-1990 were enormous for some countries, amounting to between 12% to 20% of agricultural GDP in the former countries. After the reform, the agricultural surplus siphoned out of agriculture remained extremely high in Ecuador and Uruguay but fell significantly in Argentina and the Dominican Republic. Again the lesson that can be drawn from the above study is that relatively similar reform packages can be influenced by countervailing exogenous variables and can lead to significantly different outcomes in different settings.

Returning to the issue of the impact of liberalization on market integration, there is additional evidence at the country level relating to its impact on real producer prices and price variability. Barrett (1997) uses an Arch model, where the observed price for a given region and month is a linear function of, among other variables, the real exchange rate and contemporaneous border price parity. The estimation covered five different foodstuffs in Madagascar. Essentially what he found was that both the mean of the price and its variance increased for all five commodities in the long term. In the light of increasing transaction costs following liberalization in Madagascar, Barrett interpreted these results as being consistent with increasing segmentation between some markets. In a parallel study of rice market integration in Madagascar, he had previously identified pockets of improved market interlinkages and another set of markets that became more segmented following market-oriented reforms (Barrett, 1995). He concludes that where infrastructure is poor and distances are great, as in Madagascar (conditions that are typical of many African countries), a number of agricultural commodities are non-tradable across substantial parts of the country, and thus that if "the food price distributions generated by liberalized markets are to be stimulative, one must confront the micro-economic and structural tasks of improving marketing institutions, infrastructure, production and processing technologies..." (p.170).

Another country case study by Shively (1996) of Ghana, addressing the same issues and using a similar Arch methodology, found that after an initial increase in maize price variability following adjustment, variability subsequently fell. Interestingly, Shively hypothesizes that this (unexpected) reduction in price volatility might have been influenced by government spending that atypically increased after the adjustment program began and included investments in rehabilitating roads and physical infrastructure--emphasizing again the absolutely crucial role of complementary measures for a limited "pricist" minimalist liberalization to succeed.

It has been argued that high price volatility hurts particularly the poor (both the net producers and net consumers) who have very few options in terms of seasonal storage or stocking of food. To the extent that price volatility rose following liberalization--and the evidence is somewhat mixed--it is likely to have had negative equity affects. Another measure that is part and parcel of liberalization is the elimination of pan-territorial prices. Seppälä (1997) has argued that the elimination of guaranteed pan-territorial prices is likely to increase the differences among regions within a country and affect the most vulnerable households who typically are those living in peripheral areas producing simple staple foods. This measure may tend to further marginalize areas that are already marginally endowed.

Even Sahn et al. (1997), who concluded on the basis of a long-term and comprehensive research program of SSA that the most important adjustment policies have had small beneficial effects for the majority of poor households, qualify this finding. The following quote is revealing:

> "We are not necessarily asserting that adjustment is good... Our evidence that adjustment policies do not have adverse distributional outcomes highlights the need to address an arguably more pertinent question--why adjustment policies have not succeeded in fully achieving their macro-economic and growth objectives... Given the failure of the state in propelling growth, more rapid growth in Africa depends on productive private sector investment... (There is)

> need to understand the role of the state as both the provider of public goods and an arbiter of incentives that will attract domestic and foreign investors... Better roads, port facilities, communications, and investment in human resources by improving access to higher education and health services will surely figure in such an analysis." (pp.253-254).

In short, Sahn et al. (1996) make a strong case that appropriate complementary policies to adjustment per se that improve agricultural technology and extension, infrastructure and financial markets would have a major impact on poverty alleviation in Africa.

de Janvry et al. (1997) echo much the same message within the very different context of the reform of the ejido system in Mexico initiated in 1991. The reforms ranged from implementing a package of macro-economic reforms--including trade liberalization and descaling of the role of the state in the economy--to sectoral policies redefining the role of public institutions in agriculture and eliminating subsidies to the peasants. As of the mid 1990's, the authors observed an institutional vacuum created by the decline in the state's role in agriculture, with symptoms such as a general scarcity of credit, the virtual disappearance of insurance and technical assistance and a strong fall in the use of fertilizers, hybrids seeds and chemical inputs. In order to succeed, the second agrarian reform in Mexico would require, according to de Janvry et al. (1997), in addition to a favorable macro-economic environment, institutional reconstruction, promotion of organizations and public investment in irrigation and education, as well as a comprehensive program of rural development in support of the land reform beneficiaries.

A similar type of vacuum was reported following the government's withdrawal from the fertilizer market and the elimination of fertilizer subsidies in Nigeria and Malawi. Private traders proved very reluctant to enter the fertilizer market (Gibbon et al. 1993). Clearly, the government has an essential role to create or enhance an enabling environment for traders with such measures as improving the physical infrastructure, access to commercial credit and access to market information.

Ultimately the most critical criterion of liberalization success in all of its dimensions (efficiency, equity and food security) is its impact on food production. Seppälä (1997) ran a number of tests based on regression analysis and multi-factorial analysis of variance to explore whether the extent of liberalization affected per capita food production. He claims that the results clearly show that per capita production growth is not affected by the liberalization score in SSA (p.44).

Table 2.1 shows the annual growth rates of food production by major developing regions for two periods, 1980-89 and 1989-95. If, somewhat arbitrarily, the first period is taken as representative of the pre-liberalization regime and the second period of the post reform period, Table 2.1 suggests that the growth of food production during liberalization worsened in SSA and South Asia from 2.57 percent annually in 1980-89 to 2.00 percent in 1989-95 and from 4.11 percent to 2.71 percent, respectively. The growth of food production remained the same in Latin America (about 2.4 percent in both periods), while it shot up in East Asia from 4 percent annually to a most remarkable 5.58 percent – largely reflecting the tremendous performance of China whose food production annual growth jumped from an already remarkable 4.24% a year in 1980-89 to an amazing 6.4% in 1989-95.

*Table 2.1 Annual Growth Rates of Food Production by Region, 1980-89 and 1989-95*

| Region | Annual Growth Rate (%) | |
|---|---|---|
| | 1980-89 | 1989-95 |
| East Asia & Pacific | 4.04 | 5.58 |
| -- of which China | 4.24 | 6.41 |
| Latin America & Caribbean | 2.42 | 2.40 |
| Middle East & North Africa | 3.68 | 3.81 |
| South Asia | 4.11 | 2.72 |
| Sub-Saharan Africa | 2.57 | 2.00 |

*Source: Calculated on basis of World Bank, World Development Indicators, 1997*

It is particularly relevant to ask in the context of this paper why China appeared to perform so much better than other developing countries. The liberalization process started earlier in China than it did in other developing countries. In a first wave of reforms, policy makers raised agricultural procurement prices by 20% in 1979, significantly reduced procurement quotas and formally reopened rural markets. Under the stimulation of these policy measures and the increased availability of advanced technology (such as hybrid rice) and manufactured inputs, production of grain literally took off (Carter & Rozelle, 1997). Under the 1979 reform, the commune system was gradually replaced by the production responsibility system, which started a process of decentralization of economic decision making in China's rural economy.

This was followed by two subsequent waves of reforms affecting the whole length of the marketing chain, from rural producers to urban consumers. Rozelle et al. (1996) found that the marketing and price reforms in the early 1990's led to a striking increase in the integration of markets. Some more recent and formal tests of market integration provided additional support for these results; they showed that rice and maize markets became increasingly integrated after the last wave of liberalization policies, and that this integration continued through the present (Carter & Rozelle, 1997).

Probably the main factor accounting for the ability of liberalization programs to improve the performance of agricultural markets in China, in contrast with most other developing regions, is that China went way beyond a minimalist "getting the prices right" liberalization to a very broad and complete set of liberalization measures and major institutional changes (Huang, 1998). This comes out clearly in the following quote by two experts on Chinese agriculture:

> "Market activity also requires infrastructure, including transportation and communication networks, and the development of human capital by traders with the know-how to act as intermediaries between producer and consumer.

One important aspect of China's agricultural reform success has been the successful establishment of market networks for agricultural products." (Carter & Rozelle, 1997, p.148).

## 6 SUMMARY AND CONCLUSIONS

The main argument of this paper is that it makes very little sense to discuss in general terms the impact of liberalization on agricultural markets. There is a myriad of different agricultural markets world-wide, characterized by their own specific settings. Hence the first task is to come up with an operationally useful definition of an agricultural market. The concept of exchange configuration appears to provide this missing link. Clearly, the functioning of a specific market depends on the characteristics of the item being exchanged, the characteristics of the actors involved and the characteristics of the environment (in its various dimensions such as physical, technological, socioeconomic, cultural and policy). Specific characteristics of these three elements (item, actors and environment) provide the building blocks of distinct exchange configurations. In turn, each distinct exchange configuration gives rise — through the endogenous interaction of actors and their decisions — to specific transactions.

Likewise, liberalization is a catch all word that can be used to describe a broad range of different packages of policy measures and institutions. In its minimalist sense, it is tantamount to a limited set of measures (such as a devaluation and elimination of subsidies) designed for "getting the prices right". In its broader sense, it can be interpreted as a comprehensive set of mutually reinforcing measures and institutions more closely akin to structural adjustment including, among others, public investment in physical infrastructure and research. If one is to explore the impact of liberalization on agricultural markets, both of these concepts need to be specified more concretely and operationally. Hence in this paper we examined the impact of relatively specific liberalization packages on relatively specific exchange configurations.

Among the myriad of exchange configurations that can be identified, we chose to focus somewhat on the farm household configuration in different settings in the developing world. This was motivated by the belief that if the success of different types of liberalization packages is to be evaluated on the basis of the criteria of equity and food security in addition to efficiency, the farm household in much of the Third World plays a crucial role as a prime mover at the beginning of the marketing chain.

One important implication of the methodological approach to agricultural markets taken in this paper is that the scope for generalizations and blanket policy recommendations to improve marketing functioning is limited by the context-specific nature of the particular characteristics of the elements shaping given exchange configurations. Changes in any elements—such as a liberalization regime implemented by governments—are likely to lead to differential effects, first, on the various exchange configurations within the same region or country; and, second, on the same general type of configuration across different regional or country settings. The response to liberalization depends both on the comprehensiveness of the liberalization package and on the initial conditions prevailing in the specific exchange configurations and settings under consideration.

The response of the agricultural sector to liberalization, particularly in Africa, has been disappointing. The supply response has been almost negligible. Food production per capita is still falling in many SSA countries. There is little evidence of improved market integration and the rural poor, while typically not hurt by liberalization and adjustment measures, may only have benefited marginally from these measures. Thus, market performance in terms of improved food security, efficiency and poverty alleviation leaves much to be desired.

The response of agricultural markets in SSA may be attributed, in part, to very unfavorable initial conditions such as 1) the relatively lower population density and greater spatial distribution of population in Africa compared to Asia, which represent major obstacles to the provision of an adequate rural infrastructure network, contributing, in turn, to high transportation costs, segmented agricultural product markets and very limited market orientation on the part of the small African farmers who produce largely for subsistence with low marketable surpluses; 2) no standard technological package—similar to the Green Revolution high yielding varieties for rice and wheat that have been so successful in Asia—was available and could succeed in an agronomically heterogeneous, essentially rainfed African setting. Compounding the negative effects of physical and technological factors on rural and agricultural development, it was seen that the governments in SSA almost universally followed policies and institutions that discriminated against agriculture. A large surplus was squeezed out of agriculture, and this contributed directly to agricultural stagnation. In contrast with Asia, the adverse effects of indirect taxation and direct taxation on agriculture in Africa were not counterbalanced by a reverse flow into agricultural development, particularly into rural infrastructure and irrigation.

However, an equally, if not more, important factor contributing to the poor performance of agricultural production in SSA in the post-reform era has been the very limited nature of liberalization, which largely took the form of a minimalist "getting the prices right" package. While it is clear that the various structural adjustment programs followed by African countries have substantially reduced the anti-agricultural bias in the heavy direct and indirect taxation of agriculture, contributing thereby to a rise in the prices of agricultural tradables, getting the prices right is at best only one blade of a pair of scissors. The essential lack of complementary measures — such as public investment in physical infrastructure (roads and irrigation), in research and in institution-building — meant that the second blade of the scissors needed to render price incentives effective, and thereby to increase the supply responsiveness of African farmers, was missing. There is much evidence that the growth of total factor productivity in SSA is highly correlated with public expenditures on agricultural research, and that the latter tended to be further curtailed during the adjustment process. Likewise, an inadequate transportation and distribution network can raise the marketing and other transaction costs so much that, even in the presence of attractive prices, farmers' incentives to increase output vanish and they revert largely to producing for their own subsistence.

The main counter example to Africa is China, which, through waves of reform starting about 20 years ago, was able to establish and implement a comprehensive set of complementary and mutually reinforcing policies and institutions that gave rise to more integrated markets and a sustained high growth of food output. The fact that the growth of food production in South Asia and Latin America in the 1990's was only slightly better than that of SSA—although they benefited (particularly South Asia) from more favorable initial conditions — can also be attributed to relatively partial liberalization programs. Here again, the real success stories, such as the rice intensification program in Indonesia between the mid-seventies and the mid-eighties that transformed the largest rice importing country in the world into a net exporter in less than a decade, suggest that, in most settings, only a comprehensive combination of complementary measures and institutions can yield a significant supply response.

The dilemma faced by adjusting countries is how to balance short-term cuts in public expenditures with a long-term need for improving the physical infrastructure and financing agricultural research. This is the dilemma that was referred to as the "market-relaxation-state compression hypothesis". Given the very limited public resource base that these countries can tap, it appears that one partial solution to this dilemma lies in a change in the composition of external funding. Specifically, increasing the share of agricultural Sector Adjustment Loans (SECALs) while reducing that of generalized program loans (SALs) in the World Bank lending portfolio and in the portfolios of bilateral donors suggests itself. The main advantage is that transfers imbedded in agricultural SECALs contribute directly to the building of physical infrastructure projects and the funding of an agricultural research network (i.e. a tangible productive investment) instead of taking the form of undifferentiated program and balance of payments support. When properly designed, agricultural sector loans need not reduce the conditionality leverage, but rather allow these requirements to be expressed in much more concrete and specific terms (Thorbecke, 1995).

## NOTES

[1] This paper relies heavily on Thorbecke (1992), Thorbecke (1993) and Ali and Thorbecke (1998).

[2] A commodity system approach encourages analysts to think in terms of the value added at each stage of the marketing process.

[3] For a thorough discussion of this issue and the anatomy of agricultural product markets in general, see Thorbecke (1992).

[4] For an extended discussion of these elements, see Chapter 2 of Thorbecke and Cornelisse (1991).

[5] The distinction between what is referred to market and non-market configurations comes out clearly in the following quote from Newbery, 1989,

> "Elementary economics argues that free exchange between informed individuals is mutually beneficial--there are gains from trade. Organized markets emerge to facilitate exchange, and it is a basic proposition that if agents are individually insignificant, the market structure is complete, and its operation is costless, then the resulting competitive equilibrium is one in

which there are no remaining unexploited gains from trade. The equilibrium will be Pareto-efficient, meaning that there is no alternative allocation in which some agents are better off and none worse off. An organized market is a particular type of institution for facilitating the exchange of goods and services via a medium of exchange (money), with the distinguishing features that the terms of exchange (prices) are public knowledge, and access is open to anyone wishing to trade. In reality, markets are not costless to organize and operate, and the potential gains from trade on a particular market may not be large enough to justify the emergence of an organized market. Even when the gains are sufficient, it may not be possible for the marketing agent to capture a sufficient fraction to cover his operating costs, and again the market will fail to emerge. In such cases, alternative institutional arrangements may be able to provide similar services at lower cost. A restaurant is a market for meals, but a family may be able to provide similar services with lower transactions costs."

[6] This issue is discussed in more detail in section 4.2.

[7] This section relies heavily on Ali and Thorbecke (1998).

[8] These factors have been systematically discussed in a paper by Thorbecke (1992) on "The Anatomy of Agricultural Product Markets and Transactions in Developing Countries" where comparisons were drawn between Africa and Asia.

[9] Much of the evidence comes from the excellent paper by Ahmed and Rustagi (1984). Both Thorbecke (1992) and Platteau and Hayami (1996) rely extensively on the Ahmed and Rustagi (1984) paper. The summarized evidence that follows is based on Thorbecke (1992) supplemented by more recent evidence unearthed by Platteau and Hayami (1996).

[10] Quoting from more recent sources, Platteau and Hayami (1996) mention that in the early 1990's, Africa (i.e. a group of 18 countries) had only one-sixth the rural roads density per square kilometer of land than India.

[11] See Johnston and Kilby (1975) and Pyatt and Thorbecke (1976) for a thorough discussion of unimodal vs. bimodal agricultural structures.

[12] Surprisingly, many new developing countries continue to follow an essentially bimodal agricultural development strategy after independence.

[13] TFP is defined as the difference in the growth rate of real product and the growth rate of real factor input. Therefore, a positive growth rate of TFP means that resources are used more efficiently in agriculture but not necessarily that total real output actually increased. However, it is fair to state that many studies have found very low or even negative TFP rates for SSA. Once again the earlier warning relating to the very low quality of African statistics and particularly agricultural output statistics should be borne in mind.

[14] For a synthesis and lessons of these studies, see Lecaillon, Morrisson, Schneider and Thorbecke (1987).

[15] The description which follows is based on Thorbecke (1993).

[16] This subsection is based on section 4.2 of Thorbecke (1992).

[17] Their results are summarized in their Table 2, p.3.4 and cover the following countries: Nigeria, Malawi, Tanzania, Kenya, Sudan, Indonesia, India,

Bangladesh, and the Philippines. Depending on the country, the following food grains were used: maize, rice, sorghum, and wheat.

[18] Though this analysis excluded Latin America, it is probable that the prevailing conditions fall somewhere between Asia and Africa. In much of Latin America--particularly the Andean countries--the topography is very hostile to moving commidities and transportation costs between the subsistence agriculture in the Andes and the Coast where most urban consumers reside tend to be very high.

[19] A World Bank and UNDP (1989) study of the adjustment process in Africa indicated that countries with strong reform programs improved price incentives, both for export crops and food crops, more than countries with weak reform programs, and had higher growth rates of agricultural production between 1985 and 1987. See also the discussion of this whole issue in Thompson (1991), pp. 27-28.

[20] The author of this paper was a major contributor to this evaluation.

**REFERENCES**

AHMED, R. & N. RUSTAGI (1984) *Agricultural Marketing and Price Incentives: A Comparative Study of African and Asian Countries*, International Food Policy Research Institute, May.

ALI, A.G.A. & E. THORBECKE (1998) *The State of Rural Poverty, Income Distribution and Rural Development in SubSaharan Africa*, Paper prepared for a conference organized by the African Economic Research Consortium on Comparative Development Experiences in Asia and Africa, Johannesburg, November 3-6, 1997.

BARRETT, C.B. (1995) 'Madagascar: An Empirical Test of the Market Relaxation-State Compression Hypothesis.' *Development Policy Review* 13(4):391-406.

BARRETT, C.B. (1997) 'Liberalization and Food Price Distributions: ARCH-M Evidence from Madagascar,' *Food Policy* 22(2): 155-173.

BATES, R. (1981) *Markets and States in Tropical Africa*, Berkeley, University of California Press.

BLOCK, S.A. (1994) 'A New View of Agricultural Productivity in SubSaharan Africa,' *American Journal of Agricultural Economics*, Vol. 3, No. 76, August.

CARTER, C.A. & S. ROZELLE (1997) 'How Far Along is China in Developing Its Food Markets?' In: *China's Economic Future: Challenges to US Policy* edited by the Joint Economic Committee Congress of the United states, M.E. Sharpe, Armonk, New York.

CROW, B. (1989) 'Plain Tales from the Rice Trade: Indications of Vertical Integration in Food Grain Markets in Bangladesh,' *The Journal of Peasant Studies.*

deJANVRY, A., M. FAFCHAMPS & E. SADOULET (1991) 'Peasant Household Behavior with Missing Markets: Some Paradoxes Explained,' *The Economic Journal*, Vol. 101, November.

deJANVRY, A. G. GORDILLO, AND E. SADOULET (1997) *Mexico's Second Agrarian Reform*, Center for U.S.-Mexican Studies, University of California, San Diego, La Jolla.

FOOD AND AGRICULTURE ORGANIZATION (FAO) (1992) *Agricultural Price and Marketing Policy: Government in the Market*, Policy Analysis Division, ESPT, Rome.

GIBBON, P. K.J. HAVNEVIK & K. HERMELE (1993) *A Blighted Harvest: The World Bank and African Agriculture in the 1980's*, Trenton, NJ, Africa World Press Inc.

HARRISS, B. (1982) 'Food Systems and Society: The System of Circulation of Rice in West Bengal,' *CRESSIDA Transactions*, Vol. 2, Nos. 1 and 2, pp.159-248.

HUANG, Y. (1998) *Agricultural Reform in China: Getting Institutions Right*, Cambridge University Press, Cambridge, UK.

JOHNSTON, B.F. & P. KILBY (1975) *Agriculture and Structural Transformation: Economic Strategies in Late Developing Countries*, London: Oxford University Press.

KHAN, A.R., *Reversing the Decline of Output and Productive Employment in Rural sub-Saharan Africa*, International Labour Office, Geneva, Issues in Development Discussion Paper 17, 1997. KONÉ, S. & E. THORBECKE, (1996) 'Sectoral Investment Priorities for Renewed Growth in Zaire,' Chapter 10 in: D.E. Sahn (Editor), *Economic Reform and the Poor in Africa*, Clarendon Press, Oxford.

LECAILLON, J., C. MORRISSON, H. SCHNEIDER & E. THORBECKE (1987) *Economic Policies and Agricultural Performance of Low-Income Countries*, OECD Development Center, Paris.

LIPTON, M.(1990) 'Requiem for Adjustment Lending?' *Development Policy Review* 8:437-443.

LIPTON, M. (1991) *Market Relaxation and Agricultural Development In States or Markets? Neoliberalism and the Development Policy Debate*, Eds. C. Colclough and J. Manor. Clarendon Press, Oxford.

MONKE, E. & T. PETZEL (1984) 'Market Integration: An Application to International Trade in Cotton.' *American Journal of Agricultural Economics*, Vol. 66.

NEWBERY, D.M. (1989) 'Agricultural Institutions for Insurance and Stabilization,' in: P. Bardhan, *The Economic Theory of Agrarian Institutions*, Clarendon Press, Oxford.

NGUYEN, T.T. & J. WHALLEY (1986) 'Equilibrium Under Price Control with Endogenous Transactions Costs.' *Journal of Economic Theory* 39 (August).

OYEJIDE, T.A., (1998) 'What can SubSaharan Africa Learn from the Taiwanese Development Experience?' chapter in Erik Thorbecke and Henry Wan, Editors, *Government and Market: The Relevance of Taiwan's Experience for Development Theory and Policy*, (in process).

PLATTEAU, J.-P. (1996) 'The Evolutionary Theory of Land Rights as Applied to SubSaharan Africa: A Critical Assessment,' *Development and Change*, Vol. 27.

PLATTEAU, J.-P. & Y. HAYAMI, (1996) 'Resource Endowments and Cultural Endowments in Agricultural Development: Africa vs. Asia,' Paper presented at a Round table conference on the Institutional Foundation of Economic Development in East Asia by the International Economic Association, Tokyo, December.

PYATT, G. & E. THORBECKE (1976) *Planning Techniques for a Better Future*, ILO, Geneva.

ROZELLE, S., A. PARK, J. HUANG, & H. JIN (1996) 'Dilemmas in Reforming State-Market Relations in China's Agricultural Sector,' Working Paper, Food Research Institute, Stanford University, California.

SAHN, D.E., P. DOROSH, & S. YOUNGER, (1996) 'Exchange Rate, Fiscal and Agricultural Policies in Africa: Does Adjustment Hurt the Poor?', *World Development* 24(4):719-747.

SAHN, D.E., P.A. DOROSH, & S.D. YOUNGER, (1997) *Structural Adjustment Reconsidered: Economic Policy and Poverty in Africa*, Cambridge University Press, Cambridge and New York.

SEPPÄLÄ, P. (1997) *Food Marketing Reconsidered: An Assessment of the Liberalization of Food Marketing in Sub-Saharan Africa*, UNU World Institute for Development Economics Research, Research for Action 34, Helsinki, Finland.

SHIVELY, G.E. (1996) 'Food Price Variability and Economic Reform: An Arch Approach for Ghana,' *American Journal of Agricultural Economics* 78(1):126-136.

SMITH, L.D. & A.N. THOMPSON, (1991) 'The Role of Public and Private Agents in the Food and Agricultural Sectors of Developing Countries,' FAO Economic and Social Development Paper 105, Rome.

TERANISHI, J. (1997) 'Sectoral Resource Transfer, Conflict, and Macro Stability in Economic Development: A Comparative Analysis,' Chapter 10 in N. Aoki, Y.-K. Kim, and N. Okuno-Fujiwara, *The Role of Government in East Asian Economic Development, Comparative Institutional Analysis*, Clarendon Press, Oxford.

THOMPSON, A.M. (1991) *Institutional Changes in Agricultural Product and Input Markets and Their Impact on Agricultural Performance*. FAO, Economic and Social Development paper No. 98, Rome.

THORBECKE, E. (1992) 'The Anatomy of Agricultural Product Markets and Transactions in Developing Countries,' Paper prepared for the Institute for Policy Reform, Washington, DC. The content of this paper is to become a chapter in a book written in collaboration with P.A. Cornelisse on *Markets and Transactions in Developing Countries* largely under the auspices of the Institute for Policy Reform. Chapters 1-3 of this volume appeared as IPR Working Papers in 1991.

THORBECKE, E. (1993) 'Impact of State and Civil Institutions on the Operation of Rural Market and Non-Market Configurations,' *World Development* 21(4):591-605.

THORBECKE, E. & P. CORNELISSE, (1991) 'Markets and Transactions in Developing Countries,' book being prepared for Institute for Policy Reform, Washington, DC.

THORBECKE, E. & S. KONÉ (1995) 'The Impact of Stabilization and Structural Adjustment Programs on Performance in SubSaharan Africa,' Chapter 1 in: J.-C. Berthélemy, *Wither African Economics?*, OECD Development Center, Paris.

THORBECKE, E. & C. MORRISSON, (1989) 'Institutions, Policies and Agricultural Performance: A Comparative Analysis,' *World Development*, Vol. 17, No. 9, September.

VALDÉS, A. (1996) *Surveillance of Agricultural Price and Trade Policy in Latin America during Major Policy Reforms*, World Bank Discussion Paper No. 349, World Bank, Washington, DC.

WORLD BANK (1989) *The World Bank and Senegal, 1960-87*, Operations and Evaluation Department, Report No. 8041, Washington, DC.

WORLD BANK (1997) *Development in Practice, Taking Action to Reduce Poverty in SubSaharan Africa*, Washington, DC.

WORLD BANK AND UNDP (1989)

# PART 2

## AGRICULTURAL MARKETS IN TRANSITION ECONOMIES

# 3 'WHITE GOLD' VERSUS 'FOOD SELF-SUFFICIENCY' IN FORMER SOVIET CENTRAL ASIA

Max Spoor

## 1 INTRODUCTION

The transition from a command to a market economy which has been taking place in the former Soviet Union (FSU) since 1991, with former Union republics having become independent nation states, raises fundamental questions regarding the development of their important agricultural sectors and sustainable resource use. During Soviet times, a strict 'division of labour' for commodity production and specialisation ignored transport and environmental costs, and used a variety of transfers, such as subsidised inputs and low official procurement prices [as compared with border prices]. In several Central Asian republics of the FSU (Uzbekistan, Turkmenistan, Tajikistan and southern Kazakhstan), this forced specialisation was translated into a near-monoculture of cotton, known as 'white gold', which led to overall pressure on the natural resource base, in particular on land and water.

Within the current context of slowly developing regional and domestic markets, recent changes in agricultural resource use in two of the Central Asian states (CAS) are discussed below. This chapter shows that the former specialisation in cotton is being replaced, or at least complemented by a new 'food self-sufficiency' strategy. Interestingly enough this is most visible in the countries that have undergone the least reforms, Turkmenistan and Uzbekistan. The economic and environmental impact of this strategy is analysed here, using sectoral and micro-economic data. Sustainable resource use, focusing on cotton versus grain, within a context of transition and missing markets (in which states have kept a substantial stake in the marketing of these two commodities), is therefore the main issue in this analysis. At the same time we take into account the socio-political reality of these two specific transitional economies. The chapter will discuss the impact of the current strategy, by confronting the costs [looking at aggregate and micro-economic data] with a rationality analysis within the current stage of transition.

The chapter will depart from the proposition that both countries still have a comparative and competitive advantage to producing cotton [in the case that both domestic and regional agricultural markets will actually work]. This can already be seen from various estimates that have been made of gross margins per hectare [and per unit of water] for this crop. Cotton remains the main foreign exchange earner for these economies. Forced moves towards food self-sufficiency that can be observed since [and even before] independence are nevertheless understandable from a socio-political perspective, and are not dissimilar to many other post-colonial situations. At a national level, governments tend to move towards 'economic nationalism', in particular when regional markets are fragmented, relations with neighbouring countries are tense and their insertion in world markets is still 'under construction'. At the farm level, the tendency to choose for grain - in spite of the fact that cotton can be more remunerable - is rational when markets are 'missing', and while the marketing is still heavily influenced by the state and its remaining parastatals. Furthermore, the reduction of the water consuming and environmentally damaging 'white gold' in favour of less water intensive grains [such as wheat, but not rice] should be welcomed in view of the desiccation of the Aral Sea.

However, a substantial expansion of areas under irrigation took place in both countries [but in particular in Turkmenistan] during the post-1991 period. This was mostly used for grain production, though it remained within the regionally established water allocation quota. This indicates that there has been a shift in water use, but no overall water savings have been caused by the change in crop-mix, as will be shown in this paper. The market environment in which this change took place was still heavily state-influenced, as in both countries reforms were only implemented gradually. More nominal changes [in particular in the field of privatisation] were observed than real ones (Spoor, 1998).

Finally, while not even reaching the targets set for food production, the import substitution effect has been largely outweighed by the loss in cotton-based export earnings, while -measured at the micro-economic level - grains have been produced in an inefficient and costly manner, with negative gross margins per unit of land and water in Turkmenistan and some areas of Uzbekistan.

This chapter proceeds in the following manner. The second part will provide the historical setting of the above-mentioned process of specialisation and dependency is analysed. This process grew following the inclusion of the region within the sphere of influence of the Tsarist empire, and later in the USSR, which meant that cotton became "king" in Turkmenistan and Uzbekistan. This chapter points out that during the post-1991 period the economic and political reforms, or the absence thereof, relate to the imposition of resource controls such as privatisation of land and liberalisation of agricultural markets, causing an environment of 'missing markets' in which farming units have to operate.

In the third part, a simple model is used to analyse efficiency and non-efficiency goals in order to understand the possible outcome of a shift towards national food self-sufficiency, taking into account water-related environmental costs. Although the model presupposes a 'market situation', the current changes can be partly analysed by making adjustments to the model. In the fourth section, this mostly 'state-led' move towards food self-sufficiency in Turkmenistan and Uzbekistan is further analysed. It has not provided the expected import substitution effects, as targeted output levels were not realised, in spite of optimistic official statistics.

Micro-economic data on the 'comparative advantage' of cotton versus grain are presented in the fifth part. Early estimates of a World Bank (1994) study on Uzbekistan supported the now observed shift towards grain [and particularly wheat], although these were based on much too optimistic assumptions of water savings. A more recent field-based study (WARMAP, 1997), forming part of a regional EU-sponsored project, calculated gross land & water factor margins that underline in large part the main argument of this paper: that full economic and environmental costs have to be accounted for, and a much more differentiated production and marketing strategy is warranted. However, it is argued that the governments in Turkmenistan and Uzbekistan need to relinquish their strong grip on agricultural markets, and improve the [still weak] position of farm units. Only then can the current tendency towards 'food self-sufficiency' be adjusted.

## 2 'WHITE GOLD' IN SOVIET CENTRAL ASIA

The states in the Central Asian region of the former Soviet Union (FSU) include the republics of what in Soviet times was referred to as 'Middle Asia': Kyrgyzstan, Tajikistan, Turkmenistan and Uzbekistan, and finally Kazakhstan. These largely rural societies formed a raw material producing periphery of the Tsarist empire and later the USSR, with a low level of industrialisation and relatively high levels of poverty and underdevelopment (Khan and Ghai, 1979).

Cotton was grown particularly in areas along the Syr Darya and Amu Darya rivers, already an important crop in the late 19th century since American cotton belt exports had been reduced after the civil war (Rumer, 1989; Spoor, 1993). It was to become a quasi-monoculture in the Soviet era.

The Central Asian oasis economies - in the midst of rather inhospitable deserts - were seen as having the potential to make the empire independent of imports of this crucial commodity, in the context of centre-periphery colonial relations. Subsequent Soviet rule would bring full inclusion of Central Asia within the sphere of Muscovite power, which meant that the region became part and parcel of the Soviet inter-republican division of labour, with a forced specialisation in the production of raw materials for the centre.

Traditionally, a certain balance has existed between crop production and animal husbandry in Central Asia, the former having a more or less sustainable crop-mix of cotton, grains, fodder crops, vegetables and fruits, and supported by underground water reservoirs and surface irrigation. However, cotton was to become exemplary of the process of forced specialisation within the Soviet system, with the Uzbek Soviet Socialist Republic (UzbSSR) as its main exponent. Prior to the First World War, cotton had covered about 440,000 hectares there, but by 1940 -after the process of collectivisation and the full introduction of a command economy had been completed - this doubled to more than 1 million hectares (GosKomPrognostat, 1991). The 'cotton at any price' policy directives from Moscow were steadily implemented by the local Party elite, who themselves connected their fate to cotton (Rumer, 1989; Carley, 1989; Khazanov, 1990; Spoor, 1993; 1995).

Over the following decades, irrigated cotton expanded to more than 2 million hectares, or nearly half of the country's irrigated acreage by the mid-1980's. Grain, on the other hand, shrunk in area and output in a manner that mirrored cotton's growth during this period. The World Bank (1995:5) estimates that between 1965 and 1986 the area sown with cotton increased by 1.2 percent per annum, and grain decreased by 1.3 percent, while there was a shift away from non-irrigated- towards irrigated production. By the end of this period, around 70 percent of the harvested grain was produced on irrigated land, resulting in increased tension between the two in terms of land and water resource use.

At the time that Gorbachov came to power, but particularly in the aftermath of the collapse of the Soviet Union, the tendency again reversed in favour of grain. This was due to a variety of causes. Firstly, although the trend had already begun during the perestroika reforms, regional (oblast) authorities were given much more authority over food and agricultural policies soon after the collapse of the USSR (Wehrheim, 1997). This led to a tendency to favour local-level food self-sufficiency, that can currently be observed in all of the CAS, particularly in Turkmenistan, Uzbekistan, and southern parts of Kazakhstan.

Secondly, the breakdown of the previously dominant state-order system, and the absence of new efficient food and agricultural markets in and between oblasts (regional administrative units), has strengthened this inward-looking trend. There is however still heavy state influence on the marketing of grain and cotton, considered to be strategic commodities. These 'missing markets' furthermore present a high risk of transportation problems, Mafia practices, rent-seeking behaviour of officials [in particular by police], with producers sometimes withdrawing into barter trade.

Thirdly, as the main players in the new Central Asian power game - Uzbekistan and Kazakhstan - struggle for regional hegemony, the establishment of a Central Asian market (known as the 'Common Economic Space') has not yet resulted in an agricultural development based on comparative and competitive advantage.

This would lead to more 'outward-looking' agricultural production and regionally oriented trade strategies. In most of the new independent states along the ancient silk road, important elements of the previous economic structure and its inherent power relations are still predominant even after the seven years of economic transition since the collapse of the FSU (Spoor, 1997). Contrary to the economic reforms in Kyrgyzstan, major changes in the market structure and its institutions have taken place in much of Uzbekistan, Turkmenistan and even Kazakhstan, yet farm and market restructuring is mostly absent or still rather superficial. Newly established enterprises (joint-stock companies, peasant associations, cooperatives, etc.) are still closely linked to remaining large-scale [state-dominated] trading and processing companies, maintaining their previous forms of operation, and having little influence on marketing and the operation of markets (Lerman et al., 1996; Spoor, 1997).

When it occurs, privatisation most often proceeds through the establishment of usufruct or leasing rights, while physically remaining within the perimeters of the former kolkhozy or sovkhozy, and still depending on the farm manager of the latter for inputs and sales (Lerman et al., 1996). This reflects in general the situation of Turkmenistan and Uzbekistan, which have been transformed the least in terms of political systems and economic structure. The same pattern has also been recently observed in southern Kazakhstan (Field notes of the author, Kazakhstan September 1997). In Turkmenistan, the area of these peasant (dekhkan) farms, as a percentage of total arable land, can practically be ignored in spite of official discourse on 'privatisation'. In Uzbekistan, official statistics provided by the Statistical Committee in Tashkent (GosKomPrognostat) revealed that this was not more than 2.6 percent in

1995, with marginal growth in 1996 towards 2.8 percent (Author's field notes, Uzbekistan September 1997).

There was even a reduction visible in the total landholdings of this incipient sector in some important production areas such as the Fergana valley, which indicates a stagnation of the farm restructuring process. However, the influence of household plot production is gradually increasing, partly because of some populist land distributions to households by the Turkmen and Uzbek governments in the early 1990's, that improved access to household plots and dacha gardens.

In spite of public discourse within the regimes, the slowness of transformation to adhere to economic reforms, privatisation and market liberalisation is also connected to the difficulty of splitting up former state and collective farms, as most agricultural production in Turkmenistan, Uzbekistan and southern Kazakhstan is under irrigation. The surface irrigation schemes are designed to sustain large-scale farms, and there is a genuine fear that the break-up of these large production units into small peasant farms will lead to a rapid deterioration of the existing irrigation structures (Lerman et al., 1996; Spoor, 1998). The development of water rights and water markets is still very much in an embryonic stage, though a beginning has been made with water pricing.

For these reasons, agricultural markets in Turkmenistan and Uzbekistan are still highly inefficient, monopolistic, segmented and in some cases even 'missing'. The previously dominant state order system has been largely eliminated, but with substantial monopoly power of the ruling elite still in force, newly emerging markets represent high transaction costs and substantial failures. Rent-seeking activities by market agents as well as officials often make it unprofitable to trade, both domestically as well as externally, while in the financial sphere, very few real institutional reforms have been implemented (See also De Janvry, Fafchamps and Sadoulet, 1991).

With a dominant - politically and ideologically motivated - strategy of 'economic independence', at both regional and national levels, impetus has been given to a shift away from the quasi-monoculture of cotton, towards a greater emphasis on grain production. While strongly present at the central government level (EIU, 1997a), one can now also observe that decentralised powers within the oblasts are being used to implement inward-looking regional development strategies. From the perspective of newly established farming units [whether large or small] choosing for a larger share of food in their production seems quite rational in the current 'missing market' environment.

In an interview in September 1997 with the Deputy Akim (oblast governor) in the southern Kazakh oblast of Djambul, close to the Uzbek border, this tendency was clearly indicated. Oblast food self-sufficiency levels were seen as crucial planning indicators. In the interview, it was stressed that 'they' [the officials] planned to increase the area of land under grain. In Uzbekistan and Turkmenistan the influence of these Akims is even bigger, and there are reports that political scores are settled around the issue of regional food production targets (Field notes of the author in Kazakhstan, September 1997). However, after President Nyasov of Turkmenistan had ordered a larger grain harvest for the 1996-97 season, he publicly sacked some regional governors for failing to deliver their planned targets. More recently, in March 1998, the President threatened to sue those governors who would not comply with the imposed food production targets.[1]

## 3 RESOURCE USE AND EFFICIENCY TRADE-OFFS IN THE AGRICULTURAL SECTOR

At the national level, changes in the crop-mix such as cotton versus grain, are traditionally analysed in economic theory for a situation where domestic markets are efficient, fluently linked to external ones, and where relative prices are crucial for agricultural producers to decide in advance on the planned output. This can be theoretically understood by using a simple diagram such as that given in Figure 3.1. This depicts a production possibility frontier for cotton and grain at a given technological level (curve AB). At point C this curve touches the consumption possibility curve (with slope -PC/PG, the ratio of border prices).

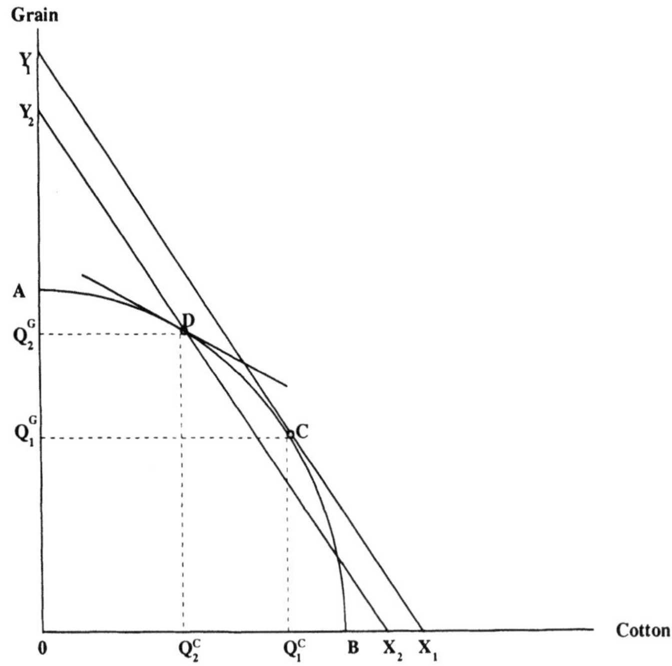


*Figure 3.1 Cotton-grain (non-) efficiency trade-off*
*Source: Adapted from Monke and Pearson (1989:8)*

If a government decides to influence the domestic relative price ratio of grain and cotton through subsidies and taxation, for example in order to promote a food security policy, a shift in crop-mix can be the consequence, leading to the production of less cotton and more grain. New quantities of grain and cotton are given at point D where the adjusted domestic price ratio is equal to the first derivative of the production possibility curve. The consumption possibility curve shifts to the left (inwards), while the purchasing power [at border prices] of the quantities Q2 [grain+cotton] is less than in the previous situation. The loss - in terms of social costs of the strategy to pursue these social 'non-efficiency goals' - can for example be measured by (Y1-Y2)*PG (Monke and Pearson, 1989:8-9).

Using the above tool to analyse the observed shift in crop-mix in Turkmenistan and Uzbekistan is however problematic for several reasons, and the model needs adaptation. The following issues have to be taken into account.

First, there are 'missing markets', as cotton and grain marketing is still largely within the control of the remnants of the former state order system. Imports and exports [of for example grain and cotton] are also still controlled by the states. Various forms of rent-seeking behaviour and substantial economic losses in production, harvesting and processing are typical for these markets.

Second, the environmental costs of producing these crops are not internalised, such as the cost of water [the scarce resource of the region]. This was already the case before the transition, and has hardly changed. Third, although the influence of border prices is now much larger than before [as food has to be imported at world market prices] and cotton has become a main hard currency earner, relative prices are still less significant than in the theoretical abstraction of Figure 3.1 in explaining the change.[2]

Cotton as well as grain is still heavily taxed in the two countries. Even if price ratios did change, the above observed shift in cropping pattern from cotton towards grain was not market induced, but state-led, as there were still sufficient policy instruments to force this upon producers.

Fourth, it remains to be seen whether in reality point C was the efficient one in terms of allocation of resources. Taking into account that grain [except rice] generally is less water intensive than cotton, inclusion of water related environmental costs will change the picture. The trade-offs between non-efficiency [food-self sufficiency] and efficiency [cotton export] goals, while including environmental [water related] costs, are shown in the adjusted Figure 3.2. Further changes have to be made in making the model more adapted to the 'missing markets' environment, but this is to be seen as a first step.

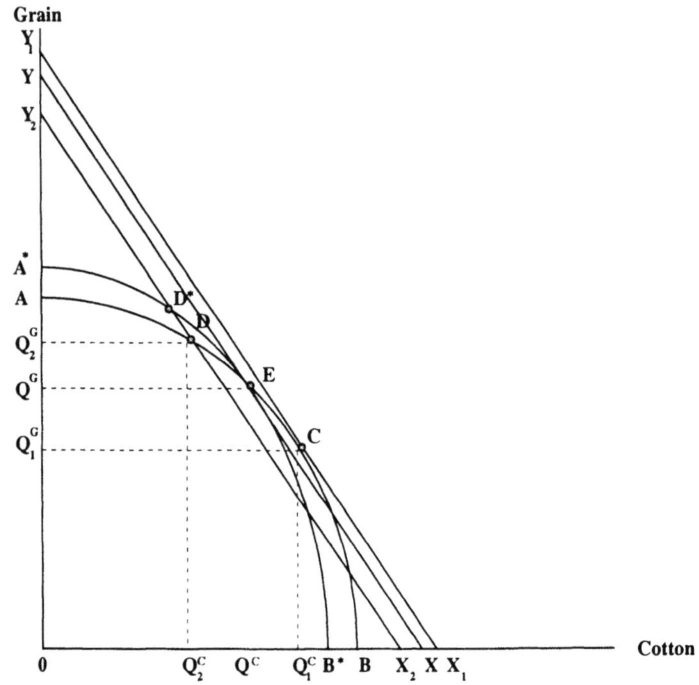


*Figure 3.2 Adjusted cotton-grain (non-) efficiency trade-off*

The adjustment in Figure 3.2 comes about by the clockwise 'pivoting' of the production possibility curve AB towards A*B*. This would result in point E actually being more efficient than C, while representing less 'social costs' than the adjusted D*. Contrary to the original curve AB the adjusted one (A*B*) represents fully incurred economic and environmental costs.

That would then identify an economically efficient and environmentally acceptable crop-mix of cotton and grain (at point E), with less cotton and more grain than point C would indicate. However, the forced move towards point D (now replaced by D*) overshoots its target, although the final real change will depend on the possibility of producing more cotton with less water (through increased yields using less land acreage) in a situation where water efficiency is extremely low (Lerman et al., 1996).

## 4 COTTON VERSUS FOOD SELF-SUFFICIENCY IN TURKMENISTAN AND UZBEKISTAN

Similar to other post-colonial independence situations, a shift is taking place in Turkmenistan and Uzbekistan, away from the overall dependency on 'white gold,' towards a national and local food self-sufficiency policy. Although the tendency to increase grain areas in Central Asia's main cotton producers, while reducing their acreage under cotton, was already visible before independence of the CAS, no dramatic shifts can be observed between 1986-1990. Table 3.1 (further illuminated by Figures 3.3 and 3.4) shows that in terms of acreage in Turkmenistan there was a decline in cotton production of only -1.1 percent per annum, with a somewhat larger percentage increase of 3.6 percent per annum in grain acreage.

*Table 3.1 Crop-mix grain versus cotton: Turkmenistan and Uzbekistan (1986-96)*

**Turkmenistan**

| | 1986 | 1987 | 1988 | 1989 | 1990 | 1991 | 1992 | 1993 | 1994 | 1995 | 1996 |
|---|---|---|---|---|---|---|---|---|---|---|---|
| (x 1000 ha) | | | | | | | | | | | |
| Grain | 165 | 189 | 198 | 179 | 187 | 240 | 330 | 435 | 428 | 649 | 661 |
| Wheat | 59 | 63 | 60 | 56 | 60 | 116 | 197 | 260 | 262 | 437 | 584 |
| Rice | 15 | 18 | 19 | 20 | 16 | 19 | 28 | 39 | 37 | 29 | 32 |
| Cotton | 650 | 633 | 636 | 634 | 623 | 602 | 567 | 579 | 559 | 560 | 537 |
| (x 1000t) | | | | | | | | | | | |
| Grain | 293 | 324 | 408 | 379 | 449 | 516 | 737 | 974 | 1130 | 1109 | 600[*] |
| Wheat | 89 | 93 | 92 | 86 | 130 | 206 | 377 | 509 | 675 | 878 | .. |
| Rice | 38 | 45 | 50 | 45 | 42 | 54 | 64 | 88 | 92 | 89 | .. |
| Cotton | 1138 | 1272 | 1341 | 1382 | 1457 | 1433 | 1300 | 1341 | 1283 | 1293 | 450[*] |

**Uzbekistan**

| | 1986 | 1987 | 1988 | 1989 | 1990 | 1991 | 1992 | 1993 | 1994 | 1995 | 1996 |
|---|---|---|---|---|---|---|---|---|---|---|---|
| (x 1000 ha) | | | | | | | | | | | |
| Grain[**] | 702 | 1005 | 1052 | 874 | 1008 | 1080 | 1212 | 1280 | 1522 | 1664 | 1690 |
| Wheat | 302 | 468 | 456 | 346 | 433 | 487 | 627 | 597 | 965 | 1165 | 1329 |
| Rice | 127 | 155 | 166 | 161 | 147 | 160 | 182 | 181 | 167 | 168 | 185 |
| Cotton | 2053 | 2103 | 2017 | 1967 | 1830 | 1720 | 1667 | 1695 | 1538 | 1491 | 1547 |
| (x 1000 tons) | | | | | | | | | | | |
| Grain[+] | 1186 | 1738 | 2083 | 1555 | 1899 | 1908 | 2257 | 2142 | 2467 | 3215 | 3562 |
| Wheat | 241 | 495 | 563 | 342 | 553 | 610 | 964 | 876 | 1362 | 2347 | 2742 |
| Rice | 399 | 506 | 581 | 484 | 503 | 515 | 539 | 545 | 498 | 328 | 450 |
| Cotton | 4989 | 4858 | 5365 | 5292 | 5058 | 4646 | 4128 | 4234 | 3938 | 3934 | 3350 |

[+] *Originally StatKom SNG (1997) estimated grain output for 1996 at 3.549 million tons. This was adjusted downwards in late 1997. However, EIU (1997a) puts the final figure at 2.786 million tons (well below the target of 4 million tons). Similarly, the wheat harvest was adjusted to 2.435 million tons (with EIU estimating 2.000 million tons). Smaller adjustments were made for rice (from 445 thousand tons) and cotton (from 3.450 million tons).*

[*] *These disastrous results are estimates from EIU (1997b), because of very adverse climatological conditions in Turkmenistan during 1996.*

[**] *There are discrepancies between the StatKom figures for grain and those published by EIU (1997a). Only in 1994 is there a substantially different estimate, the reason is left unexplained (see Table 3.2).*

*Source: Statistical Handbook for States of the Former USSR, 1993 (Washington, 1994); Statistichkiy Ezhegodnik SNG v 1996 godu (StatKom SNG, 1997)*

For Uzbekistan, the changes were greater as its grain acreage grew by 11.6 percent per annum in the same period, while cotton [after an initial rise] decreased by -2.8 percent per annum. Possibly this was also because of internal pressures to reduce food imports and become more independent from Moscow, while population growth was high and total food demand per annum strongly rose.

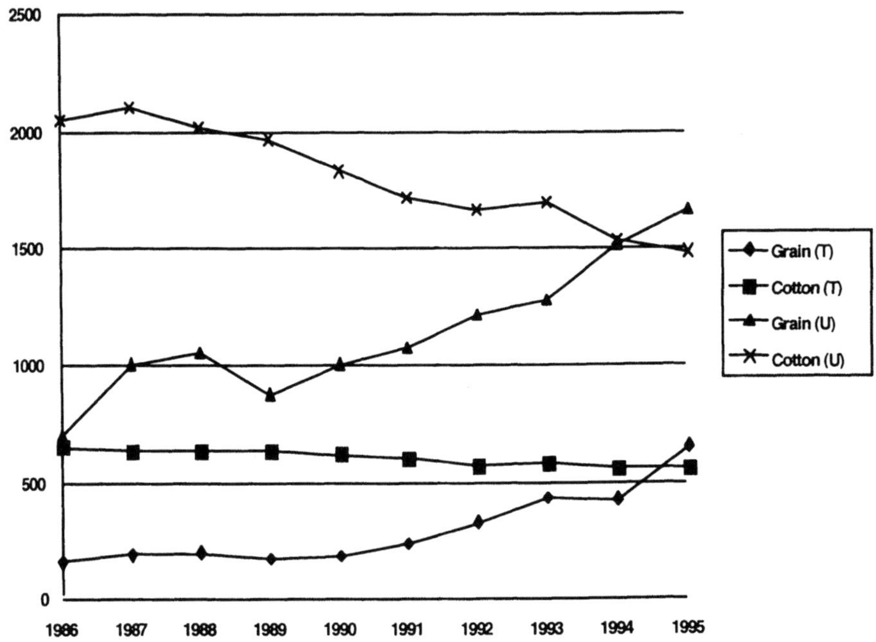


*Figure 3.3 Land use in Turkmenistan and Uzbekistan (1986-95): Grain versus Cotton (x 1,000 Hectares)*
*Source: See Table 3.1*

However, as there was also a shift towards producing more grain under irrigation, and therefore a strong increase in yields, the increase in grain production was substantially higher during the same period, namely 12.0 percent per annum in Turkmenistan and 15.8 percent per annum in Uzbekistan (with wheat output rising even faster).

In the first half of the 1990's, land-use shifted more strongly in the direction of grain, by gradually diminishing the cotton acreage and expanding the total cultivated area. After 1991, regional transport and marketing structures disintegrated as horizontal and vertical links in the former Soviet economy suddenly broke. World market prices soon had to be paid for food imports [while also receiving more hard currency for cotton]. However, in the regional context of fragmented markets and sometimes tense political relations, the emerging independent states combined their political independence with 'economic nationalism', choosing for the self-sufficiency variant of food security policies. At local levels this tendency was replicated, while newly established farm units behaved rationally in choosing to cultivate more food, in the context of 'missing agricultural markets', still heavily influenced by the state. Finally, it is unclear whether environmental concerns have played an important role in this trend, as in some cases [such as in the Aral deltas of the Syr and Amu Darya], cotton has been replaced by much more water-intensive paddy production.

Taking 1990 as the base year, being the last 'normal former' Soviet year, dramatic changes took place with regard to the crop-mix of cotton and grain (see Table 3.1; Figures 3.3 & 3.4). This is particularly evident within the scope of this paper's focus on cotton versus grain, and abstracting from other changes in other sub-sectors such as animal husbandry, fruits and vegetables.

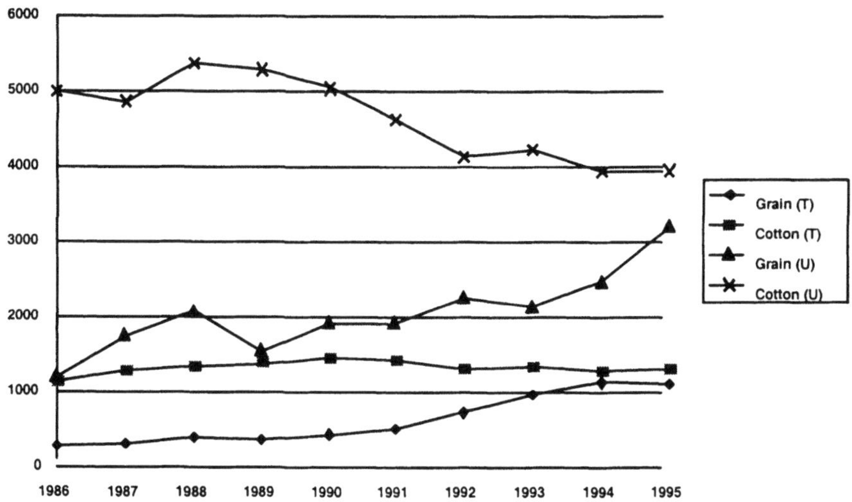


*Figure 3.4* *Crop output in Turkmenistan and Uzbekistan (1986-95). Grain versus Cotton (x 1,000 Tons)*
*Source: See Table 3.1*

Soon after independence, Turkmenistan would declare its non-inclusion in any form of economic co-operation with the other Central Asian States. It then pushed its grain acreage from 240,000 ha in 1990 to 649,000 ha in 1995, its land under wheat production rising to a level more than five times that of the last pre-independence year. Turkmen cotton acreage diminished at a rate of 2.0 percent per annum (Table 3.1).

This situation could only be realised through strongly expanding the areas under irrigation, within the existing limits of regional water quota allocation, but in a 'technically sub-optimal way' (World Bank, 1994:22). Land cultivated with grain expanded from just over 1 million hectares in Uzbekistan in 1990 to nearly 1.7 million ha in 1995, a yearly increase of 11.4 percent. During the same period, the cotton acreage dropped from more than 1.8 million ha to around 1.5 million, which is a -4.5 percent drop per annum.

The emphasis on wheat within the grain acreage actually shows the food self-sufficiency strategy of the Karimov regime in Uzbekistan, as wheat acreage more than doubled from 433,000 ha in 1990 to 1,165,000 ha in 1995 (or 24.0 percent per annum, with output growing at a rate of 41.5 percent according to official statistics). The change in Turkmenistan under President Nyasov was reported to be even greater, with a 25.4 percent per annum increase of grain acreage and 45.0 percent for wheat.

If we go back for a moment to the official data presented in Table 3.1, it is possible to estimate the benefits and costs of the strategy followed, by comparing the import substitution effect of producing more grain with the losses incurred by producing less cotton (FOREX and water savings). In order to do so, in Table 3.2 non-official [and more realistic] estimates are also presented.

*Table 3.2 Contested data on cotton & grain output (1991-1996)*

| (x1,000 tons) | *StatKom* 1991 | *StatKom* 1995 | *EIU* 1995 | *StatKom* 1996 | *EIU* 1996 |
|---|---|---|---|---|---|
| **Turkmenistan** | | | | | |
| Grain | 516 | 1109 | 850 | 600 | 400 |
| Wheat | 206 | 878 | *[673]* | *[475]* | *[317]* |
| Cotton | 1433 | 1293 | 1305 | 450 | 430 |
| (x1,000 tons) | *StatKom* 1991 | *StatKom* 1995 | *EIU* 1995 | *StatKom* 1996 | *EIU* 1996 |
| **Uzbekistan** | | | | | |
| Grain | 1908 | 3262 | 2000 | 3562 | 2743 |
| Wheat | 610 | 2347 | *[1439]* | 2742 | *[2112]* |
| Cotton | 4646 | 3934 | 3934 | 3350 | 3450 |

*Note: Data between square brackets are estimated by using given ratios*
*Source: EIU (1997a, 1997b); StatKom SNG (1997)*

The official data of StatKom SNG (Statistical Committee of CIS) substantially differ from those of the Economist Intelligence Unit during the years 1995-96 (see Table 3.2). If import and export parity prices were used for imported wheat and exported cotton, the data would point to a small FOREX-gain (74 million US$) for Turkmenistan between 1992-95, and a substantial loss for 1996 (450 million US$), when using the more realistic EIU data.[3] However, it must be noted that the year 1996 is exceptional, because the harvest failure was largely caused by extreme climatological circumstances of extended drought.

For Uzbekistan, this loss is estimated to be respectively 64 million US$ and 14 million US$. With a lower wheat import parity price (such as was used by the World Bank, 1994), the picture obviously becomes more negative than that stated above. For Uzbekistan the overall target was 4.5 million tons of grain. In spite of optimistic official data, it seems that even by 1996 [which was a relatively good year], only 2.7 million tons was reached, necessitating substantial grain imports. The EIU has calculated even greater losses caused by the import substitution strategy. The EIU (1997a) has become highly critical of the Uzbek leadership in the past years, because of the observed lack of reform policies and recent stagnation in relation to initial plans.[4]

This loss apparently excludes the environmental costs of producing more cotton than grain in the areas dedicated to the latter, which would certainly make this observation less dramatic. However, import substitution contributed to the rapidly emerging balance of payment crisis in late 1996, which postponed the announced convertibility of the Uzbek Som, and led to the reintroduction of budgetary finance, a spur in inflation, and the suspension of an IMF credit. The effects are more difficult to estimate in Turkmenistan, as the country has a buffer against external shocks with its natural gas exports. Nevertheless, it repeatedly failed to meet the 1.2 million ton target of grain output, and was struck by a disastrous 1996 harvest for grain and cotton (EIU, 1997b). However, as we will see in the next section, costs there have possibly been relatively even higher because gross margins per unit of land and water on the production of wheat [at economic prices] are mostly negative for Turkmenistan.

## 5 COMPARATIVE ADVANTAGE AND PRODUCTION SYSTEMS

It is important to also look at the micro-economic level to see whether it is efficient to produce grain or cotton at all, and where. Although a study of gross margins assumes that 'markets will work,' it remains interesting to study the economic returns of these [and other] crops, pricing inputs and output at border prices, in particular to form a future perspective of agricultural production in the region. There are, however, only few serious studies that can provide a micro-economic foundation for sectoral agricultural strategies to be followed in the CAS. The World Bank (1994), focusing on quantifying transfers to and from the agricultural system, concluded that the agricultural sector in Uzbekistan was being heavily taxed, and recommended the following:[5]

> as long as quotas determine levels of production, economic returns will not be maximised and farmers will not grow the crops with the highest economic return. State orders bypass any reference to market price signals as a guide to resource allocation. Farmers could receive much higher prices if they were allowed to sell their production freely. A change in crop pattern could increase production replacing grain imports (World Bank, 1994:52).

This policy recommendation was presented with aggregate estimates of economic costs of producing wheat in comparison with other main crops for Uzbekistan (divided into five regions).[6]

However, quite optimistically the estimates were based on the important assumption that water savings would be made in other crops 'providing new farm land with adequate water to produce the additional wheat'. In practice, as we have seen above, there was an increased trade-off between irrigated land for alternative use, while the assumed water savings were hardly realised, caused by deteriorating irrigation systems, lack of maintenance and new investments.

Gross margins are compared for various crops, including wheat, maize, rice and cotton in Table 3.3. The World Bank (1994) estimated that on average the gross margin [at economic prices] for cotton was US$ 657 per hectare, US$ 407 for wheat, and US$ 680 for rice. In terms of water - the most scarce factor - the outcomes were US$ 47 for cotton, US$ 64 for wheat, and US$ 58 for rice, per 1,000 $m^3$. The last result is rather surprising as rice produced in the delta areas of the Amu and Syr Darya rivers (that flow into the Aral Sea), needs about three times as much water as cotton.

*Table 3.3 Gross margins at economic prices per factor (land and water) Turkmenistan and Uzbekistan*

| (US$) (Per ha) | WARMAP (1997) | | World Bank (1994) |
|---|---|---|---|
| | Turkmenistan | Uzbekistan | Uzbekistan |
| Cotton | 395 | 508 | 657 |
| Wheat (winter) | -29 | 151 | 407 |
| Rice | .. | 127 | 680 |
| Maize | -176 | -49 | 283 |
| (Per 1000 $m^3$) | Turkmenistan | Uzbekistan | Uzbekistan |
| Cotton | 80 | 147 | 47 |
| Wheat (winter) | 7 | 64 | 64 |
| Rice | .. | 9 | 58 |
| Maize | -108 | -63 | 23 |

*Source: WARMAP (1997), WUFMAS Annual Report for 1996 Agricultural Year (Tashkent, EU Office). World Bank (1994: Annex 2), Uzbekistan: Economic Memorandum Subsidies and Transfers (World Bank, Washington)*

Finally, for irrigated maize, gross margins per ha and per unit of water used were still positive (23 US$ per 1000 m3), suggesting that double cropping with wheat and maize would be preferable to cotton in most regions of Uzbekistan. The state-led strategy towards 'food self-sufficiency' was therefore at least in the case of Uzbekistan backed by the cited World Bank study, which was undertaken in preparation for an agricultural sector analysis that was never completed (World Bank, 1995).

The Water Management and Agricultural Production Project of the EU recently produced a report of which the data is compared with the findings of the World Bank (1994) in Table 3.3. The farm survey, which was completed in June 1996, presents a more complex and at the same time less rosy picture of the benefits of the diversification process that has taken place. According to WARMAP (1997), wheat in Turkmenistan is generally grown with negative economic returns per unit of land, and only marginally positive for water. The gross margin for maize grain is substantially negative in both cases, as in Uzbekistan. In the latter country there are positive returns for wheat, while for rice the gross margin per unit of water again is only barely positive [taking into account the very intensive water demand of rice], while most likely negative for Turkmenistan. However, the small number of farms per country [6 in Turkmenistan, 16 in Uzbekistan] influence the data, and cautious interpretation of the results is warranted. For example, the average sample yield for wheat in Turkmenistan was 1.80 tons/ha (nationally in 1995: 2.01 tons/ha), which would mean a slight adjustment upwards of the gross margins, albeit still remaining negative. For Uzbekistan the average sample yield for wheat was 2.52 tons/ha (nationally in 1995: also 2.01 tons/ha), which would indicate a substantial overstatement of the estimated gross margins in this case (see Table 3.1).

Finally, the shift towards the production of more grain and less cotton would normally provide substantial water savings, a point which was also made in our theoretical presentation of the trade-off between efficiency and non-efficiency goals. However, the extensive growth in grain production - increasingly on irrigated lands - outweighs the volume of water savings per hectare. Following official estimates for the acreage sown with wheat and cotton during the period 1992-96, an accumulated increase in water use for grain & cotton production of 1.6 $km^3$ is shown for Turkmenistan and 1.2 $km^3$ for Uzbekistan. This was calculated by using the aggregated data on acreage in Table 3.1, using indicators of water intensity 6,000 $m^3$ water/hectare for grain and 13,500 $m^3$/hectare for cotton (World Bank, 1994; Lerman et al., 1996).[7]

If a comparison is made of wheat versus cotton, the increase is slightly higher as the growth of wheat acreage is faster than grain in general. Hence, this would not lead to an improvement, but to a stagnation if one relates these figures to the yearly water deficits in the regions that are responsible for the desiccation of the Aral Sea (World Bank, 1996; Spoor, 1998).

## 6 CONCLUSIONS

The two CAS that have been analysed here (Turkmenistan and Uzbekistan) have chosen to move away from the extreme dependency of cotton, and to take on an agricultural strategy with a perspective of food self-sufficiency. This is an understandable strategy from various points of view, although it can also prove to be a costly one. In a situation where regional markets are still fragmented, and tensions between countries make it difficult to move towards a regional common market, national governments tend to strive for 'economic independence'. The existence of 'missing (domestic) markets' [where states still maintain strong control over marketing of grain and cotton], means that whenever there is room for newly established agricultural production units, they are likely to choose for the 'safe' option of producing food, for which there is demand in a variety of market channels. The fact that producers [and even trading companies] can neither directly export cotton nor import grain, makes it rather difficult to pursue a strategy based on comparative [and competitive] advantage.

The food self-sufficiency strategy of Turkmenistan, under which the cultivated and irrigated area was expanded in order to grow wheat, is not based on comparative and competitive advantage. Actually, wheat and maize grain are stimulated through domestic protection representing the 'non-efficiency' goals of the Nyasov regime that wishes for political reasons to reduce its dependency on food imports. The costs of this strategy are to be financed by the [large] receipts of natural gas exports. It is however questionable how long and to what extent such a policy can be prolonged. The situation is more complex for Uzbekistan. Wheat expansion has in part continued at the cost of cotton, and there are certainly regions where wheat [and maize grain] can be produced at positive gross margins when measured at border price levels, while also contributing to sustainable resource use, in particular of water (World Bank, 1994; WARMAP, 1997). Nevertheless, the proceeds are often lower than those for cotton, even on a per unit of water factor basis. In this case, an across the board expansion of grain (and particularly wheat) is not economically viable from an 'efficiency goal' point of view either.

Concentration of grain production in those areas where soil and climatological conditions are able to provide optimal yields, will contribute to a viable national food security strategy which takes into account limited food availability and difficulties of [external] provision in times of scarcity. The WARMAP (1997) study also indicated that there is a great need to overcome the cotton versus grain type of dichotomy, and that there are substantial gains made in focusing on fruits and vegetables, a lesson which recalls a nationalist Uzbek slogan heard in Tashkent as early as the 1970s: 'down with cotton, long live the orchards' (Khazanov, 1990:25; Spoor, 1993). This would however necessitate substantial structural reform and investments.

It can be concluded that the current food self-sufficiency strategy of Turkmenistan and Uzbekistan did not actually deliver the expected results, and if one focuses on the period 1995-96, this strategy has been economically costly. The 'food self-sufficiency' strategy of the past years was 'state-led' and has therefore taken place in an institutional environment where necessary reforms towards greater influence of peasant producers and the development of internal agricultural markets [and their linkage to external ones] are very slow indeed. Although one could have suspected this, such a process does not represent the outcome of a necessary policy debate on the future role of cotton [in terms of its foreign exchange earning capacity and contribution to rural employment] versus its negative impact on the rural environment of Central Asia.

A rethinking of strategy as the result of both the 1996 balance of payments crisis in Uzbekistan, and the continued non-compliance to the overly-ambitious domestic food production targets set by both the Uzbek and Turkmen governments, is however still questionable. This is especially so if we take into account the limited amount of market-oriented and political reforms that have been undertaken up until now. Nevertheless, there is an urgent need for an agricultural diversification programme that will base itself on regionally defined agro-ecological comparative and competitive advantage, and last but not least, decisions by agricultural producers themselves. This cannot be realised unless governments of these CAS relinquish their substantial controls over agricultural markets of grain and cotton, and actually implement policies that will promote the construction of new agricultural markets.

## NOTES

[1] SWB/SUW/0445/WC/3, 23 August 1996 (Summary of BBC World Broadcasts). Also this was more recently reported by a RL/RFE OMRI-report of 20 March 1998.

[2] From the two studies that are being cited in the fourth part of this paper (World Bank, 1994; WARMAP, 1997), it can be calculated that taxation was more or less similar for cotton and grain in 1993, while farm gate prices [weighted averages] in both cases rose in real terms by around 200-300 percent in the period following. Nevertheless, grain became fundamental for home consumption needs. In the absence of data, this particular effect could not be measured here.

[3] We used the parity prices from WARMAP (1997), i.e. for wheat 270 US$/ton and for cotton 404 US$/ton

[4] EIU (1997a:28) states:

> The policy [of shifting from cotton to grain, MS] was a major contributing factor in the 1996 economic crisis. In 1995 and 1996, the target was not reached, necessitating large, expensive and unforeseen grain imports. In 1995-96 Uzbekistan effectively spent $ 1.6 billion, taking opportunity costs into account, trying to save $ 700 million in grain import costs. The methodology of calculation is unfortunately left unexplained by the source.

[5] For 1993 it was estimated that US$ 1.6 billion was transfered through the official procurement system, while the farm enterprises received US$ 0.7 billion in return through subsidized inputs. The net outflow [from agriculture] was therefore estimated at US$ 0.9 billion, the largest share being represented by underpricing of cotton (World Bank, 1994:50-1). The report uses 'international prices' for cotton (366 US$/ton) and wheat (228 US$/ton), which are somewhat less than in WARMAP (1997), but with a comparable ratio between them.

[6] No details were provided on how the 'regional analysis' was performed, or -if a sample was used- how representative the sample was.

[7] The water efficiency of crops in Soviet Central Asia is very low indeed. This is caused by very high losses of water [transport, evaporation, leakage].

**REFERENCES**

CARLEY, P.M. (1989) 'The Price of the Plan. Perceptions of Cotton and Health in Uzbekistan and Turkmenistan.' In *Central Asian Survey*, vol.8, no. 4, pp.1-38.

DE JANVRY, A., M.FAFCHAMPS AND E.SADOULET. (1991) 'Peasant Household Behaviour with Missing Markets: Some Paradoxes Explained.' In *The Economic Journal*, November, pp.1400-17.

EIU (1997a) Country Profile Uzbekistan 1997-78. London: *The Economist Intelligence Unit.*

EIU (1997b) Country Report Kyrgyz Republic, Tajikistan, Turkmenistan. London: *The Economist Intelligence Unit*, 4th Quarter 1997.

GOSKOMSTAT (1987) Narodnoe Khozyaistvo Uzbekskoy SSR zo 70 Let Sovetkoy Vlasti. Tashkent: State Committee of Statistics.

GOSKOMSTAT (1991) Narodnoe Khozyaistvo Uzbekskoy SSR. Tashkent: State Committee of Statistics).

GOSKOMSTAT RESPUBLIKI KAZAKHSTAN (1996) Selskoe Khozyaistvo Respubliki Kazakhstan. Almaty: GosKomStat.

KHAN, A.R. AND D.GHAI (1979) *Collective Agriculture and Rural Development in Soviet Central Asia.* London: MacMillan.

KHAZANOV, A. (1990) 'The Ecological Situation and the National Issue in Uzbekistan.' In *Environmental Policy Review*, vol. 4. no. 1, pp.20-27.

LERMAN, Z., J. GARCIA-GARCIA AND D.WICHELNS (1996) 'Land and Water Policies in Uzbekistan.' In *Post-Soviet Geography and Economics*, vol. 37, no. 3, pp.145-74.

MONKE, E.A. AND S.R. PEARSON (1989) *The Policy Analysis Matrix for Agricultural Development.* Ithaca & London: Cornell University Press.

RUMER, B.Z. (1989) *Soviet Central Asia "A Tragic Experiment".* Boston: Unwin Hyman.

SPOOR, M. (1993) 'Transition to market economies in former Soviet Central Asia: dependency, cotton and water.' In *The European Journal of Development Research*, vol. 5, no. 2, pp.142-56.

SPOOR, M. (1995) 'Agrarian Transition in Former Soviet Central Asia: A Comparative Study of Uzbekistan and Kyrgyzstan.' In *The Journal of Peasant Studies*, vol. 23, no. 1, pp.46-63.

SPOOR, M. (1996) 'Upheaval Along the Silk Route: the Dynamics of Economic Transition in Central Asia.' In *Journal of International Development*, vol. 9, no. 4, pp.579-88.

SPOOR, M. (1997) 'Agrarian Transition in the former Soviet Union: the Case of Central Asia.' In M. Spoor (Ed.) *The 'Market Panacea'. Agrarian transformation in developing countries and former socialist economies.* London: Intermediate Technology Publications, pp.95-111.

SPOOR, M. (1998) 'The Aral Sea Basin Crisis: Transition and Environment in Former Soviet Central Asia.' In *Development and Change*, vol.29, pp.409-35.

STATKOM SNG (1996) Statistichiskiy Ezhegodnik SNG v 1995 Godu. Moscow: StatKom SNG.

STATKOM SNG (1997) Statistichiskiy Ezhegodnik SNG b 1996 Godu. Moscow: StatKom SNG.

WARMAP (1997) Water Use and Farm Management Survey (WUFMAS) Annual Report for 1996 Agricultural Year. Tashkent, Uzbekistan: EU Delegation.

WEHRHEIM, P. (1997) 'Agrarian Reform in Russia: The Case of Pskov, Orel and Rostov Oblasts.' In M. Spoor (Ed.), *The 'Market Panacea', Agrarian transformation in developing countries and former socialist economies.* London: Intermediate Technology Publications, pp.142-55.

WORLD BANK (1993) Statistical Handbook 1993, States of the former USSR. Washington: World Bank.

WORLD BANK (1994) Uzbekistan, Economic Memorandum: Subsidies and Transfers. Report 12934-UZ. Washington: World Bank.

WORLD BANK (1995) *Republic of Uzbekistan: Selected Issues in Agriculture*. Internal Country Memorandum, July 1995. Washington: World Bank.

WORLD BANK (1996) *Developing a Regional Water Management Strategy: Issues and Workplan*. ASBP Technical Paper Series, April. Washington, Tashkent: World Bank.

# 4 RUSSIAN GRAIN MARKETS 1991-1998: TRANSITION WITHOUT A FUNCTIONING DOMESTIC MARKET?

Michael Kopsidis

## 1 INTRODUCTION

Russian agricultural production has declined continuously since 1990 except in 1997. Gross Agricultural Output (GAO) in 1998 reached little more than 55% of the initial level of 1990 (own calculation, OECD, 1998, p.49 and OECD, 1999a, p.244). The slowdown was further sustained in 1999 (East Europe, July 1999, p.6). An important reason for the poor performance of agriculture during the previous years has been the disintegration of the economic area not only of the former Soviet Union but also of the Russian Federation itself. One central question that arises in this critical situation is whether the creation of a functioning domestic market would form an important support mechanism for inducing a process of sustainable agricultural growth. The findings of recent research in economic history as well as the experience of many developing countries, show that this is indeed the case. A positive relation does exist between the creation of a domestic market and growth of agricultural production (e.g. Johnston & Kilby, 1975; Wrigley, 1987; Grantham, 1989a).

We pose the following questions, which arise from economic history and development policy, with the focus on Russia:

1) What are the consequences for the efficiency of domestic markets of a weak central state and regional governments that aim at extensive control of agricultural trade?
2) How important are sufficient infrastructure and the lowering of transport costs for the development of an integrated agricultural domestic market, as compared to the abolition of trade- restricting measures at the regional level?
3) What role do economically and politically powerful urban centres play in the development of agricultural marketing institutions?

Findings from economic history and development economics can help to understand and overcome the difficulties and obstacles that the Russian Federation is facing in the course of transition. Consideration of historical integration processes is useful to understand the preconditions which have to be fulfilled in order for an efficient, competitive and growth-inducing domestic market to emerge. In addition, a comparative analysis of the Russian case helps to answer the question, why the disintegration of Russian agricultural markets has taken on such outstanding proportions.

Corresponding to the comprehensive nature of this paper, a very general definition of market integration is useful. According to the "law of one price", the integration process has to be seen as the spatial extension of a uniform market price, only differentiated by transportation and transaction costs. Isolated local markets are merging to form a larger market with uniform price formation. The factor allocation changes according to the integration process in both developed and developing market economies. Reductions in the transportation and transaction costs accompanied, by a growing market demand, lead to the expectation of higher profits by agricultural producers. These increasing profits are often the result of growing specialisation. Both the production and, within developing market economies, the marketed share increase. In a situation where market-oriented structures are only just developing, it is necessary to investigate to what extent market orientation of producers and consumers within the market area increases, rather than to restrict the study to an allocation analysis.

This chapter focuses on agricultural trade and production related to grain. First, the Russian grain trade is of considerable importance for all questions concerning food security because of the patterns of consumption and low incomes of consumers. This is was also historically true for Europe, as well as for many developing countries. Second, under better conditions the most important Russian grain producing areas would be able to meet the requirements to compete on international grain markets. Thus, a functioning domestic market and the related introduction of market mechanisms serve as a preliminary stage for integration into the world market.

Following an overview of the disintegration of the Russian grain market the third, fourth and fifth sections contain a summary of the relevant research findings in economic history and development economics for Russia. These results are used to explain the present Russian situation in section 6. A short discussion of the prospects for the creation of a nation-wide Russian grain market concludes this chapter.

## 2 THE DISINTEGRATION OF THE RUSSIAN GRAIN MARKET 1991-1998

Most studies on Russian food markets confirm the non-existence of an integrated national market for the majority of agricultural food products including bread and wheat flour (Gardner & Brooks, 1994; Loy & Wehrheim, 1996; Kopsidis & Peter, 2000). Studies reveal a severe decline in interregional grain trade within the Russian Federation. At farm level the share of grain production that was marketed fell from 38% in the early 1980's to less than 25% in 1992/93 (Stockbridge, 1997, p.200). Between 1990 and 1997 the volume of annual long-distance grain transport within the Russian Federation fell by much more than the amount of grain available per year on the Russian domestic market.[1]

If transport performance deteriorates more than the available amount of grain, the share of grain used for interregional exchange declines. This is what has happened in the Russian Federation. The transport volume has fallen to a quarter of the 1990 level, whereas the supply has dropped to one half. Even if a part of this significant decline in interregional grain transport is an adjustment to real transport costs (which were subsidised during Soviet times) by far the largest share appears to be caused by the fragmentation of Russian grain markets during transition (Kopsidis, 2000a).

There are a number of important reasons for the declining domestic grain trade and the lasting fragmentation of the Russian grain market.

1. The growing independence of the political entities of the Russian Federation and a simultaneous decrease in central power have given considerable scope for restrictive local and regional administrative regulations concerning agricultural trade.
2. Transaction costs have increased considerably in inter-regional trade due to the absence of an institutional framework, the demonetisation of the Russian economy, growing barter trade, a lack of private marketing institutions as well as currently paralysed and under-financed public buying and trading organisations.
3. The continuous deterioration of the infrastructure during the previous decades and the aggravating crisis of the Russian transport sector since the beginnings of the 1990's, as well as a sharp rise in fuel prices, have led to a strong increase in transport costs. The latter has resulted in a significant deterioration of the terms of trade of agricultural products compared to fuel (Gardner & Brooks, 1994; Wegren, 1996; Sedik, Foster & Liefert, 1996; USDA, 1996; OECD, 1995, p.111, p.137; OECD, 1996, p.95, p.118; OECD, 1998, p.267; OECD, 1999a, p.171; Poser, 1998).

It seems that political developments following the decline of central government structures after 1990 are the main causes for the fragmented grain markets in Russia (Gardner & Brooks, 1994; Melyukhina & Wehrheim, 1996; Stockbridge, 1997; OECD, 1998, p.267; OECD, 1999a, p.177; Kopsidis, 2000a). In recent years the responsibility for agricultural policy has been increasingly transferred from central government to regional and local authorities. This development is a result of the central government's tight fiscal policy aimed at achieving a balanced budget, as well as a consequence of its increasing loss of power. The federal government in Moscow is no longer able to enforce basic rules such as the free circulation of goods within the national territory, which is absolutely necessary for the existence of a domestic agricultural market. Although, the presidential decree of December 1993 prohibited any regional grain trade restrictions, it was completely ignored by the regional administrations. The various regions have adopted very different agricultural policies. In most cases it would appear realistic to assume a conscious persistence of planned economy oriented marketing structures on the part of the local authorities. Where this is not the case, this is not necessarily the result of a conscious market orientation. It is more likely to be the consequence of a lack of financial funds or of a local government in disarray. It could thus be said that:

> "Although the termination of the command economy and recent reforms in the federal grain monopoly have greatly reduced the central state's role in cereals production and distribution, the state apparatus that evolved to facilitate central planning continues to exert major influence over production. The state today is not the unified state that existed in the pre-reform years. Much central power has devolved to the regions, and it is here that the administrative controls of the state apparatus persists" (Stockbridge, 1997, p.219).

The local and regional grain trade in the Russian Federation is subject to far-reaching restrictions and regulations. Among other things, local authority regulations contain regional export restrictions, which allow exports only when all agricultural producers have fulfilled their often very extensive delivery duties to the local public funds. Additionally, the danger of total export bans in situations of impending food shortages, local price controls, and strict regulation of any kind of grain trade by regional authorities have put a heavy strain on interregional grain trade. The impact of these regulations on the domestic grain market were revealed clearly after the bad grain harvest in 1998 - the worst in 40 years - which coincided with the strong Rouble depreciation in the wake of the financial crisis. Many regional administrations strictly prohibited any outflow of grain and other staple foods when suppliers from Moscow and other cities started to look for domestic producers in other regions to replace imports. The condemnation of regional embargoes as anti-constitutional by the federal government and its threat to punish disobedient regions had no effect and was totally ignored (OECD, 1999a, pp.177).

Russian private trade, which is growing despite all these problems, can only work in a more or less lawless environment and is often limited to the tasks which public authorities allow.[2] The old structures are often only workable to a certain degree, but they are still so powerful that they can severely obstruct the change to a more market-oriented agricultural trade (Stockbridge, 1997, 210).

In addition to the growing independence of regions, it seems that the fragmentation of Russian grain markets has also been aggravated by the privatisation process (Kopsidis, 2000a). Controlling the privatised public procurement and trade organisations was a precondition for regional administrations to establish their dominant position on local grain markets; it has also facilitated the enforcement of local trade restrictions. The public procurement and trade organisations for grain, were only formally privatised in 1994. The newly founded joint stock companies are still controlled by public shareholders, mainly from the regional administrations as well as by former directors. Following Stockbridge it can be concluded that for grain marketing a large number of regional monopolies have partially replaced the former central government monopoly and these regional level monopolies "exert considerable influence over the flow of commodities through the marketing chain and between regions" (Stockbridge, 1997, 219).

The decrease of domestic agricultural trade and market disintegration are also closely linked with the growing importance of barter, which is a result of the demonetisation of the Russian economy. Among other things demonetisation means that Russian enterprises are increasingly cutting out the banking sector and settling their claims and liabilities without involving financial intermediaries. In the literature this process is referred to as "disintermediation" and is attributed to public regulations of the payment systems and institutional defects of Russian banks, which actually encourage fraud. Enterprises have consequently resorted to cash and barter transactions:

> "This also meant breaking ties with client or supplier enterprises in far-off regions which payments reached only after a long lapse of time and where the transaction costs of barter were higher" (Poser, 1998, 168).

Additionally, the poor performance of the transport sector impedes interregional grain trade. The structural weakness of the railway-dependent Russian grain transport, which is a Soviet heritage, has been aggravated since 1990 due to the economic decline (Holt, 1993, p.3, p.73; Jaehne & Penkaitis, 1987, p.47). Three main reasons can be identified for this:

a) In Soviet times, transport costs and the optimisation of logistics played no role in the choice of location for processing plants of the milling industry. An increasing shortage of freight capacity and growing transport costs were the consequences.
b) During the last decades, investments to modernise the infrastructure or even to maintain the given level of the railway network and number of trains, which were sufficient at the time, have increasingly declined (Wegren, 1996, p.171, p.175; USDA, 1996, p.5; OECD, 1996, p.129). The critical public finance situation has led to an accelerated decline of investment in the transport sector (OECD, 1996, 73).
c) The inefficient Russian transport system, characterised by a continuous reduction of its performance and growing transport costs, can only be maintained by high subsidies for its tremendous energy consumption. Even in 1992 public subsidies for transport amounted to 2% of the GDP. At the end of the 1980's transport costs per ton per kilometre were six times higher in Russia than in the US (comparison on a US dollar basis; Holt, 1993, p.3). Since that time public financial support for energy consumption has been significantly reduced.

The fragmentation of the domestic grain market has had two main consequences for grain and agricultural production in general:

1) The already weak regional specialisation of grain production diminished further after 1990. Many regions seem to follow a highly expensive policy of self-sufficiency concerning grain and other agricultural products (Melyukhina & Wehrheim, 1996; Ioffe & Nefedova, 1997, p.87; Kopsidis, 2000b). The stabilisation of grain production has best been achieved in regions not belonging to the Russian supply areas, which have favourable natural conditions for grain cultivation, but in those with a profitable energy and industrial sector, which is able to bear the cost of an expensive output-maximising agricultural policy. The tremendous costs of an expensive self-sufficiency policy in the latter regions are borne by a flourishing energy sector. The autonomous republics of Tatarstan, Bashkortostan, Udmurtiya, the Tyumenskaya oblast in the Volga, the Urals and the Western Siberian region are good examples of this kind of "socialist islands" preserving the old centrally planned economy at a regional level (Kopsidis, 2000b).[3]
2) Until the financial crisis and the sharp Rouble depreciation during August and September 1998 agricultural imports - mainly from Western countries - almost totally crowded out Russian suppliers from the most promising markets such as Moscow and St Petersburg.

The manifold problems described above make it difficult and time consuming to trade Russian agricultural products, even over short distances. Russian domestic trade is generally a very risky business. Thus, it is cheaper and more secure for urban centres like St Petersburg and Moscow to obtain supplies through exports from Western countries rather than depend on their traditional supply areas within the former Soviet Union (OECD, 1995, p.143; OECD, 1996, p.128). In 1997 Russia was one of the world's largest importer of agricultural food products.[4] Since the main imports are concentrated in the urban agglomerations and regions with the highest incomes, Russian agriculture and the food industry have lost some of their most promising domestic markets during the last years, especially for meat, butter and food products with high added value. This development has also had an impact on Russian grain production through a decline in domestic feed demand. However, the main reason for the falling feed grain demand is the nation-wide drop in real income. Annual Russian consumption of meat and meat products per head declined within only eight years from 75 kg (1990) to 48 kg (1998 preliminary estimate) (OECD, 1999, p.260). Under these circumstances the loss of some important demand centres such as Moscow and St Petersburg aggravates the crisis, but it is not the only reason for the reduced feed grain demand.

In contrast to food products with high added value the export statistics reveal a drastically reduced influence of grain imports on Russian grain markets since 1993. The net import share in overall Russian grain consumption decreased from 20.1% (1991/92) to 5% (1995/96).[5] Furthermore, despite the worst grain crop since 40 years and its classification by a FAO report as a nation which faces an actual or potential situation of food shortage Russia became a net exporter of grain in the second half of 1998 (East Europe, No. 199, OECD, 1999a, 184). This is partly due to the strict administrative control of grain prices and trade in most Russian regions leading to grain prices far below the world market levels. This indicates that the reasons for the nevertheless substantial marketing problems of the main grain producing areas can be found within Russia itself, and they are linked to the disintegration of the Russian domestic market.

The description sketched above shows that only a pessimistic picture of the creation of a functioning agricultural market can be given at present. On the one hand, the central government institutions controlling the agricultural markets are increasingly unable to work, even to solve crucial issues concerning food security because of their lack of liquidity. On the other hand, the vacuum thus created is not adequately filled by private marketing institutions. Because of the declining power of the Moscow central government and the growing influence of regional authorities in the field of politics and economics, the nation-wide monopolistic grain marketing structures have been weakened, despite the fact that their survival at a regional level has been encouraged. Until now no effective efforts on the part of the central government to create a functioning nation-wide domestic market can be observed.

The question needs to be asked now which political, institutional and infrastructural preconditions have to be met for a well-functioning domestic market to emerge. Western European economies in their historical development, and many countries in developing countries have faced similarly difficult integration problems. Thus, economic history and development economics can help to find solutions for the Russian Federation.

## 3 URBAN CENTRES AND INFRASTRUCTURE. THE DEVELOPMENT OF INTEGRATED AGRICULTURAL COMMODITY MARKETS IN WESTERN EUROPE AND PRUSSIA, 1600-1880

Starting in the early modern age, the formation of highly integrated regional markets, followed by national and later international agricultural markets in Europe, has been closely linked with the development of high-income and fast growing urban centres (Wrigley, 1987, Perren, 1989, Grantham 1989). Only in urban centres were the preconditions met (institutions, capital, knowledge) that made the expansion of free enterprise operations in a continuously growing area possible. Sustainable market-induced agricultural growth would not have taken place without the fundamental improvement of the traffic infrastructure, accompanied by falling transport costs. Together with improved infrastructure, information costs decreased and dense communication networks emerged. In the case of the Russian Federation, it would be interesting to settle the question how important urban centres like St Petersburg and Moscow are to develop domestic agricultural markets.

All over Western Europe and Germany the merging of many local markets into one large national market with uniform prices was closely linked to the abolition of traditional internal tariffs and numerous local tariff privileges by the developing central state (for Germany Ohnishi, 1973; Berding, 1980). Looking at the German case empirical studies prove clearly that the strengthening of central state structures accompanied by the disappearance of particularism since the beginning of the 19th century were essential for the development of highly integrated agricultural markets between 1810 and 1830 according to the law of one price (Kopsidis, 1996, pp.260). Only with the Napoleonic and Prussian customs reforms did a nation-wide Prussian domestic market emerge. Even a little increase of the grain trade volume was sufficient during the first decades of the 19th century to create a large Prussian domestic market according to the "law of one price" (Kopsidis, 1996, pp.337). However, the impact on agricultural trade volume and production by the customs reforms was not significant, due to the lack of infrastructure.

High transport costs and a lacking transportation infrastructure prevented larger investments necessary to effect a continuous increase in production in fertile but remote regions. Thus, those regions lacked a secure perspective for long-term, market-oriented production planning. Many similarly structured remote Russian agricultural regions are facing the same difficulties. Hence in Prussia, the first phase of market integration, characterised by an insufficient infrastructure and high transport costs, was not accompanied by a significant acceleration of agricultural growth (Kopsidis, 1996, p.260). Reviewing the historical evidence leads us to address the question whether total liberalisation of the domestic market, without accompanying measures like improvement of the infrastructure and creation of market institutions, would be sufficient to shape an efficient Russian grain market.

In Prussia the inefficiency of markets described above only changed with the rapid extension of the railway network after the mid-1840's. In the 1850's a decrease in price fluctuation is clearly recognisable, continuing for several decades, although wide harvest fluctuation continued until 1880. In the second phase of market integration, beginning in 1850 the amounts of inter-regionally traded goods increased considerably and food security improved due to the extension of railway networks (Kopsidis, 1996, p.260). Until about 1870, railway construction mainly led to an increasing intensification of domestic farming, but not to an increase in imports. This could not have been caused by customs and agrarian reforms alone. Due to secure marketing prospects and the strong drop in transport costs with constantly increasing demand in industrial regions, it seemed to be economically justified to make cost-intensive efforts to achieve permanent surplus production through intensifying farming. Following the historical experience, decreasing transport costs due to an improved infrastructure and traffic technology are essential for inducing market-oriented agricultural development. This also seems to be true for the Russian grain economy in transition.

## 4 PUBLIC GRAIN MARKET POLICY VERSUS PRIVATELY ORGANISED TRADE: "THE GREAT TRANSFORMATION" IN AGRICULTURAL MARKETING IN WESTERN EUROPE AND PRUSSIA[6]

The main stages in the establishment of a privately organised grain market will be considered, using the examples of England between 1600 and 1850 and Prussia between 1780 and 1850. England, the homeland of liberalism, provides a good example illustrating the long process of establishing market economy principles, even under relatively suitable conditions concerning the sensitive issue of food supply. The development of Prussia, however, reveals the important role of a liberal reform-oriented bureaucracy and central power for establishing free markets in a backward society.

Around 1500 agricultural products in England were traded in public market places, subject to legal restrictions which closely monitored the actions of sellers and buyers. The strong control of private intermediate trade was the aim of all local authority regulations. The profit-oriented speculation and hoarding of private merchants rather than bad harvests were considered by the population and the public officials to be the most important reasons for dramatic price increases and the resulting famines (Overton, 1996, p.135; Huhn, 1987, p.39; Roscher, 1852, p.72). A deep mistrust of private trade based on similar arguments is also widespread in modern Russia.

Efforts to enforce traditional regulations were significantly diminished in England after 1680, although they were still officially law. The increasing liberalisation of agricultural trade and the rise of large private grain trading enterprises operating in increasingly large areas was closely connected with improvements in infrastructure. As a result, market towns faced increasing competition. They risked losing the entire trade to their neighbours if they did not liberalise and adapt their market organisation to the needs of farmers and merchants, for example by opening commodity exchanges. Local trade monopolies were increasingly difficult to maintain, since merchants were now in a position to avoid regulated markets. Growing urban-industrial centres established free markets and put pressure on reluctant market centres to open up their markets. Furthermore, innovations in private trading led to considerable decreases in transaction costs. Due to the improved infrastructure and thus the increase of profitable marketing channels and marketing possibilities (direct sale, sample markets, deals on futures markets), the monitoring of trade was rendered increasingly futile (Overton, 1996, p.143; Chartres, 1985, p.496; Stevenson, 1985; Thwaites, 1985). This aspect of historical development is of particular interest for the investigation of grain markets in transition, which are controlled by regional monopolies, such as those in the Russian Federation.

In addition to the economic development in England, deep-rooted traditions and the formation of a liberal philosophy also played an important role in the development of increasing trust in markets. These comprise on the one hand the special early modern sense of justice, and the English legal system, as manifested in Common Law, the basis of English legislation, and, on the other hand, the belief in a self-regulating market, based on the principles of the Enlightenment. From the beginning of the 17th century, English courts increasingly protected the rights of private individuals to dispose absolutely freely of their property against state demands. As a result of the English revolution, the trade monopolies of the Crown were abolished by Parliament after 1640 and an extensive deregulation policy was enforced (Hill, 1980, 56, pp.124). The belief in the free market as the best of all possible economic worlds started with the Enlightenment. Liberal economic ideas experienced almost unchallenged support in English society in the course of the 18th century. The ideological pressure of "laissez faire" was one of the most effective tools to enforce the liberalisation of the English agricultural market. Strong regulation of the domestic market was increasingly considered counterproductive for supply, since it was assumed to result in the reduction of trade and production.[7]

The general public's and the social elite's lack of trust in market mechanisms is probably a considerable obstacle to the development of market structures in the Russian Federation. However, even in England the development towards a well-established free agricultural market was not without problems. Despite the improved market efficiency, after crop failures considerable price increases occurred in England, which disproportionately burdened the low-income classes and led to a surge in the so-called "food riots" until the end of the 18th century. Interestingly enough, the greatest protests against rising prices took place in the first European country to be no longer threatened by starvation (Stevenson, 1985; Thwaites, 1985). For the 18th century it has to be noted that local authorities enforced regulating measures at the local level only after massive riots, quickly removing them again after the crisis. Thus, private agricultural trade could operate under calculable conditions in the long and medium term.

The state bureaucracy played a greater role in the modernisation of state and society in Prussia and Germany as a whole.[8] The ideas of Adam Smith had a strong influence on large parts of the Prussian bureaucracy, already in the last decades of the 18th century. The book "Wealth of Nations", published in 1776, was regarded by the reform-minded parts of the administration as a guide for modernising their backward society. The liberalisation of the internal and external grain trade was one of the most important and contentious issues in economic policy (Huhn, 1987; Mooser, 1988). However, before Napoleon's conquest of Germany, none of the efforts of the reform-oriented parts of the administration were successful. The so-called enlightened absolutism in Prussia and some other German states was not powerful enough to carry out substantial reforms against the resistance from a backward society. It was not possible to introduce a long-term calculable liberal grain market policy. Reform opponents were repeatedly able to gain the upper hand within the state administration, partly because liberal ideas had only little support among the population. This situation was very similar to the present inconsistent grain market policy in Russia.

The situation changed radically with Prussia's defeat by Napoleon in 1806. Prussia faced its dissolution as a state. The conservative part of the administration had no concept of a solution to this general crisis. This enabled the reform-minded chancellor Hardenberg to gather a small elite of liberal officials in order to elaborate and push forward the reforms. The persistent maintenance of the absolutist principles was considered the greatest danger for the further existence of Prussia. Freedom of trade, made law in 1810/11, was the central point of a general economic reform program and led to a rigorous abolition of the old grain market policy. However, it has to be mentioned that the economic reforms (Ordnungspolitik) only legitimised an economic development that had already started decades before. For the future they were only useful in slightly speeding up the abolition of potential obstacles to further modernisation. They did not generate a process of sustained growth. In Germany also, due to population growth and decreasing employment shares in agriculture, conditions had developed that allowed a sufficient food supply only through the expansion of markets. In contrast to the strong conviction of the reformers, freedom of trade did not directly lead to the accelerated economic growth and increasing trade hoped for to help solve the structural crisis in large rural areas of Prussia, which had been caused by rapid population growth. Both began three decades later in the course of industrialisation. This historical experience shows us not expect quick economic results from regulative policy (Ordnungspolitik) alone.

## 5 STATE CONTROL OF THE GRAIN MARKET AND PRIVATE TRADE

In addition to creating private enterprise institutions, the Russian Federation needs to decide on the extent of regulation of the domestic market, especially because of the parallel existence of state and private purchasing and marketing institutions, and because of the crucial impact of the former on the free market's efficiency. Moreover, since the end of 1994, leading representatives of the agro-food sector have called for a policy of national self-sufficiency based on comprehensive government control of agriculture, arguing that the market would otherwise fail in this vital respect. The argument and situation are well known from the experiences of many developing countries. The self-sufficiency policy pursued in many Asian and African countries between 1950 and 1980 was accompanied by the almost total abolition of any supra-regional private domestic trade in order to be able to establish state purchase and distribution monopolies (Staatz, 1991, p.15). Experience with such a policy leads to an extremely critical look at similar tendencies in Russia, especially with regard to their negative impacts on the efficiency of markets and on agricultural production (Stone, 1989; Staatz & Wohl, 1991).

Throughout history and until the present day, some of the most drastic measures of government-controlled agricultural trade policy in Russia and other countries have been trade restrictions and the compulsory sale of part of the production. Especially the former leads to serious negative effects (Ahmed, 1989). The balance function of markets is disrupted and regional price differences are enhanced. In the short run, the price pressure in surplus areas will ease state purchasing; in the long run, however, production can be expected to decrease. According to all previous experience, extensive black markets with high transaction costs emerge, leading to an increase in prices. Furthermore, each state regulation of the agricultural market influences the expectations of private traders. An uncertain and erratic public purchase policy results in an increased unpredictability of the market, creating a feeling of uncertainty among private traders. Under these circumstances private storage will be only sub-optimal with regard to the intertemporal balance. It is also true that a quantity-oriented state market regulation, as that attempted in many Russian regions, affects the efficiency of markets more than a price-oriented policy.

It is very difficult to maintain state monopolies. In underdeveloped economies, budget restrictions and lack of administrative efficiency used to be the greatest obstacles. Underfinanced public agencies are unable to stabilise market prices because they cannot buy up and sell sufficient amounts of grain to significantly influence market price movements. Furthermore, lacking financial resources impede more flexible operations on the market, so that prices cannot be stabilised. As a result, an "anticyclical" price policy can quickly turn into its opposite. Delayed payments endanger the realisation of an aimed for market regulation, especially where there is inflation. Moreover, there are no economies of scale regarding the costs of purchasing. On the contrary, according to previous experience, the more comprehensive public purchasing is intended to be, the more the costs per registered farm increase. In addition, official arbitrariness and corruption lead to a considerable growth in transaction costs and prevent a functioning of the market. More than four decades of development policy have shown that interventions which do not work according to the market mechanism (regulations, prohibitions, rationing etc.) lead to corruption and cannot be enforced by an ineffective administration (Streeten, 1988, 12).

The frequently found fixing of purchasing prices, irrespective of quality or transport costs, as well as their level, which is generally too low, creates further problems. A purchasing price without any consideration of quality aspects hampers quality-oriented production. Standard prices for a large area result in a concentration of purchasing on remote areas with high costs of transport, and thus in the crowding out of private trade in these regions. Such prices function as substitutes for deficient infrastructure. Purchasing prices that are too low, and the connected compulsory deliveries lead to reluctance among private entrepreneurs to invest in transport and storage, because of fear of confiscation.

Where the government attempts to take political control of an issue as sensitive as national food security, intensified state regulation of grain marketing is often justified as being necessary because of market failure. However, a significant number of market deficits certainly have to be attributed to administrative measures (Staatz & Wohl, 1991, pp.80). Therefore, if a government decides on the parallel existence of a free market and a public marketing system, not only the question of whether state interventions can really countervail market imperfections arises. The effects of the chosen form of public regulation on the functioning of the free market also need to be analysed.

## 6 THE PROSPECTS FOR THE EMERGENCE OF A UNIFIED DOMESTIC GRAIN MARKET IN RUSSIA

Three main obstacles impede the creation of a functioning nation-wide Russian domestic grain market: the far-reaching autonomy of regional political authorities in agricultural policy, an insufficient infrastructure, and the reluctance of all actors in agricultural policy to create an institutional framework consistent with the free market economy.

The historical development of the Western European economies shows that, in a first phase, integration processes according to the "law of one price" can be achieved despite insufficient infrastructure and inefficient markets characterised by an unsatisfactory level of interregional and inter-temporal exchange of goods. Thus, not every market-related phenomenon should be viewed as a success - which is frequently done in publications about transition economies. Do increasing Russian grain price fluctuations really indicate progress in transition (implicit in USDA 1996, p 5), or should they rather be interpreted as a sign of collapsing food supplies? As regards the early 19th century, it has been clearly proved that strong grain price seasonality was a result of inefficient markets. Highly fluctuating harvests and a simultaneously inelastic demand could not be balanced by sufficient interregional trade flows that would stabilise prices (Kopsidis, 1996, pp.260). The growing subsistence economy, which is manifested in a sharply increasing household production, is a striking symptom of the lack of a functioning food marketing system as well.[9]

However, to induce agricultural growth, modernisation of the Russian infrastructure is essential. In contrast to underdeveloped economies, the Russian Federation essentially has sufficient infrastructure to carry out long-distance trade according to its needs. But insufficient investments over several decades have led to reduced efficiency of the given capital stock. An effective decrease in transport costs accompanied by a cut in subsidies could only occur in the case of large investments to modernise the outdated infrastructure. Only a significant decrease in transport and information costs will enable rural Russia to integrate into a growth-inducing nation-wide domestic market. Due to the lack of well-functioning infrastructure, which results in high storage, transport and transaction costs, the potential comparative advantages in production of many agricultural regions cannot yet be realised.[10] The procurement of the necessary funds as well as the realisation of huge infrastructure projects would at least require close co-operation between regions. The effectiveness of economic reforms (Ordnungspolitik) in initiating the development of market institutions also depends strongly on the infrastructure. A functioning transport and communications infrastructure not only leads to decreasing transaction costs of market operations; it also extends the spectrum of transactions for private enterprises. This in turn complicates the control of the market by local authorities, causing conditions that are favourable for the elimination of (semi-)governmental regional market monopolies. According to historical experience, strong local regulation of agricultural markets could be neutralised in the course of infrastructure development, because of the initiation of competition between market locations or regions. Without large investments in the transport and communications sector in rural areas, there is a real danger of ambitious economic reforms, such as privatisation, becoming ineffective.[11] Modernised infrastructure is a necessary prerequisite to induce private investment in agriculture. The question arises whether a market-oriented change of the institutional framework in agriculture would have any effects worth mentioning without considerable financial support for infrastructure investments. In this situation a "Marshall Plan" for the development of rural infrastructure in Russia would definitely be plausible.

In addition to deficiencies in infrastructure, the attempt on the part of many local authorities to prevent the export of agricultural products, or to restrict it by administrative measures in order to secure the local food supply, poses another great problem for the internal Russian agricultural trade. According to the historical experience of continental Europe it can be argued that, with inadequate infrastructure, the development of a domestic market can be accelerated by the central power exerting massive pressure to overcome the resistance of local rulers who create trade restrictions through regulations.[12]

The economic history of Western Europe has shown that economic development and market integration processes start from some dynamic regions and gradually extend over a great area. Urban agglomerations like London, Paris or the German Ruhr were of central importance for a market-oriented agricultural development, and especially for the emergence of private agricultural trade enterprises operating over a large area and supported by a suitable institutional framework. Against this background the serious consequences of the loss of high-income urban agglomerations like Moscow and St Petersburg as markets for the further development of Russian agriculture and domestic marketing channels have been obvious. The supply of these cities with highly subsidised imports from the West has made it unnecessary to develop trade structures consistent with the free market economy such as private enterprises, or to reduce transport and transaction costs within Russia. It also has reduced the pressure to break up inefficient monopolies within the production regions. This would only happen if Moscow and St Petersburg had to provide for their populations from their traditional supply outlets. Before the Russian crisis wealthy regions could easily circumvent the malfunctioning Russian food market and harsh conflicts with conservative administrations to open their markets by importing food from abroad. Thus, high Western imports and a reform-hostile and amorphous Russian agricultural policy are two sides of the same coin.

Since food imports contracted sharply after August 1998 due to a lack of hard currency (OECD, 1999, p.181), Russian regions are now forced to find new supply channels within the country. This pressure is an opportunity to improve the emergence of a domestic market, at least in the medium term. But now, after the severe crop failure of 1998 and a low harvest in 1999, the tendency to isolate regional markets by stopping any outflow dominates. One can therefore hope that, despite a weak central power, the more liberal-minded regions such as Moscow and St Petersburg will gain influence over an increasing number of underdeveloped regions, and put pressure on them to develop competitive domestic markets. Economic modernisation that starts from single regions would indeed correspond to the historical European pattern of development. The historical development of Western European agricultural sector and agricultural markets can be described as a process that first concentrated on particular regions, but in the course of time accelerated to extend over an increasingly large area. At least the first phase of this "regional transition approach" could be realised, even under a weak central power.

According to this view, the crucial problem for the development of Russian agricultural markets is the reluctance to accept reforms, and the backward orientation of the whole agricultural sector. All measures needed to create a competitive agriculture have been obstructed until now. It could be argued that, because of a totally disrupted agricultural sector with nearly all farm enterprises facing bankruptcy, a temporary protectionist policy would be useful in a first stage of reconstruction. But this would only hold if serious efforts were undertaken to indicate absolute determination to modernise the whole agricultural sector and to introduce market mechanisms. Contrary to the Prussian case a market-oriented group within the government, which has a comprehensive reform strategy for the agricultural sector and the power to realise it, is still missing. In such a situation, a temporarily protectionist policy would certainly involve the danger of keeping encrusted structures alive[13]. Therefore, temporary protection according to the infant-industry argument only makes sense if it is part of a reform programme of essentially market-oriented forces.

According to historical experience, one possible strategy to create a domestic market under the difficult Russian conditions is to strengthen the co-operation between economically powerful, reform-minded regions, to put pressure on reform-hostile regions to open their markets, and to enable free trade within the Russian Federation. The chance of such an alliance of reform-minded forces occurring, at least in the long run, has grown since the currency crisis in 1998.

## NOTES

[1] For the data and the calculation as well as a brief discussion about the reliability of these both figures see Kopsidis (2000a).

[2] An ideal example of this is the Saratov region, which is located on the Volga. Saratov has the highest grain surpluses of all Russian regions. Even in this export-oriented region private traders are strictly controlled and they can only work if they have good relations with the regional authorities. Besides this, between 1994 and 1998 in Saratov only 15-20% of grain sold by agricultural producers was bought by private traders and 80-85% by the regional public grain procurement organisation which was founded in 1994 and is the successor of the Soviet procurement agency (Tourbabine, 1999, pp.37).

[3] The link between preservation of the centrally planned economic system at the regional level and the relatively good performance of agriculture has also been observed by the OECD:

> "... it seems on the first sight, that regional administrations that have maintained Soviet style agro-food policies, like large-scale state procurement, input subsidisation, and zonal pricing, have been rather successful in terms of preventing a substantial decline in agricultural output in their regions" (OECD, 1998, 278).

[4] OECD (1999b, 70); Until the sharp Rouble depreciation the share of imports in the total Russian demand for agricultural food products increased to more than

30%, and for cities like St Petersburg and Moscow was higher than 75% (East Europe, No. 184).

[5] Own calculations based on data from Stockbridge (1997, 198) and an Agra Europe Special Report (1996, 4). This development can be mainly explained by a fall in imports rather than by an increase in exports.

[6] Concerning the term "The Great Transformation" see the classical work of the same name by Polanyi, about the market- driven transformation of premodern societies in Western Europe, published in 1944.

[7] For further references see Overton (1996, p.143), Kopsidis (1993, pp.34, 48).

[8] For further references concerning the Prussian reforms see Kopsidis (1993, p.51).

[9] OECD, 1995, p.140; The share of household production within the total Russian agricultural production nearly doubled between 1990 and 1998 and has reached a level of more than 50% (OECD, 1998, 14; OECD 1999a, 174). It should be mentioned, however, that a part of this production is marketed.

[10] Koester deals with the essential role of transport costs for the comparative advantages in foreign trade, for the agriculture of continental, low-infrastructure developing countries. His comments are also valid for the regional trade within a country:

> "Transport costs are an important determinant for a country's comparative advantage. This is especially true for bulky staples with relatively low production costs but high transport costs per unit. Hence, in identifying a country's comparative advantage, information is needed, not only about domestic costs of production for individual products but also about the distance to the market where the products can be sold abroad and the transport costs from the location of production to the destination of exports;" (Koester, 1986, p 53).

[11] These experiences occurred in the course of the 19th century agrarian reforms in Prussia (Kopsidis, 1996).

[12] Gardner and Brooks argue in the same direction: "It appears that a central state strong enough to force local political reform is a prerequisite for effective liberalization in the sense of prices determined by competitive market forces" (Gardner & Brooks, 1994, p.645).

[13] For a similar argumentation see Wegren (1996, p.176).

**REFERENCES**

*Agra Europe Special Report* No. 88 (1996) The Russian Grain Market. London.

AHMED, R. (1989) 'Pricing Principles and Public Intervention in Domestic Markets.' In R. Ahmed & J.W. Mellor (eds.) *Agricultural Price Policy for Developing Countries*. Baltimore and London, pp.55-80.

AHMED, R. & J.W. MELLOR (1989) 'Agricultural Price Policy-the Context and the Approach.' In: R. Ahmed & J.W. Mellor (eds.) *Agricultural Price Policy for Developing Countries*. Baltimore and London, pp.1-10.

BERDING, H. (1980) 'Die Reform des Zollwesens in Deutschland unter dem Einfluß der napoleonischen Herrschaft.' In *Geschichte und Gesellschaft*, 6. Jg., 523-537.

CHARTRES, J.C. (1985) 'The marketing of agricultural produce.' In: J. Thirsk (ed.) *The agrarian history of England and Wales 1640-1750: agrarian change*. Cambridge, pp.406-502.

*East Europe Agriculture and Food* (1996) no. 169, no. 171, no. 199, no. 202.

GARDNER, B.L. & K.M. BROOKS (1994) 'Food Prices and Market Integration in Russia: 1992-93.' In *American Journal of Agricultural Economics*, vol. 76 (August), pp.641-646.

GRANTHAM, G. (1989) 'Agricultural Supply During the Industrial Revolution: French Evidence and European Implications.' In *Journal of Economic History*, vol. 49/1, pp.43-72.

HILL, C. (1980) *The Century of Revolution 1603-1714*. London.

HOLT, J. (1993) *Transport Strategies for the Russian Federation*. Washington D.C.

HUHN, M. (1987) 'Zwischen Teuerungspolitik und Freiheit des Getreidehandels: Staatliche und städtische Maßnahmen in Hungerkrisen 1770-1847.' In H.J. Teuteberg (ed.) *Durchbruch zum modernen Massenkonsum*. Münster, pp.37-89.

IOFFE, G. & T. NEFEDOVA (1997) *Continuity and Change in Rural Russia. A Geographical Perspective*. Boulder.

JAEHNE, G. & N. PENKAITIS (1987) *Getreidevermarktung in der Sowjetunion*. Berlin.

KERRIDGE, E. (1986) 'Early modern English markets.' In B.L. Anderson & A.J.J: Latham (eds.) *The market in history*. pp.121-154.

KOESTER, U. (1986) 'Regional Cooperation to improve Food Security in Southern and Eastern African Countries.' In *Research Report 53*, IFPRI. Washington.

KOPSIDIS, M. (1993) 'Liberale Wirtschaftspolitik im Zeitalter der Industrialisierung.' In R.H. Tilly (ed.) *Geschichte der Wirtschaftspolitik. Vom Merkantilismus zur sozialen Marktwirtschaft*. München, pp.34-68.

KOPSIDIS, M. (1996) *Marktintegration und Entwicklung der westfälischen Landwirtschaft 1780-1880*. Münster.

KOPSIDIS, M. (2000a) 'Disintegration of Russian Grain Markets in Transition: Political and Economic Dimensions.' In *Communist Economies & Economic Transformation*, vol. 12, no. 1 (in print).

KOPSIDIS, M. (2000b) 'The Effects of Disintegration: Regional Grain Production and Marketing in Russia 1992-1995.' In J. v. Braun, E. Serova & K. Frohberg (eds.) *Russia's Food Economy Towards Truly Functioning Markets* (forthcoming 2000).

KOPSIDIS, M. & G. PETER (2000) 'Market Integration During Transition? The Case of Regional Wheat Flour Prices in Russia 1994 - 1996.' In *IAMO-Discussion Paper*. Halle (forthcoming 2000).

KRÄMER, A. (1996) 'Politikalternativen für die russische Getreidewirtschaft.' In *Agrarwirtschaft*, Jg. 45, pp.259-267.

LOY, J.-P. & P. WEHRHEIM (1996) 'Spatial Food Market Integration in Russia.' In *The Russian Food Economy in Transition*, no. 6. Kiel.

MELLOR, J.W. & R. AHMED (1989) 'Agricultural Price Policy for Accelerating Growth.' In R. Ahmed & J.W. Mellor (eds.) *Agricultural Price Policy for Developing Countries.* Baltimore and London, pp.265-291.

MELYUKHINA, O. & P. WEHRHEIM (1996) 'Russian Agricultural and Food Policies in the Transition Period: Federal and Regional Responsibilities in Flux.' In *The Russian Food Economy in Transition*, no. 5, Kiel.

MOOSER, J. (1988) 'Gewalt und Verführung, Not und Getreidehandel.' In H. Berding (ed): *Soziale Unruhen in Deutschland während der Französischen Revolution.* Göttingen, pp.218-236.

OECD (1995) *Agricultural Policies, Markets and Trade in the Central and Eastern Europe Countries, Selected New Independent States, Mongolia and China.* Monitoring and Outlook 1995. Paris.

OECD (1996) *Agricultural Policies, Markets and Trade in Transition Economies.* Monitoring and Evaluation 1996. Paris.

OECD (1998) *Review of Agricultural Policies: Russian Federation.* Paris.

OECD (1999a) *Agricultural Policies in Emerging and Transition Economies 1999.* Paris.

OECD (1999b) *Agricultural Outlook 1999-2004.* Paris.

OHNISHI, T. (1973) *Zollpolitik Preußens bis zur Gründung des Deutschen Zollvereins.* Diss. Göttingen.

OVERTON, M. (1996) *Agricultural Revolution in England. The Transformation of the Agrarian Economy 1500-1850.* Cambridge.

PERREN, R. (1989) 'Markets and Marketing' In G. Mingay (ed) *The Agrarian History of England and Wales 1750-1850.* Cambridge, pp.190-274.

PETERSON, E.W.F. (1991) 'Mechanisms for Achieving Food Self-Sufficiency.' In F.J. Ruppel & E.D. Kellog (eds.) *National and Regional Self-Sufficiency Goals. Implications for International Agriculture.* Boulder, pp.27-44.

POLANYI, K. (1977) *The Great Transformation. Politische und ökonomische Ursprünge von Gesellschaften und Wirtschaftssystemen.* Wien (1st edition 1944).

POSER, J. A. (1998) 'Monetary Disruptions and the Emergence of Barter in FSU Economies.' In *Communist Economies & Economic Transformation*, vol. 10, no. 2, 157-177.

ROSCHER, W. (1852) *Ueber Kornhandel und Theuerungspolitik.* Stuttgart und Tübingen.

SEDIK, D., C. FOSTER & W. LIEFERT (1996) 'Economic Reforms and Agriculture in the Russian Federation, 1992-95.' In *Communist Economies & Economic Transformation*, vol. 8, pp.133-148.

STAATZ, J. M. (1991) 'Conceptual Issues in Analyzing the Economics of Agricultural and Food Self-Sufficiency.' In F.J. Ruppel & E.D. Kellog (eds.) *National and Regional Self-Sufficiency Goals. Implications for International Agriculture.* Boulder, pp.13-25.

STAATZ, J.M. & J.B. WOHL (1991) 'The Evolution of Food Self-Sufficiency Policies in West Africa' In: F.J. Ruppel & E.D. Kellog (eds.) *National and Regional Self-Sufficiency Goals. Implications for International Agriculture.* Boulder, pp.65-87.

STEVENSON, J. (1985) 'The `moral economy´ of the English crowd: myth and reality.' In A. Fletcher & J. Stevenson (eds.) *Order and disorder in early modern England.* Cambridge, pp.218-238.

STOCKBRIDGE, M. (1997) 'Strategic Aspects of Cereals Sector Reform in the Russian Federation.' In: LD. Smith & N. Spooner (eds.) *Cereals Sector Reform in the Former Soviet Union and Central and Eastern Europe.* New York, pp.196-220.

STONE, B. (1989) 'Relative Prices in the People´s Republic of China: Rural Taxation through Public Monopsony' In: R. Ahmed & J.W. Mellor (eds.) *Agricultural Price Policy for Developing Countries.* Baltimore and London, pp.124-154.

STREETEN, P. (1988) *What Price Food?.* Ithaca, New York.

THWAITES, W. (1985) 'Dearth and the marketing of agricultural produce: Oxfordshire c. 1750-1800', In *Agricultural History*, vol. 33, pp.119-131.

TOURBABINE, S. (1999) *Getreidevermarktung im Oblast Saratow.* Master Thesis (unpublished).

USDA, Economic Research Service (1996) *Former USSR Update: Focus on Russian Grain Marketing*, WRS-96-S1, March 18. Washington.

WEGREN, S.K. (1996) 'From Farm to Table: The Food System in Post-Communist Russia.' In *Communist Economies & Economic Transformation*, vol. 8/2, 1996, pp.149-183.

WRIGLEY, E.A. (1987) *People, Cities and Wealth. The Transformation of Traditional Society.* Oxford.

# 5 RURAL LAND MARKETS AND ECONOMIC REFORM IN MAINLAND CHINA

Xianjin Huang, Nico Heerink, Ruerd Ruben and Futian Qu

## 1 INTRODUCTION

With the start of agrarian reforms in mainland China in 1978, rural factor markets (i.e. markets for labour, capital, land, and technology) began to develop despite the fact that resources are still collectively owned. Since 1992, a market economic system has been introduced that has accelerated the development of the rural land market. Due to rapid industrialisation and urbanisation and the general improvement of socio-economic conditions, market functions have gained importance in the allocation of rural land.

The purpose of this chapter is to examine recent developments in rural land markets in mainland China and the driving forces behind these changes. We discuss prevailing problems in rural land markets, and present policy recommendations for the improvement of the functioning of rural land markets in mainland China. The development of rural land markets becomes increasingly important when factor markets for labour and finance, and commodity markets for inputs and outputs are subject to deregulation. It is argued that exchange conditions on these factor and commodity markets will eventually lead to further demands for institutional development of the land market.

The development of land markets is usually considered an important condition for agricultural growth. Binswanger et al. (1994) refer to frequently occurring distortions in land, labour, credit and commodity markets as major causes of reduced resource use efficiency and retarded growth. Transfers of land ownership or usufruct rights enable a more efficient allocation of resources, stimulate in-depth investments for improved land management (i.e. soil conservation measures) and reduce risk-aversive behaviour of farm households. Limitations in access to land tend to drive down the profitability for farmers and thus reduce the reservation price of labour. Land market failures also restrict farmers' supply response and make incentives for farmers' engagement in commercial production less effective.

The establishment of the land market is influenced by changes occurring on other factor and commodity markets. Carter and Mesbah (1993) demonstrate that the emergence of rural labour markets for off-farm employment enables marginal farmers to give up land rights. Moreover, viable family farms will become important when the collateral function of land is recognised for lending on the financial market. Feder et al. (1988) consistently demonstrate the importance of land titles for enhancing agricultural investment and farm productivity. Land tenure security tends to improve farm-level incentives to invest in land improvement and fixed capital, but also enables loans to be better secured. Land titling might also reduce the transaction costs on other markets and thus increase agricultural productivity by favouring factor allocation towards more efficient producers (Feder & Feeney, 1993).

Historical developments in rural land markets, especially during the socialist revolution, are briefly described in section 2 as a background to understanding the impact of the economic reforms implemented since 1978. Factors that contributed to the re-emergence of rural land markets after 1978 are described in section 3. The present structure of rural land markets is discussed in section 4. A distinction is made between the transfer of land ownership (which has remained in the hands of collectives) and the transfer of land usufruct rights (which have been assigned to individual households since the onset of the reforms). Section 5 reviews current problems in rural land markets related to insecurity of land use rights and the absence of social security provisions. In section 6 some recommendations are made for improving the functioning of rural land markets.

## 2 RURAL INSTITUTIONS AND LAND MARKETS IN MAINLAND CHINA BEFORE 1978

The independent family farm, albeit small and fragmented, was the traditional farming institution in rural China for thousands of years prior to the founding of the People's Republic. In the wake of the socialist revolution, nearly half of the farmland in rural China was owned by landlords who rented land to peasant families, for fees often as high as 50% of the main crop's value (Lin, 1997). The Land Reform Law of China, promulgated in June 1950, stipulated that cultivated land was allotted to everybody according to distance from dwelling and soil fertility, giving farmers the ownership of farmland. But socio-economic developments led to increasing differences between rich and poor farmers. Rich farmers needed more land to expand agricultural production, while poor farmers had to sell part of the land for making a living. Under these conditions, it was common to rent out land to tenants (Wang & Xu, 1996). With the development of rural land markets, labour and capital markets developed rapidly.

In order to develop socialism in rural areas and to assist poor farm households with insufficient means of agricultural production, from 1950 onwards farmers were organised into Mutual Aid Teams (MATs) by the government according to principles of voluntary participation. Each MAT consisted of 2-7 farm households. By 1954, already 58.3% of total farm households had joined these teams. In most MATs, land and other means of agricultural production were owned by private households, but production activities were carried out collectively. Rich teams bought land and other means of production from poorer teams in order to expand production.

The Chinese Central Government (CCG) considered the system of individual farming an "isolated", backward, conservative and decentralised economic system, which was not effective. To transform this small-size, individual economic system into a large-size, co-operative economic system, farm households were forced to join so-called Primary Agricultural Producers Co-operatives (PAPCs), consisting of 10-30 households. All means of production had to be converted into a joint stock, but the ownership mostly remained with the private households. In order to encourage hard work, it was decided that the payments for land by a PAPC should not exceed the payments for labour. The price of land was valued at 14-25% of the output in an average year, only about 1/2 of shadow market rent (Wang & Xu, 1996). In some areas, other means of agricultural production (even livestock) were confiscated by the co-operatives by order of the local government. Most PAPCs were more effective than individual farmers or teams because of better administration and production organisation.

The PAPCs were transformed in 1956-57 into Senior Agricultural Producers Co-operatives (SAPCs) composed of 200 farm households. Ownership of land, farm animals, and other main means of agricultural production was transferred (without compensation) from the farmers to the co-operatives. Only 5% of total cultivated land area was allotted to farmers as family plots. Family plots could not be sold or bought on the market. By the end of 1956, 96% of all farm households had joined the co-operatives.

In 1958, Rural People's Communes (RPCs) were established in order to integrate the local government agency with the farmers' co-operatives. Originally, most agricultural production activities were organised by communes, but farmers could no longer recognise any relationship between their work and income, causing lower production incentives ("free rider" problem). In 1959, most agricultural production activities were organised by Agricultural Production Brigades (APBs). Since these proved to be too large to operate efficiently, they were sub-divided into Agricultural Production Teams (APTs) in 1962. Each RPC was composed of 10-20 APBs, and each APB was composed of 10-20 APTs with about 30-50 farm households. The RPCs, APBs and APTs shared the ownership of the means of agricultural production. The APBs possessed most of them, the APTs possessed less, and the RPC possessed the fewest. Of the total cultivated land area, only 5-15% was allotted as family plots.

During this period, resource allocation and production administration were arranged by government agencies. There were no markets for land and other resources. The requisition of rural land for urban, industrial purposes was the only exception. Collectives that transferred part of their land, received compensations in kind (employment in urban industries for some members of the collectives) or a relatively low amount in cash (3000-6000 yuan/ha).

Despite the dramatic increase in the use of modern inputs in the 1960's and 1970's, the performance of agriculture in China continued to be poor. Between 1952 and 1979, the growth rate in grain production equalled 2.4% per year, i.e. only 0.4% above the population growth rate. The output growth rates of other farm products (cotton, fruit, oil-bearing seeds, etc.) were not much higher than the population growth rate either (Lin, 1997). A fundamental reform of the rural economy and of the economic planning system to establish more effective agricultural production systems became inevitable.

## 3 THE RE-EMERGENCE OF THE RURAL LAND MARKET AFTER 1978

### 3.1 BACKGROUND

Increasing poverty and hunger motivated reforms of the rural economic system in mainland China. In Xiaogang Village, Fengyang County, Anhui Province, farmers started to allocate collective land resources to individual households in the spring of 1979 (Qu, 1991). Land usufruct rights were returned to individual farmers and collective labour was changed into household management. Consequently, farmers' remuneration became better linked to their labour input and land output and the "free rider" problem was largely solved. After examining the effects of the first spontaneous reform initiated by farmers, the CCG conceded this practice and named it the Household Responsibility System (HRS). The system was first restricted to hilly or mountainous areas and to poorly functioning APBs or APTs. In practice, this restriction could not be maintained and by 1981 the HRS was given full official recognition and universal acceptance. By the end of 1983, almost all rural households in mainland China had adopted the HRS (Lin, 1997).

The HRS contracts land to individual households for a period of 15 years. After fulfilling the government procurement quota obligations, which serve as an implicit land rent, farmers can either sell their surplus production or retain it for household consumption. The CCG currently aims to stabilise the HRS as a farming system institution. From 1993, a new policy was adopted allowing land contracts to be extended for another 30 years.

The success of the agricultural reform greatly encouraged the moderate political leaders to undertake more market-oriented reforms in rural as well as urban sectors. The rural reform was the driving force behind the market-oriented reform in mainland China (Lin, 1997). Central control and state planning were gradually reduced and market mechanisms were introduced into production, distribution, and trade in the other economic sectors. Most resource allocation decisions are now made at market prices instead of at planned prices (World Bank, 1992). The government is gradually establishing a legal framework for a market economy. The economic reforms greatly promoted economic development. From 1978 to 1993, the average growth rate of the GNP was over 9 per cent per annum (Xie, 1996).

### 3.2 FARMLAND FRAGMENTATION AND THE LAND MARKET

Under the HRS, land plots were usually allocated to households on the basis of household size, soil quality and distance from the dwelling. As the person-to-land ratio was high, this method of land distribution generally resulted in small, fragmented patches of farmland, each parcel system associated with a ridge and irrigation ditches (Zen, 1991). Available information for 1984 shows that 71% of land was contracted to farmers according to family size, 8% of land according to the number of labourers in a family, and 21% of land according to both. Every family was allotted a similar proportion of each land quality type in terms of soil fertility and location. Consequently, on average each household was allotted 0.56 ha of arable land divided into 9.7 parcels, with an average land parcel size of only 0.06 ha in 1984 (Qu et al., 1995). The fragmentation of land has become even more pronounced than it was before the establishment of the People's Republic. During the 1930s, an average farm household in China cultivated of 0.34 ha of land dispersed over 5.6 parcels (Buck, 1937).

While the HRS has stimulated agricultural productivity, land fragmentation has reduced the efficiency of agriculture (Chen, 1987; Lu and Dai, 1987; Zen, 1991; Hu, 1997). A study in Jinsa County (GuiZhou Province) indicates that newly developed ridges and escape canals take up 5% of total farmland. Assuming this estimate is representative for mainland China, it means that 6.37 million ha of farmland was wasted during the land distribution in 1978-83.

The emergence of a land market is partly a response to the demand for farm consolidation. Some farm households cultivate many parcels of land that are sub-leased from absent farmers. For example, one specialised farm household in Jintan City (Jiangsu Province) cultivates 9.87 ha of land divided over 140-150 parcels (the farmer does not know the precise number of plots) from 8-9 Village Groups in 1997.

### 3.3 TRANSFER OF RURAL SURPLUS LABOUR AND THE LAND MARKET

Since the start of the rural reform, there has been a large labour surplus in rural areas because of the improvement of agricultural labour productivity. Although Chinese rural non-agricultural enterprises have absorbed 120.18 million rural labourers, it is estimated that 160.27 million persons (i.e. 35.9% of the total rural labour force) were underemployed in 1994 (Chen, D., 1996). The main problem for the farm sector is that 40% of the labourers obtain 80% of the national income from agriculture, while 60% of the labourers receive only 20% of the income (Xu, 1996). A study in Anhui Province showed that, on average, each household owned 0.44 ha of arable land with an average household size of 4.53 persons in 1994. The net income from farming was nearly 4,000 yuan RMB (1 dollar = 8.2 yuan RMB), which means that it was very difficult to support the family without additional income from the non-farm sector (Chinese Society of Agronomy, 1996).

Table 5.1 shows that differences in incomes between the farm and the off-farm sector have increased considerably over time in Xishan City, Jiangsu Province (columns (1) - (3)). To seek higher incomes, more and more underemployed labourers opt for off-farm employment (columns (4) - (8)). In 1985 only 76% of the total number of rural labourers were employed by the off-farm sector, while in 1993 this rose to 96% (column 5). The transfer of under-employed labourers makes more arable land available for remaining farmers and permits an increase in agricultural labour productivity. Farmland used per labourer increased from 0.46 ha in 1985 to 0.65 ha in 1993 (see column (9)). The development of rural labour markets clearly enhanced the dynamics of rural land markets.

*Table 5.1 Income and employment in arable farming and the off-farm sector, and land availability in Xishan City, Jiangsu Province*

| | income | | | labour and employment | | | | | land |
|---|---|---|---|---|---|---|---|---|---|
| | Arable Farming (yuan) | Off-farm Labour (yuan) | Ratio [2/1] | Rural Labour (x 1,000) | Off-farm Labour (x 1,000) | Agri-Culture (x 1,000) | Arable Farming (x 1,000) | Joint [5+6-4] (x 1,000) | Arable land p.c. (ha.) |
| Year | (1) | (2) | (3) | (4) | (5) | (6) | (7) | (8) | (9) |
| 1985 | 3,279 | 13,985 | 4.3 | 54.02 | 41.17 | 18.80 | 12.75 | 5.95 | 0.46 |
| | | | | | (76%) | (35%) | (24%) | (11%) | |
| 1986 | 4,328 | 16,220 | 3.8 | 54.46 | 43.42 | 16.77 | 11.11 | 5.73 | 0.51 |
| | | | | | (80%) | (31%) | (20%) | (11%) | |
| 1987 | 4,017 | 20,415 | 5.1 | 54.81 | 45.85 | 14.83 | 10.92 | 5.87 | 0.52 |
| | | | | | (84%) | (27%) | (20%) | (11%) | |
| 1988 | 4,119 | 23,942 | 5.8 | 55.34 | 49.66 | 12.04 | 10.74 | 6.36 | 0.53 |
| | | | | | (90%) | (22%) | (19%) | (11%) | |
| 1989 | 4,196 | 25,994 | 6.2 | 55.37 | 48.76 | 16.07 | 10.55 | 9.46 | 0.54 |
| | | | | | (88%) | (29%) | (19%) | (17%) | |
| 1990 | 4,194 | 29,503 | 7.0 | 55.39 | 48.74 | 13.31 | 11.05 | 6.66 | 0.51 |
| | | | | | (88%) | (24%) | (20%) | (12%) | |
| 1991 | 4,147 | 36,518 | 8.8 | 55.64 | 49.64 | 12.96 | 10.56 | 6.96 | 0.53 |
| | | | | | (89%) | (23%) | (19%) | (12%) | |
| 1992 | 4,907 | 72,556 | 14.8 | 55.81 | 51.00 | 11.47 | 9.10 | 6.66 | 0.61 |
| | | | | | (91%) | (21%) | (16%) | (12%) | |
| 1993 | 5,074 | 103,672 | 20.4 | 56.34 | 54.00 | 11.01 | 8.60 | 8.67 | 0.65 |
| | | | | | (96%) | (20%) | (15%) | (15%) | |

*Note: Income data in columns (1) and (2) refer to average incomes per person employed in arable farming and off-farm labour, respectively. Joint employment in column (8) refers to persons employed simultaneously in agriculture and off-farm employment*
*Source: Based on data from Xishan City Statistical Yearbook (1986-94)*

## 3.4 THE SECOND RURAL LAND CONTRACT REFORM AND THE LAND MARKET

Although domestic prices of grain have increased to levels very close to or even slightly above the international price level during the period 1984-95, the growth rate of grain production (1.24%) was lower than that of population (1.36%) over that period (Lin, 1997). Management costs for fragmented land remained very high, while the small size of landholdings formed an obstacle to further increases in productivity and efficiency. Only 6% of cultivated land in mainland China (738 million ha) belongs to "scale farming" (i.e. farms larger than 10 mu = 0.67 ha per farm labourer) (Xu, 1996). Efficiency differs greatly between different managing units, as can be seen from the information on Jintan City presented in Table 5.2. Although land productivity per ha is roughly the same, productivity per labourer and production for the market is much lower in "small-scale farming" (less than 0.67 ha per labourer).

*Table 5.2 Comparison of efficiency of different managing systems Jintan City, Jiangsu Province (1995)*

| | share in total cultivated land | productivity of labour (yuan p.c.) | grain output per labourer (kg p.c.) | land productivity (yuan/ha) | grain yield (kg/ha) | commodity share in grain production |
|---|---|---|---|---|---|---|
| small-scale farming | 89.0% | 2,509 | 5,828 | 16680 | 10,350 | 44% |
| scale farming: | | | | | | |
| family scale farming | 9.3% | 18,915 | 10,587 | 17280 | 10,725 | 65% |
| farms owned by centres for agricultural technology: | 0.5% | 19,010 | 12,707 | 15787 | 10,275 | 98% |
| farms owned by village | 1.1% | 17,280 | 12,986 | 14939 | 10,050 | 97% |
| tock co-operation farms | 0.1% | 17,750 | 11,888 | 14250 | 9,825 | 85% |

*Note: The commodity share in grain production is the total share sold to the government and at the market*
*Source: Data provided to the authors by the Jintan Department of Rural Work*

To promote sustainable development of agricultural production, the CCG initiated a new policy (the "second round reform") in 1993 when a start had to be made with the renewal of rural land contracts. Important elements of this new policy are: (1) land contracts are extended for another 30 years; (2) land contracted by households does not change hands when the household size changes in order to reduce adjustment and fragmentation of land; and (3) contracted land can be transferred to other farmers in return for compensation, with the purpose of increasing the farming scale and improving the productivity of land and labour.

In practice, local governments had already introduced different HRS initiatives and land stock systems. Two main types of HRS systems can be distinguished: (i) the system of contracted and mobile land, and (ii) the system of grain ration and responsibility land. In the former system, contracted land is allocated to persons with land usufruct rights for 30 years. Mobile land may be changed for non-agricultural use or farmland adjustment (i.e. caused by population changes, establishment of infrastructure) and is used for a shorter period (mostly 3-5 years), but can be exchanged through market mechanisms, e.g. auctions, or invited tenders. Grain ration land can be obtained by every rural household paying an agricultural tax to the government and a relatively low fee to the collective, but its use is decided by the government. Responsibility land can be obtained by farm households that pay, in addition to the agricultural tax, a higher fee to the collective, and have to sell a fixed portion of the agricultural production to the government (see section 4.2). The average proportion of grain ration land to responsible farmland is 1:2 (Central Policy Research Division and Ministry of Agriculture, 1997). In practice, different land use modalities are sometimes combined.

## 3.5 INDUSTRIALISATION, URBANISATION AND THE RURAL LAND MARKET

Since the beginning of the economic reform, industrialisation and urbanisation have accelerated in mainland China. During the period 1980-95, the proportion of the population living in urban areas increased from 13.6% to 28.9% (Chen, Y., 1996) . The share of the industrial sector in Gross Domestic Product increased from 49.0% to 73.5% during the same period (Chen, D., 1996). To meet the increasing demand for expansion of industrial and urban areas, a large share of cultivated farmland was changed into non-agricultural land by land requisitions and other market mechanisms (land stock, land lease, etc.).

## 3.6 CONCLUSION

The emergence of the market economy led to a growing demand for reform of the free land use system in the urban areas of mainland China. The 1979 Sino-foreign Joint Venture Enterprises Law marked the start of the change in the free land use policy. The price mechanism and market competition were introduced in the process of urban land allocation. This urban land use system reform also led to reforms in the non-agricultural land use system in rural areas. A non-agricultural land market, mainly aimed at rural non-agricultural enterprises and farmer housing, emerged in rural areas.

Rapid economic growth and increasing demands for market supplied commodities gave rise to substantial reforms in agricultural policies. With the introduction of the household responsibility system, fragmentation of cultivated land was recognised as an important impediment to efficiency improvement. Rising agricultural labour productivity and strong increases in off-farm employment opportunities enhanced further development of the land market. Legal measures were taken to enable land transfers and to improve security of usufruct rights. The conversion of agricultural land towards non-agricultural activities (i.e. housing, industries) proved, however, to be an ongoing process that now exercises a strong influence on the rural land market.

Major driving forces for the formation of rural land market in mainland China are shown in Figure 5.1. The dynamics of land market development are mostly enhanced by the rapid emergence of an articulated labour market. Demand for non-agricultural labour increases due to industrialisation and urbanisation tendencies. Demand for land emerges from urban areas (for industry and housing) and from rural areas (for land consolidation). The increase of non-agricultural labour requires an improvement of rural factor productivity to satisfy global food demands.

Commercial agricultural can only be further developed when land rights are adequately defined.

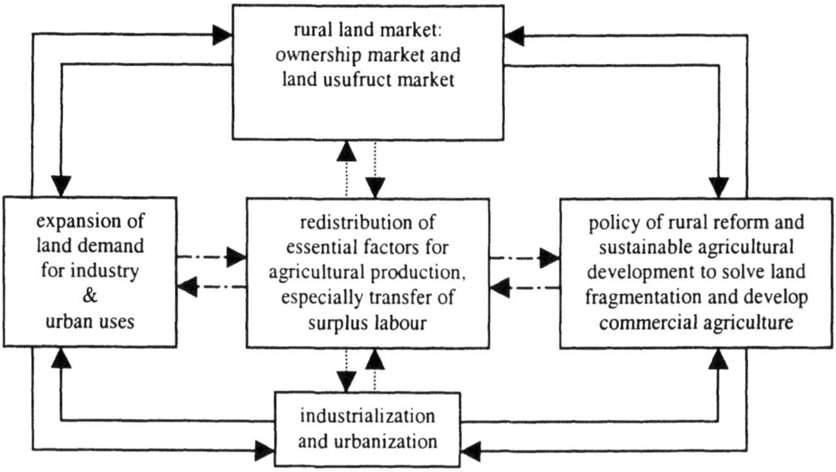


*Figure 5.1 Macro-background of the formation of rural land markets in mainland China*

## 4 RURAL LAND TRANSACTIONS IN MAINLAND CHINA

Rural land transactions can be classified into two types: transactions on the land ownership market and transactions on the land usufruct market. The land market structure is represented schematically in Figure 5.2.

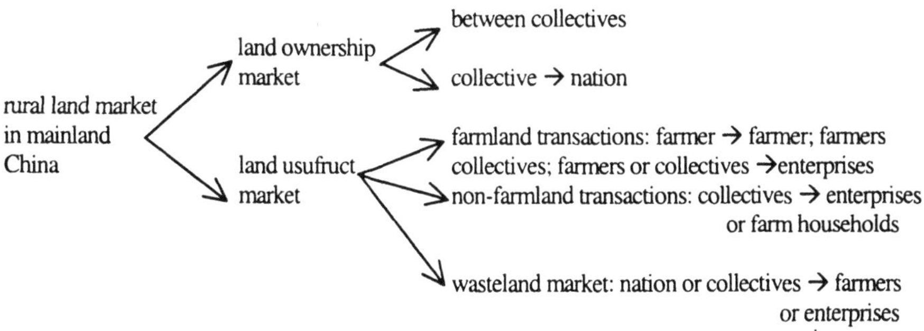


*Figure 5.2 Rural land market structure in mainland China*

In the following we will review the currently available mechanisms for transfer of land ownership or usufruct rights in each of these market segments. Major attention is given to the prevailing conditions in land contracts (compensation payments, short- or long term contracts, taxation, etc.) and the derived security for investments in agricultural intensification.

## 4.1 LAND OWNERSHIP MARKET IN MAINLAND CHINA

According to the Work Regulations of Rural People's Commune (1962), the Chinese Constitution (revised in 1988) and the Agriculture Law (1992), there are four kinds of land owners in mainland China: the nation, the town or township, the village, and the village group. The Land Administration Law of 1986 recognises only three kinds of landowners, the nation, the town(ship) and the village; the latter are the largest landowners.

It is generally believed that only land usufruct markets exist in mainland China (see e.g. China's Land Yearbook, 1995). Huang (1993), however, shows that transfers of land ownership also occur in mainland China. Two types of land ownership market can be distinguished, one between collectives and another between collectives and the nation. Some land ownership transactions between collectives were reported e.g. in South Jiangsu Province, where poor collectives sell land to richer collectives to establish infrastructure and develop specialised agriculture (e.g. for "sight-seeing" agriculture and aquaculture) (Zhou, G., 1994). Land requisitions are the main mechanism through which land ownership is exchanged. Before 1997, the government was the legal authority in land requisition at different levels (not including town and township). However, many local governments made land requisitions that exceeded their authority in order to rapidly develop local industries and urban areas. Land requisitions made during the period 1991-94 should have been legally approved by the State, but only 18.7% was actually examined (Qu et al., 1997). Recently, only the state and province-level governments can examine and approve land requisition.

Dong (1993) argues that land requisitions are governmental actions and cannot be regarded as a land market because the governments make requisitions of land from collectives at low prices and sell at higher prices on the land market. For example, in South Jiangsu Province, land requisition compensation is 750-900 thousand yuan/ha, but the market price is about 1,500 thousand yuan/ha, whereas the compensation distributed to farmers is only 150-225 thousand yuan/ha (Zhou, G., 1994). The compensation obtained by farmers is based on 3-6 times the average output over the last three years, compensation for fixed investment on farmland (i.e. agricultural infrastructure, compensation for agricultural product during maturation period) and compensation for non-working farmers because of loss farmland. In our opinion, the big price differences indicate that land requisition is still a highly imperfect market.

## 4.2 FARMLAND USUFRUCT TRANSACTIONS IN MAINLAND CHINA

Government action is still predominant in farmland usufruct reallocation. The second round land reform carried out by the CCG since 1993 has stimulated the formation of farmland usufruct markets. A study by the Ministry of Agriculture shows that 47.8% of farmland (in 31.5% of the village groups) was removed from the grain ration and the responsibility land system during the period 1990-95. After the systems reform, many farmers only obtained grain ration farmland, while responsible farmland is mostly contracted by invited tenders. This resulted in bigger farms that could achieve greater efficiency. In 1994, 20% of grain ration farmland and 60% of responsible farmland were planted under "scale farming" (Tan, 1996).

In 1990, the rate of farmland transfer was 6.3% for mainland China, and 1.7% for the East, 10.2% for the Middle, and 2.3% for the West (Qu et al., 1995). Available data for 1995 indicate that the transfer rate was much higher: 14.57% in Jintan City (East), 5% in Anhui Province, 1% in Hubei Province (Middle), and 6-10% in Inner Monggu of (West) (Central Policy Research Division and Ministry of Agriculture, 1997).

### *4.2.1 FARMLAND MARKET ESTABLISHED BY THE GOVERNMENT*

There are two kinds of farmland markets: one is established by the governments, the other is a spontaneous market. The farmland market established by the governments mainly emerged in developed areas. Typical examples include land stock co-operatives and village collective farms.

#### Land stock co-operative institutions

In Nanhai City, Guangdong Province, rural industry represents a major share of gross output value. Only 20% of rural labourers are farmers, and 70% of the net per capita income comes from non-agricultural employment (Wang and Xu, 1996). Based on successful experiments carried out in some towns or townships from 1991 to 1994, Nanhai City established a land stock co-operative institution reform in its rural area in 1994. All farmland was managed by the Land Stock Co-operative Companies established by villages. The farmers became shareholders, sharing the dividend at the end of each year. The farmers can also sell their stock at the market. Through the establishment of stock co-operative institutions, farmland can be merged and agricultural labour can be transferred to the non-agricultural sector. Consequently, rural production factors are being re-allocated through the market mechanism.

#### Village collective farms

In the South of Jiangsu Province, rapid rural industrialisation has taken place. The non-agricultural sector contributes 92.6% to the gross regional product, and 70% of rural total labourers work in the non-agricultural sector (State Economic Planning Commission, 1995). Farmland is very scarce in the area: only 0.055 ha per person, and 0.469 ha per rural labourer (Tan, 1996). Many farmers sublease their land to work in the off-farm sector. In 1994 almost 20% of grain ration land and 60% of responsible land were subleased in Xishan City. Because of the higher efficiency (economies of scale), village collectives obtain farmland from farmers working outside agriculture. Village collective farms include three types of organisation: village commission farms, rural non-agricultural enterprise-owned farms, and agriculture technology centres. In Xishan City, for example, the land area controlled by village commission farms increased from 8% in 1991 to almost 27% in 1994, while the family farms area decreased by 14.5% during the same period. Village collective farms also exist in other developed areas including Beijing City, Shanghai City, Guangdong Province, Zhejiang Province, etc. They are operated by a democratically elected village commission, responsible for infrastructure construction, establishment of enterprises and farms, collection of taxes and fees, and transfer of agricultural technology. Rural non-agricultural enterprises sometimes run farms as a means to spread investment risks. Finally, centres of agricultural technology were established by the government but are now independent institutions.

#### 4.2.2 *FREE FARMLAND MARKET*

The free market for farmland usufruct transfers between farmers is more common in both developed and developing provinces. Farmland is mainly subleased between relatives and friends for longer or shorter periods through different market arrangements. In Anhui Province, 49% of the land transfers concern the sublease of contracted land for a long period (Central Policy Research Division and Ministry of Agriculture, 1997). In other regions, most farmland is privately subleased through oral contracts for a short time (one year or a planting season). In South Jiangsu Province, the farmland is subleased mainly for short-season rice production, since the productivity of rice is about 9,000 kg/ha (compared to 4,500 kg/ha for wheat) and rice is the main food, whereas the collectives also grow wheat during the following season. Sublessees are only responsible for paying agricultural tax. Other taxes raised by collectives and local governments and for contracted grain are paid by the land owner, who maintains the benefits from the transferred land kept for social security reasons.

### 4.3 WASTELAND TRANSACTIONS IN MAINLAND CHINA

Transfers of wasteland (i.e. land with low productivity, including water gullies, bare mountains, sloping fields, etc.) have been taking place since 1992, mainly through auctions. There are two kinds of wasteland transaction markets: one is a completely open market, the other is a restricted open market in rural communities.

#### 4.3.1 *COMPLETELY OPEN MARKET FOR WASTELAND TRANSACTIONS*

In 1993 and 1994, Helongjiang Province sold one million ha of wasteland to counties and others by auction, contract, and lease (China Land Yearbook, 1995). Land usufruct obtained by auction can be kept for 50-70 years. In most areas, the income derived from wasteland transactions goes to the owners of wasteland, i.e. the state or the collectives. Every farmer, collective, or firm from a different area or country can buy wasteland. In Maodanjiang City of Helongjiang Province, 15.8 thousand ha was sold to 3,300 farm households through auction before 1994 (China Land Yearbook, 1995). Many firms or collectives buy wasteland as an investment. For example, one company in Shenzheng City bought 13.33 thousand ha, a joint enterprise of Dalian City bought 12 thousand ha, and Huaxi Village (the richest village in China, located in Jiangyin City, Jiangsu Province) bought 100 ha wasteland in Helongjiang to establish agricultural production bases of rice or beans (Xu, B., 1996).

#### 4.3.2 *RESTRICTED OPEN MARKET FOR WASTELAND TRANSACTION IN RURAL COMMUNITIES*

The auction is the main market mechanism for wasteland transactions. In Luliang Prefecture, an area with severe land erosion located in the Loess Plateau, 77.8% of the land area is mountainous and 14% of the total land area is not developed. To develop more land and to protect it against erosion, a wasteland market was established in 1992. Wasteland is only allocated to local (collective) farmers, companies, or government departments through auctions, contracts, and lease.

In 1992-93, 72,900 ha of wasteland was auctioned, of which 67% was shared by 30,856 farm households, 29% by joint-families, and 4% by 60 local government departments or local companies (Li et al., 1994). The farmers received 30 to 60 years of usufruct rights through wasteland auctions. The wasteland prices range from 15 to 525 yuan/ha, depending on land quality, duration of the land usufruct right, and the supply and demand of wasteland. Due to increasing demands for wasteland, the price increased more than 10-fold in some villages between 1993 and 1995 (Hasetaden and Li, 1996).

Farmers who bought the wasteland have a certificate for wasteland usufruct and a regular contract. The wasteland right promotes investment in wasteland development. Despite an average income of only 735 yuan per capita in 1995, more than 20 million yuan were invested for wasteland purchase in 1992-93, without any credit provided by banks (Wang, X., 1994). Wasteland buyers obtain relatively secure land usufruct rights compared to farmland contractors. The latter make less long-term investments in existing farmland. Contractors cannot transfer or inherit land usufruct of wasteland, and have no lawful contract and certificate of land usufruct. Moreover, local governments find it very difficult to provide farm households with secure land ownership rights, because frequent adjustments in the land assigned to households are required due to population changes, establishment of agricultural infrastructure, and so on.

## 4.4 NON-AGRICULTURAL RURAL LAND MARKET IN MAINLAND CHINA

In the rural area of mainland China, 16 million ha of land is used for rural housing. Between 1980 and 1995, 40,000-100,000 ha of land was used for establishing rural non-agricultural enterprises each year (Li, Y., 1996). The free and unlimited access to non-farm land in rural areas has led to a decrease of the available farmland, and increasing housing land disputes between farmers.

At the same time as the far-reaching reform of the urban land use system began, a start was made with the reform of the rural non-farm land use system in more developed areas (e.g. in Nantong City, Jiangsu Province; Shandong Province, etc.) from 1988 onwards. Two patterns of non-farm land transactions can be distinguished: indirect transactions and direct market transactions.

**Indirect transactions**
Land is first changed from collective land into state land by means of land requisitions, and then it is transferred. This indirect route is followed because, according to the Land Administration Law of mainland China, no transactions can take place with collective land. In practice, however, the number of non-farm land market transactions is increasing rapidly. Land obtained through the indirect rural land market is used for building, as an investment, as collateral for obtaining credit, and for paying debts (Zhang, 1997).

**Direct market transactions**
Collective land usufruct can be directly traded on the market. As reported for Wuxi City (China's Land Yearbook, 1995), this is being done because it is proving impossible to control rural transactions.

The two systems of land transactions have different consequences for the income received by the government and local institutions. Income obtained from the direct land market flows directly to the village groups, villages, and counties. On the indirect market, the government pays a relatively low compensation for land requisition to farmers or village group, while the full income from land lease and stock flows to different government agencies. The income received by the governments from land transfers is sometimes partly forwarded to villages and towns or townships for the establishment of infrastructure or for the purpose of developing the non-farm sector.

The duration of non-farm land lease transactions varies according to use. Short-term lease land cannot be traded on the market through sublease, mortgage, or stock. Long-term lease is up to 70 years for residential land; 50 years for industrial, educational, cultural, health, sports or mixed use, and 40 years for commercial, entertainment and tourism use (Zhang, 1997). In China, Fengshui (which literally means 'propitious place'; determined by wind and water) is a very important factor for the location of housing and graves. To allocate this scarce resource without controversy, some rural areas distribute land with good Fengshui by auctions. A long-term lease can be obtained in three ways: through auction, invited tenders, or negotiation.

## 5 PROBLEMS OF LAND MARKET OPERATION IN RURAL CHINA

Well-functioning land markets play an important role in promoting industrialisation, urbanisation, and sustainable agricultural development in the rural area of mainland China, provided that sufficient off-farm employment possibilities are available. Experiences from the pre-communist period indicate that they may also lead to a concentration of land ownership and create a class of landless people. The farmers currently interested in transferring farmland in China, however, are particularly the farmers with off-farm employment. These groups have relatively high incomes (see e.g. Table 5.1).

The number of rural land transactions through market mechanisms is still fairly limited. Lin (1987) attributes this to the absence of a credit market for farmers because lenders' rights are not sufficiently protected. While in Latin America the lack of credit markets is indeed an important constraining factor (Shearer, 1991), in India borrowing has played a negligible role in financing land purchases (Sarap, 1995). The key problem in China is the security of land rights and their transferability. The security of rural landed property is adversely affected by the current system of land assignment to farmers. Security of land rights is the foundation of any land market. Insecure land rights hamper the transferability of, both ownership and usufruct rights of, rural land. In addition, the number of land transactions is limited by the lack of complementary social security institutions in rural areas that permit rural households a fallback position in times of unemployment, illness or retirement.

### 5.1 GOVERNMENTAL ACTION AND INSECURITY OF COLLECTIVE LAND OWNERSHIP

Government behaviour is an important contributor to the insecurity of land ownership rights. The Land Administration Law provides that all urban land must be owned by the state. The collectives mainly own farmland, but some cities have insisted that land used by rural enterprises must be owned by the state. According to the Chinese Constitution and Land Administration Law, collective land can be expropriated only when it is needed for public goods. A review of 200 land requisition projects in 11 counties carried out in 1992 showed that only 21% were used for public goods (e.g. highways, schools) and 5% for government purposes (e.g. office building, housing), while as much as 74% were used for companies or real estate development (Farmers' Daily, Beijing, 19th April, 1993). Therefore, some farmers say that collective land is like meat that can be cut when the government wants to eat (Zhou, G., 1994).

Land prices increased rapidly from 1992 to 1994. Guangdong Province, Shanghai City, and other large cities made requisitions in advance to save on the compensation for land requisition. This advance requisition has been forbidden by the State Council since 1995, because it leads to a decrease of farmland area that is not advantageous to agriculture development.

### 5.2 INSECURITY OF COLLECTIVE LAND USUFRUCT AND THE LAND ASSIGNMENT SYSTEM

Although the CCG rents farmland to farmers for 15 or 30 years, land usufruct rights are frequently changed. Between 1978 to 1995 user rights had been adjusted at least once for 95 percent of total farmland, 3.1 times on average (Central Policy Research Division and Ministry of Agriculture, 1997). Adjustments are made as a result of changes in family size or labour transfers, but also for other purposes including setting up scale farming, consolidation of fragmented farmland, establishment of infrastructure, and so on.

In many areas, no formal land contracts were signed with households from the beginning of the HRS. Other farmers lost their first land contracts, while no new contracts were signed. Sometimes, local governments or villages unilaterally change or cancel land contracts. In Hubei Province, there were about 15,000 disputes concerning land contracts in 1995, caused by villages or local governments who change or cancel at will (Central Policy Research Division and Ministry of Agriculture, 1997). In Huaiyin City, Jiangsu Province, a township government used police to take back farmers' contracted land in 1996 in order to establish a centre for agriculture technology farm owned by the township.

Because of the frequent adjustment of farmland, many farmers do not want to make long-term investments (e.g. Dorner and Saiba, 1981; Feder et al., 1988). In order to obtain immediate benefits from the land before an anticipated change takes place, farmers have been "mining" the soil and land, without taking care to leave resources for the future (Hu, 1997). Some farmers are even reported to throw salt or alum on the farmland instead of fertilisers when they know that their land usufruct will be adjusted, with the purpose of obtaining quick, short-term benefits. Consequently, soil organic matter content and nutrient balances deteriorate strongly (Qian, 1997;

Chinese Society of Agronomy, 1996). In mainland China, 59.1% of total farmland is already short of phosphate and 22.9% is short of potassium (Chen, Y., 1996).

Land rights are also important for farmers to obtain loans (e.g. Shearer et al., 1991; Sarap, 1995; Vogelgesang, 1996). In mainland China, only water and land usufructs of rural enterprises can be mortgaged, while collective land usufructs cannot be mortgaged. If farmers do not have a certificate of farmland usufruct, they have no collateral for borrowing.

Local governments or collectives levy many kinds of fees on farmland and farm households to raise funds or make up departmental spending. Examples are family planning fees, rural education fees, road establishment fees, irrigation establishment fees, fees given to soldiers' families, and training fees for militia. Between 1994 and 1995, lawful taxes and fees on farmers increased 14.3%, while net income per farmer increased by only 8.3% (Zhao et al., 1997). Taking into account the use of unlawful fees, the increase may often have been higher. These frequent government levies increase the insecurity of land rights.

### 5.3 LACK OF A SOCIAL SECURITY SYSTEM IN RURAL AREAS

Most households consider farmland as insurance, especially in developing areas. Although there are many kinds of social security systems for the urban population in mainland China, there is no such system for rural households. Results from a survey in Jiangsu Province (a high-productivity area in mainland China) show that, if farm households earned enough non-agricultural income, only 14.6% would return the contracted land to collectives. 63.7% of the households prefer to keep the contracted land, even though half of these consider it very difficult to make a living on their contracted land (Qian, 1997). At present, the contracted land is the best and only security system for rural households. In some instances, farmers abandon their land to pursue other business, but pay cash for renting the land that remains uncultivated. This evidently represents a waste of the scarce land resources in China (Hu, 1997).

To consolidate the fragmented land and develop sustainable agricultural systems, a social security system ought to be established in rural areas. In developed rural areas, many farmers work in rural enterprises if these enterprises are in good condition, and sublease contracted farmland to collectives or farmers. If the rural enterprises are no longer in good condition, they return to their contracted farmland again. So, the land transfer is mostly for a short period of time. Land transfers would be more secure if the rural enterprise institution were reformed, e.g. by establishing an unemployment insurance system for the farmers working in rural enterprises.

## 6 DISCUSSION AND CONCLUSIONS

As a result of the economic reforms and the development of agricultural input and output markets, rural land markets will become increasingly important for land allocation in mainland China. With respect to land ownership, few land sales by collectives still take place. Market functions, however, gain importance in land requisition by the government. With respect to land usufruct rights, the wasteland market already performs as a competitive market.

The urban land market is developing rapidly in developed areas or cities, as a result of the increasing demand for land and the far-reaching reforms in the urban land use system. But the current state policy on non-agricultural land use has a negative influence on non-agricultural land markets in rural areas. About 10% of farmland is now transferred through the market. The prevailing rural farmland market is characterised by temporary transfers and the absence of market prices.

The establishment of rural land markets in mainland China can be considered the result of a particular combination of government policies and market development. The state enables land transfers of wastelands for land consolidation purposes, and tries to increase land security through more prolonged usufruct rights. Nevertheless, land requisition for non-agricultural purposes results in an increasing pressure on agricultural land use. While population growth and urbanisation require a substantial increase in agricultural productivity, fragmentation of farmland and limited access to modern inputs (seeds, fertilisers) reduce the efficiency of agricultural production. Through the developments in the labour market (off-farm employment), rural surplus labour is transferred to other sectors and labour productivity has increased substantially. Further growth requires, however, an equally high increase in land productivity based on farm-level investment. Reinforcing property rights is considered an important condition for encouraging farmers to adopt more intensive land use and more commercial production.

Further development of the rural farmland market in mainland China is generally considered of primary importance to improve the effectiveness of economic policies aimed at stimulating farmers to increase productivity. To stimulate the functioning of rural land markets, the following institutional reforms are required:

1. Ensuring the security of land rights. The collective land ownership is equal to state land ownership in an economic sense. A free market for land use rights can be established, except for public land use.
2. Making more stringent regulations of governmental behaviour by law in order to protect land rights security and land market operation.
3. Establishing a social security system in rural areas, especially in developed areas.
4. Reforming the household registration system to allow more rural people to work and live in urban areas.
5. Using the value of farmland as the basis for assigning usufruct rights.
6. Establishing an efficient land contract enforcement system and a land market information service organisation.
7. Establishing a rural land usufruct mortgage system to promote land development.

**REFERENCES**

BUCK, J. L. (1937) *Land Utilisation in China.* London: Oxford University Press, pp.181.

BINSWANGER, H.P., K. DEININGER & F. FEDER (1994) 'Power, Distortions, Revolt and Reform in Agricultural Land Relations.' In: J. Behrman & T.N. Srinivasan (eds.) *Handbook of Development Economics*, vol. IIIB. Amsterdam: Elsevier, pp.2661-2772.

CARTER, M.R. & D. MESBAH (1993) 'Can Land Market Reform Mitigate the Exclusionary Aspects of Rapid Agro-export Growth?' *World Development*, vol. 21, no. 7, pp.1085-1100.

CENTRAL POLICY RESEARCH DIVISION AND MINISTRY OF AGRICULTURE (1997) 'The Investigation and Analysis of the Second Round Land Contract.' [in Chinese] *Rural Economic Papers*, no. 4, 4-13.

CHEN, D. (1996) *The Boost of China's West* [in Chinese]. Shanghai: Far-East House, pp.197.

CHEN, Y. (1996) *On Sustainable Development* [in Chinese]. Beijing: China's Planning House.

CHEN, X. W. (1987) 'The Rural Economy in China is Transferring from Supra-Conventional Growth to Conventional Growth.' [in Chinese] *Economic Research*, no. 12, pp.47-50.

CHINA'S LAND YEARBOOK 1994-95 [in Chinese] (1995) Beijing: China's People House, pp.331.

CHINESE SOCIETY OF AGRONOMY (1996) *On Chinese Grain of 2000* [in Chinese]. Beijing: China's Agricultural Science & Technology House, pp.333.

DONG, R. (1993) 'Do Not Take Public Land as a Commodity.' [in Chinese] *Housing & Real Estate*, no. 3, pp.56-58.

DORNER, P.& B. SALIBA (1981) 'Interventions in Land Markets to Protect the Rural Poor.' Land Tenure Centre Research Paper no. 74. Madison: University of Wisconsin.

FEDER G., T. ONCHAN & Y. CHALAMWONG (1988) *Land Policies and Farm Productivity in Thailand.* Baltimore and London: Johns Hopkins University Press.

FEDER, G.T. & D. FEENEY (1993) 'The Theory of Land Tenure and Property Rights.' *World Bank Economic Review*, vol. 5, pp.135-153.

HASETADEN & LI. (1996) 'Land Institution Reform in Chinese Rural Area: Wasteland Usufruct Auction.' [in Chinese] *Chinese Rural Economy*, no. 4.

HU, W. (1997) 'Household Land Tenure Reform in China: its Impact on Farming Land Use and Agro-Environment.' *Land Use Policy*, vol. 14, no. 3, pp.175-186.

HUANG, X. (1993) 'On Land Ownership Market.' [in Chinese] *Problems of Agricultural Economy*, no. 11, pp.34-36.

HUANG, X. (1995) 'On Land Rights Discrimination.' [in Chinese] *Problems of Agricultural Economy*, no. 7, pp.46-49.

HUANG, X. (1997) *On Farm Land Value* [in Chinese]. Beijing: China's Agriculture Press.

HUANG, X, F. QU & N. HEERINK (1997) 'Valuation of Arable Land in Mainland China: Comprehensive Estimates for the Agricultural Subregions.' In: P.W.J. Uithol and J.J.R. Groot (eds.) *Proceedings workshop Wageningen-China, Wageningen, May 13th, 1997.* Wageningen: Research Institute for Agrobiology and Soil Fertility (AB-DLO).

LI, Y. (1994) 'On Wasteland Auction in Luliang Prefecture.' [in Chinese] *Chinese Rural Economy, no.* 5, pp.27-30.

LI, Y. (1996) *Who Gives Land to Us?* [in Chinese]. Beijing: China's Land Press.

LIN, J.Y. (1987) 'Rural Factor Markets in China After the Household Responsibility System Reform.' Yale: Yale Economic Growth Centre Discussion Paper.

LIN J.Y. (1997) 'Institutional Reforms and Dynamics of Agricultural Growth in China.' *Food Policy*, no. 3, pp.201-212.

LU, M. & X.J. DAI (1987) 'An Analysis of Peasant Households' Economic Behaviour at Present.' [in Chinese] *Economic Research*, no. 7, pp.34-37.

QIAN, Z. (1997. 'A Positive Study into Farmers' Attitudes to Rural Land Reform and Implications for Policy.' [in Chinese] *Chinese Rural Economy, no.* 4, pp.66-70.

QU, F. (1991) *On Land Institutions in China's Rural Area: 1949-1990* [in Chinese]. Nanjing: Jiangsu People's Press.

QU, F., N. HEERINK & W. WANG (1995) 'Land Administration Reform in China: Its Impact on Land Allocation and Economic Development.' *Land Use Policy*, vol. 12, no. 3, pp.193-203.

QU, F. & X. HUANG (1997) *On China's Land Institution* [in Chinese]. Xuzhou: Chinese University of Mining & Technology House, pp.76.

ROSE, J. (1994) 'The Role of Philosophy and Politics in Urban Development and Control.' In . D. Chiddick, A. Millington & F. N. Spon (eds.) *Land Management: New Directions*. New York.

SARAP, K. (1995) 'Land Sale Transactions in an Indian Village: Theories and Evidence.' *Indian Economic Review, vol.* 30, no. 2, pp.223-240.

SHEARER, B. et al. (1991) *The Reform of Rural Land Markets in Latin America and the Caribbean: Research, Theory, and Policy Implications*. Madison: University of Wisconsin.

STATE ECONOMIC PLANNING COMMISSION. (1995) *'95 Population, Resources and Environment in China* [in Chinese]. Beijing: China's Environmental Science House, pp.28-35.

TAN, L. (1996) 'Scale Farming and its Adaptable Cropping System in Southern Jiangsu.' [in Chinese]. Nanjing: Nanjing Agricultural University, pp.6.

VOGELGESANG, F. (1996) 'Property Rights and the Rural Land Market in Latin America.' *CEPAL Review*, no. 58, pp.95-113.

WALKER, T. (1997) 'From Shadow to Spotlight.' *Financial Times*. September, pp.20-21.

WANG, X. (1994) 'Wasteland Development: Institution, Policy and Farmer's Behaviour.' [in Chinese] *Chinese Rural Economy, no.* 11, pp.3-7.

WANG, Z. & B. XU (1996) *On Land Institution of China's Rural Areas* [in Chinese]. Beijing: Economy Administration House. pp.32-50.

WORLD BANK. (1992) *China: Reform and the Role of the Plan in the 1990's*. Washington, DC: World Bank.

WORLD BANK. (1994) *China: Internal Market Development and Regulation*. Washington, DC: World Bank.

XIE, J. (1996) *Environmental Policy Analysis: A General Equilibrium Approach*. Gateshead, Tyne & Wear: Athenaeum Press.

XU, B. (1996) 'The Key to Ensure Self-sufficiency in Food is Policy.' [in Chinese] *Rural Socio - economic Papers* no. 3, pp.5-8.

ZEN, D. (1991) 'Collectives' Land in Countryside Should be Used under Taxation [in Chinese]. In *Study on China's Land Problems*. Beijing: China's Economy House, pp.78-81.

ZHANG, X. Q. (1997) 'Urban Land Reform in China.' *Land Use Policy*, vol. 14, no. 3, pp.187-199.

ZHAO, M. et al. (1997) 'On Counting of Farmers' Burden.' [in Chinese] *Problems of Agricultural Econo*my, no. 3, pp.43-46.

ZHOU, G. (1994) 'The Situation and Reformation of Rural Land Institution.' [in Chinese] *Real Estate Forum*, no. 3, pp.18.

ZHU, F. et al. (1996) *On China's Poverty Alleviation* [in Chinese]. Beijing: Chinese Planning Press, pp.8.

# 6 PERFORMANCE OF MARKETS IN A CONTEXT OF LIBERALISATION: THE HOG INDUSTRY IN NORTHERN VIETNAM

Katell Le Goulven

## 1 INTRODUCTION

Vietnam is now listed amongst the countries which are « in transition » from a centrally planned to a « market oriented » economy. Since 1986 the Vietnamese government has implemented a process of liberalisation called « Renovation » . The Renovation has led to a complete reorganisation of the agricultural commodity chains. From the establishment of the first agricultural cooperatives in Northern Vietnam in 1959 up until the late 1980's, agricultural production was part of a hierarchical chain of command from the Ministry of Commerce down to farmers who were organised into production brigades. The agricultural outputs were then marketed by State-run companies in charge of collection, transportation, processing and retail sales in State-run stores. Renovation policies were adopted during the fourth Vietnamese Communist Party congress in 1986. A change to a market-oriented economy was voted in and the government's first declaration was to recognise the household as the basic unit of production and to legalise private ownership.

Regarding the organisation of commodity chains, the most important change was the 1988-1989 decision to abolish the systematic State enterprise subsidy and to legalise private business for households and firms. These measures occurred at the same time as price controls were abolished and the Dong currency was devalued: two key signs of a market economy

Through this wave of reforms, the goal of the Vietnamese government was to enhance sustainable development and industrialisation. As far as agriculture was concerned, the objectives were: to stimulate agricultural growth and its contribution to income, to promote food security, and to succeed in switching to a market economy linked to the rest of the world (through the export of agricultural products) (Vu Tuan Anh, 1994). In this paper, we attempt to analyse the ability of Vietnamese agricultural markets to achieve those goals. Within this perspective, we focus attention on the hog subsector, which carries special weight in Vietnam. First, hog production accounts for 71% of the Vietnamese livestock (General Statistical Office, 1996) and pigs are almost raised in every rural household. Second, pork represents 83% of all meat consumed by the Vietnamese (General Statistical Office, 1996). And third, from the 'collective' period until the present, hog raising and selling has been one of the only ways for Vietnamese peasants to save money and earn cash income. Hog production is as important to the rural Vietnamese household economy as pork is to the Vietnamese meat diet.

This chapter is divided into three main parts, which relate to results extracted from data collected in the field in 1996 and 1997 during PhD research. The first part sketches the present organisation of the subsector on the basis of surveys carried out for every actor in the marketing chain, from producers to retailers. The second part provides some insights into the performance of the organisation previously described in an Industrial Organisation (IO) framework. The third part discusses this performance within a New Institutional Economics (NIE) framework.

The case study of the hog subsector in Northern Vietnam allows us to illustrate the necessity of considering institutions when analysing the performance of markets in transitional economies.

## 2 THE PRESENT ORGANISATION OF THE HOG MARKETING CHAIN

Northern Vietnam comprises 17 provinces, which account for 48% of the national hog production. Of this production, 26% is located in the 11 provinces of the North mountains and midlands (mainly in the provinces of Vinh Phu and Ha Bac), and 22% of hogs are raised in the Red River delta, where Hai Hung and Ha Nam Dinh are the primary production areas in the North of the country. The hogs of the mountains and midlands remain in the areas where they are produced, for local consumption. Vinh Phu, Ha Bac and the Southern districts of Bac Thai, as well as the provinces of the Red river delta, supply the rural surrounding areas and the large consumer basins: Ha Noi, Hai Phong and Quang Ninh. We can identify two different marketing chains: (i) one supplying the rural markets where hogs are collected, slaughtered and sold by a butcher, and (ii) one supplying the large consumer markets such as the capital of the country (Ha Noi), the first port of the country (Hai Phong) and the tourist area of Quang Ninh. In this paper, we focus on the latter chain through which 50% of the production from the North is marketed as meat demand rises in urban areas.

This section provides a general overview of the structure of the hog marketing chain, from the Northern countryside to the municipalities of Ha Noi and Hai Phong. The data used was collected from 200 surveys carried out among hog producers and traders in eight locations in Northern Vietnam.

### 2.1 GENERAL FEATURES OF THE HOG AND PORK MARKETING CHANNEL

The hog marketing chain that we consider goes from fattened hog production down to pork retail, see Figure 6.1.

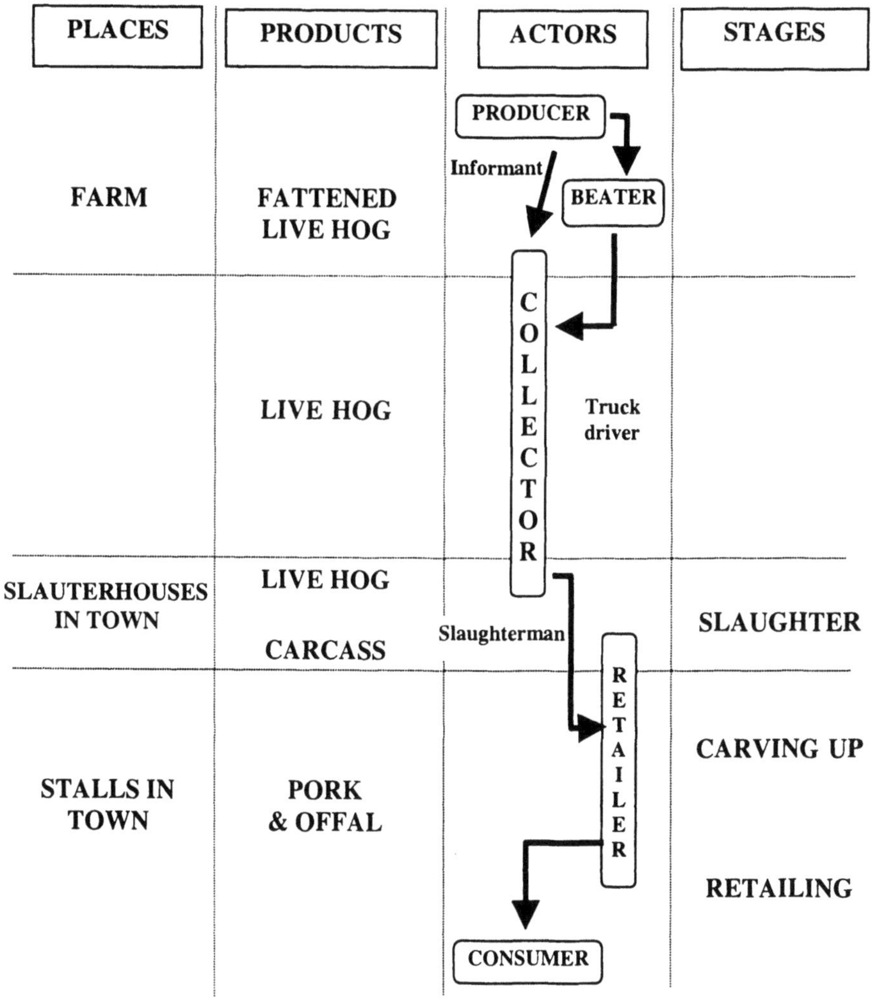


*Figure 6.1 Structure of the hog marketing chain*

Vietnamese hog production is spread throughout the rural areas, as 83% of producers raise one or two hogs a year (Le Ba Lich, 1996). Nevertheless, we can distinguish two different kinds of producers:

1. Traditional producers who only raise hogs for cash income and manure, and to make sensible use of crop residues. Production costs can be equated with opportunity costs, even if sometimes the hogs are fattened up intensively (using industrial feed) just before selling.

2. Market-oriented producers who raise 15 to 20 hogs at once. These farmers use the by-products of their extra activities (rice alcohol production, rice husking,

tofu production) to feed the animals. They represent 3 to 5% of all producers in rural areas of the Red River delta.

At the other end of the chain, we find two different types of pork consumers: fat meat eaters in the countryside versus lean meat consumers in the big cities where living standards are much higher. As a consequence, animals are sorted by the traders who supply the cities.

Between the consumers and producers, there are various actors who coordinate the marketing of the animals. The first stage of the marketing process is the selling of hogs from the farmyard. In the Vietnamese countryside, home consumption is almost non-existent, except for weddings or traditional ceremonies, and 80 to 90% of the total production is sold on rural or urban markets. Farmers can either sell their hogs to a local butcher or to a collector who supplies city markets. The latter is in charge of the transport of the animals to the city, as well as the slaughter and the selling of carcasses to urban meat retailers.

As there is no refrigeration within the chain, hogs are transported live and are slaughtered close to the place where pork is sold and consumed on the same day as purchase. At the longer and more complex urban end of the chain, collectors might sometimes buy hogs as a result of information from « beaters » or informants to increase the scale of their collection. Further down the chain, collectors use the services of "slaughtermen". These actors organise slaughter, connect supply and demand, and provide credit to both parties, as will be seen later on.

Slaughtering plants in the cities, even in the capital, are very basic operations. The standard private facility frequently consists of a convenient sidewalk or a paved forecourt inside a private house. The tools used are usually big knives. Slaughter requires no specific investment as animals are killed and dressed manually, using a knife only. Bristles are removed using hot water. All these jobs are done by pieceworkers hired by the slaughterman who owns or rents the « slaughterhouse ».

Finally, carcasses are sold to retailers who carve up the meat (into cuts on and off the bone) and sell it in markets or from small stalls in the street. Meat buyers are called retailers as there is no specialisation in meat wholesale; so any buyer can occasionally wholesale at least part of the carcass she[1] bought, and retail the rest. Offal retailers are more specialised as they sell heads and ears, feet or tripe.

Before deregulation all meat products belonged to the State. Nowadays, changes in ownership occur along the marketing chain. When exchanging, actors are linked together in a variety of combinations, some of which will be analysed in the following section. Some of the participants in the operations of the marketing chain exercise more influence than others even if they have no ownership titles to the product exchanged.[2]

## 2.2 RURAL COLLECTORS

Hog production is spread over a wide area, which generates high transaction costs in the first step of the marketing chain. Collectors have to spend a lot of time searching for potential sellers in their village and in the neighbouring ones (an average of four hours for one to two hogs). Besides collection costs, collectors have to cope with uncertainties on hog yields (ratio carcass weight/live weight), on hog health (sanitary aspect) and on carcass fat content (percentage of meat), as they are buying live animals and selling carcasses. These uncertainties can lead to opportunistic behaviour on the part of the sellers, who may try to feed the hogs before the transaction or to hide the real state of health of the pigs from buyers. On the one hand, this risk can partly be reduced if one is able to evaluate the quality of the carcass from a live hog. Each buyer can acquire this kind of specific knowledge by being taught and/or by learning-by-doing. This knowledge is gained individually, even if common criteria can usually be referred to. On the other hand, specific contracts can be adopted as an institutional mechanism to regulate exchange and reduce the risk of ex-ante opportunism (as we will see later on).

In order to bring down these costs collectors work with go-betweens: beaters or informants. Usually, collectors are aware of all the imminent sales in their villages and the way each peasant raises his own livestock. This information helps them assess the real value of the hogs. For instance, the collectors are aware that hogs fed on rice alcohol, tofu cake residues or industrial feed are leaner than those raised in a traditional way.

The information gathered by the collector is also based on his experience as a social member of his village, the habits and the reputation of the farmers, their reasons for selling, peasant knowledge of marketing mechanisms and so on. The collector uses these elements to gain greater insight, which helps him reduce risk and improves his bargaining position when buying hogs from the peasants. However, this insight is only valid at the village level. Gathering data on a greater scale would imply huge costs in time and personal involvement on the part of the collector in each village where he would like to buy hogs. This is why, when working in other villages, collectors resort to informants' services. The informant[3] has his personal database for his own village and sells it to the collector when required. Informants are usually simple farmers or occasional collectors, and their wages are related to the number of hogs sold (at a commission of 1% of the total value of the animal). Informants have no property rights to the animals finally purchased by collectors from peasants. Beaters have exactly the same function as informants but they only work for one collector and are not expected to take any initiative. In this vertically coordinated relation, beaters are paid according to the number of hogs bought and their presumed quality (which provides a high incentive to buy hogs with lean carcasses). In addition to reducing the risks linked to the quality of hogs, informants and beaters lower transaction costs, as collectors can concentrate on buying the hogs rather than scouring the countryside[4] to find potential sellers, thereby extending their collection area. Finally, informants and beaters make peasants from their own village sell to collectors on credit. In such cases, a go-between acts as a guarantor for collectors.

Collectors, informants and beaters form a rural network of buyers and mediators, within which information on prices and farmers' behaviour circulates efficiently. Most of the time they work alone or in small groups of two or three members of the same family or very close neighbours. Even if they compete to buy hogs, great solidarity prevails between people doing the same job. In addition, they are connected to urban marketing networks, which are the key to selling livestock in towns.

For peasants involved in the first transaction, their knowledge of prices and market mechanisms is, most of the time, restricted to the village or the district area. Even when some of the producers are aware of the way hogs can be sold in the cities, they do not have the opportunity to go and market them themselves, as they are not connected to specific networks. The only way farmers can keep buyers under pressure, if they do not agree to the terms of the contract, is to not sell their hogs. Such an alternative will be in vain if the purpose of the sale is to raise cash income in order to pay agricultural taxes, provide crop inputs or cover unforeseen expenses. Even if it is not that imperative to sell, producers can hardly afford to keep on fattening hogs once they have reached 80 or 90 kg, as feeding them is no longer profitable.

To recapitulate, in the first transaction, the two sides are not on an equal footing. The buyer is integrated into a network where each has the others' support and knows market rules, whereas the seller is faced with no alternative but to sell his product. Stopping production is not an alternative for farmers as hog raising is often the only way to save money and get cash income. In such a context, producers are totally dependent on buyers.

Collectors are in a position of strength in the countryside, but they no longer control the exchanges in the city, where they need slaughtermen's services to contact retailers and organise the slaughter.

## 2.3 SLAUGHTERMEN IN THE CITIES

Pork yields remain uncertain until the second transaction is carried out. The owner of the slaughterhouse where hogs are killed orchestrates[5] the transaction between the seller and the retailer, without buying the animals. His role is to connect supply and demand, to provide all the facilities for slaughtering the animals and funds for buyers and sellers. Collectors sell the animals according to carcass weight, although they bought them alive. However, as there are no cold storage facilities, retailers must choose the hogs alive and discuss the price of the carcass before slaughter. Contracting this way, the risk is shared between both traders, while the slaughterman just takes a percentage (2% of the carcass value) plus a payment in kind.

In Ha Noi, an estimated 150 slaughtering plants operate, either on a small scale (less then 10 hogs per day) or on a larger one (from 20 to 150 hogs per day). The smaller ones are supplied by small collectors who transport hogs by motorcycle from the nearby countryside. The owners of the larger ones directly deal with big collectors who transport the hogs (30 to 100 per trip) to the capital by truck.

Slaughtermen not only connect rural collectors with retailers scattered in the cities but they also provide credit to both forms of traders. They even pay cash to collectors in the morning, although some of the animals will not be sold, to allow the collectors to return home on the same day. Retailers buy all the carcasses on credit. In the afternoon, when they visit all the markets, slaughtermen recover money from retailers. They make enquiries about the state of demand and their competitors' supplies at the same time.

Through this advance/credit system, slaughtermen keep both collectors and retailers in a position of dependency. Moreover, being connected with both sides, they are able to concentrate information on supply and demand. They regulate the hog supply. Indeed, a collector is affiliated to only one slaughterman through a contract, which defines credit and hog delivery on a daily basis. In contrast, a slaughterman works with an average of 5 to 6 collectors.

The current state of the marketing chain as described above, is far removed from the previous one, as only private actors are concerned with the domestic market and only some of them seem to have taken advantage of the new system. In the following sections we analyse the performance of the hog marketing chain within two different economic frameworks.

## 3 MARKET STRUCTURE AND PERFORMANCE

« In the field of Industrial Organisation, we seek to ascertain how market processes direct the activities of producers in meeting consumer demands, how those processes can break down, and how they adjust, or can be adjusted, to make performance conform more closely to some ideal standards » (Scherer, 1990). In the Vietnamese context, ideal standards are related to the aims of the government, i.e. the switch to a market economy, achieving food security and stimulating agricultural growth.

### 3.1 THE ENHANCEMENT OF AGRICULTURAL PRODUCTION AND FOOD SECURITY

The Vietnamese Gross Domestic Product (GDP) has increased continuously since the first steps of liberalisation were taken. The World Bank estimates that the programme of reforms has yielded growth rates of 9.3 percent GDP in 1996, boosted industrial output growth by 15.6 percent, and increased exports by 25.6 percent. Such an evolution is partly due to the increase in agricultural production (+ 4%) but increasingly to the sectors of services and industry (+14,5% and +11,8% respectively) (van Potten & al., 1996). Nevertheless agriculture still represents one third of the GDP and employs 70% of the national labour force.

Thanks to impressive gains in productivity and despite the importance of domestic needs[6], Vietnam is now the world's third rice exporter. Considering meat, even if the average rural food intake remains deficient in protein (Ha Huy Khoi, 1996), the production of pork meat for 1992 (591,400 tons according to raw data from General Statistics Office) covers domestic needs (579,700 tons[7]).

Even though the rapid expansion of the Vietnamese economy and agriculture has led to the achievement of food security, a gap remains between rural and urban areas. Whereas the GDP per capita for the whole country is 300 $, the average is 800 $ in Ho Chi Minh City, 500 $ in Ha Noi and less than 200 $ in the countryside (van Potten & al., 1996). When trying to understand the origin of this differentiation, one is tempted to scrutinise the relations between agricultural production, processing and marketing in an attempt to evaluate the performance of the hog subsector.

### 3.2 STRUCTURE OF THE HOG SUBSECTOR AND CONCENTRATION IN THE CHAIN

In Table 6.1 we present the Gross Marketing Margin (GMM) of each actor. This is the share of the final selling price captured by a particular agent in the marketing chain.

*Table 6.1 Actors' gross marketing margins in the hog chain between Nam Thanh and Hai Phong*

| | Actors' Gross Marketing Margins (GMM) | | | |
|---|---|---|---|---|
| | Per hog (80 kg of liveweight, 60 kg of carcass) | | Per year | |
| | Percentage of the total value of the animal | In thousands dongs[1] | Number of hogs marketed | In thousands dongs |
| Producer's share | 70 | 760 | 2 | 1,520 |
| Informant's share | 1 | 10 | 700 | 7,000 |
| Collector's share | 4 | 44 | 3,000 | 128,000 |
| Slaughterman's share | 5 | 56 | 7,300 | 408,800 |
| Offal retailer's share | 3 | 38 | 170 | 6,460 |
| Pork retailer's share | 16 | 170 | 170 | 28,900 |
| Taxes, transport... | 1 | | | |

*[1] 1 US$ = 11 600 dongs*
*Source: personal data collected through surveys conducted in 1996*

The shares in the hog marketing chain in Vietnam are similar to those found in most developing countries (Abbott, 1987) and in France in the early 1980's (Boutonnet, 1995). As far as marketing margins are considered, the structure of the marketing chain does not reveal the emergence or the appropriation of major rent.

Second, there is no concentration at the production level: 82.6% of the 16 million hogs produced in 1994 were spread amongst households raising one or two animals. When focusing on slaughterhouses in the two biggest cities of Northern Vietnam, the results of the surveys show that the concentration is not important in Ha Noi, but may tend towards a monopoly in Hai Phong. To estimate the level of concentration, we use the market concentration ratio CR, defined as the percentage of total industrial sales (carcasses as the outputs of the slaughterhouses) contributed by the few largest firms, ranked in order of market share. The most common variant is the percentage of total industry sales originating from the four leading firms: CR4 (Scherer, 1990).

On the one hand, the largest plant in the capital slaughters 6% of the total amount of meat sold, the CR4 is 15% and the CR8 is 24%, thus we cannot infer a situation of either monopoly or oligopoly.[8] On the other hand, the largest plant in Hai Phong slaughters 20% of the carcasses marketed in the city, whereas each of the other slaughterhouses produces 1 to 2% of the global output. The aim of the largest plant in Hai Phong is to increase its production in order to capture 60 to 70% of the carcasses sold (Le Goulven & al., to be published). The Herfindahl-Hirshman Index[9] would have been of great help as another suitable measure of market concentration in the cities. However, the surveys did not provide an exhaustive list of slaughterhouses or of their market shares.

The basic index used to evaluate the level of firm concentration does not reveal any tendency towards monopolistic power amongst slaughterhouses in Ha Noi. We could go further in the classical analysis of market structure and resulting performance. More sophisticated and subtle tools would allow us to highlight where market power is effective. However, data for individual firms, such as price elasticity, demand function and return on capital is not available in Vietnam because of the lack of specific studies and the difficulties of collecting relevant data in a country where the informal economy accounts for 75% of the national economic activity (Madesclaire, 1994).

The structure (concentration index) as well as the performance (GMM) reveal that the hog market in North Vietnam is closer to perfect competition than to oligopoly.

### 3.3 TRANSPARENCY OF INFORMATION AND ENTRY BARRIERS

The description of the marketing chain illustrates the central position of slaughtermen in the organisation and regulation of the market. Slaughterhouses in both Ha Noi and Hai Phong adjust the prices of carcasses every day according to information on fluctuations in supply and demand. Every morning before the slaughter starts, slaughtermen visit several other plants to evaluate the supply and every afternoon they estimate the demand in the city markets.

From the description of the marketing chain structure, it appears that there are some entry barriers for professions related to hog trading. As defined in Industrial Organisation theory (IO), entry barriers are the production costs which must be borne by the potential entrant to an industry, but not by existing firms (Scherer, 1990). In the Vietnamese hog market, production costs are negligible at each stage of the marketing process, as none of them requires important investment The costs that shape entry barriers are not production costs but transaction costs, which are not taken into account within the IO framework.

Using IO tools to analyse hog marketing in Northern Vietnam, the market appears efficient enough to allow for food security, to stimulate agricultural growth and be competitive. Nevertheless, if we consider the aims of the Vietnamese government as performance criteria, but within a New Institutional Economics (NIE) framework, our findings become different.

## 4 MARKET AND INSTITUTIONS

Going further than considering production and distribution costs, NIE focuses on the costs involved in making transactions, and the social rights and obligations (institutions) which people devise to reduce them (Hubbard, 1997). Terms of transactions are costly to measure and to enforce (North, 1990). Contracts and institutions either evolve to reduce those costs or fail to do so. In the hog marketing chain we attempt to identify such contracts and institutions to evaluate their impact on market efficiency.

### 4.1 THE MISSING LEGAL INSTITUTIONS

#### *4.1.1 THE BASIC FUNCTIONS OF THE STATE*

The withdrawal of the State from the economy and the liberalisation of trade (new tax policies, private property rights) have resulted in inefficient basic functions of the government. The adoption of new rules by all the actors has not been immediate. In addition, the existence of corruption and abuse by representatives of the authorities affects the trust that actors should have in the impartiality of legal institutions.

In Vietnam, the State remains relatively weak in assuming basic functions such as enforcing the new laws that have been passed since the beginning of the process of liberalisation (Fforde & de Vylder, 1996). The economic context is also characterised by restrictions on capital markets (Dao Thê Tuân, 1997) and shortcomings in the enforcement[10] of private property rights and contractual agreements between private actors (Schwarz, 1995).

### *4.1.2 MEAT INSPECTION AND HEALTH CONCERNS*

In the Ha Noi inner-city area, 97 slaughterhouses (of which 79 were for hogs only) were legally registered in 1997 (Courrier du Vietnam, 1997). However, most of those we surveyed were not registered and were not subject to veterinary inspection before or after slaughter. Indeed, the Department of Animal Health and Production's (DAPH) animal quarantine and meat inspection centre in Ha Noi estimates that only 20 or 30% of the total number of carcasses in the capital are actually controlled.

We interviewed 20 slaughterhouse tenants/managers and were present at slaughter several times in each location during a period of more than three months both in Ha Noi and in Hai Phong. All the hogs that died either during transport from the countryside to the city or in the stockyard were sold to meat retailers. Dead animals are sold at very competitive prices and are thus reserved for special retailers, according to the clientelization strategy adopted by the seller.

Laws concerning health and veterinary regulations do exist in Vietnam but are not efficiently enforced even when slaughterhouses are legally registered. In fact, the situation at times threatens human health. In 1994-1995, for instance, 93% of the tiêt canh (coagulated blood) tested by the Centre for Hygiene and Epidemiology did not match the required Vietnamese norms. The number of human deaths due to food poisoning rose continuously between 1990 and 1995 (Courrier du Vietnam, 1996).

Not only is public health threatened but also potential meat exports. Domestic needs are now met but even if Vietnamese hogs were to conform to international standards, export would be impossible as long as veterinary laws are not enforced. The lack of properly enforced sanitary regulations to prevent contamination of herds by swine fever or foot and mouth disease[11], as well as contamination of meat in the slaughterhouses constitute insuperable barriers to trade.

### *4.1.3 TAX COLLECTION*

When it comes to taxes, the State is hardly more efficient than it is with veterinary inspection in a subsector where most slaughterhouse are unofficial. But, even when slaughterhouses are properly registered by local authorities, slaughter and commercial taxes are hard to collect. We have shown, for instance, that only 3% of those taxes were collected in 1996 in the second biggest city of Northern Vietnam, Hai Phong (Le Goulven, 1996). Moreover, the tax bureau of Hai Ba Trung district in Ha Noi city revealed that, in 1996, taxes were paid on 4,437 hogs, which represents only 26% of what is officially slaughtered in this district.[12] We can thus expect the actual percentage to be lower.

The situation is much the same in the countryside where slaughter taxes are also collected. In fact, these taxes are usually collected twice (in rural and in urban areas) although the law calls for only one collection because taxes constitute local government funds. The tax, which amounts to one US dollar per hog sold or slaughtered, is regarded as very high by most Vietnamese[13]. Moreover, tax officials usually associate poor tax collection with people's poor awareness of any laws, and mismanagement (Phuong Van, 1996). We would add to the previous explanations the corruption of taxmen by collectors that we witnessed during our stay in the field.

As a consequence and at the national level, the Minister of Finance, Ho Te, declared in 1996 that even if taxes collected in the first quarter of 1996 were 42.1 percent more than the same period the year before, they had still only achieved 20.1 percent of the total planned for that year.The inefficiency of the tax system affects the fiscal balance of the government and has resulted in an increase in informal trade. When tax rates are too high, there is no incentive for businesses to invest, free-riding is enhanced, and illegal trade also increases. As a result, the State has to invest in costly strategies to detect and sanction tax dodgers. When only a minority of people operate within the formal sector, the State is tempted to increase tax rates in order to cover its needs. However, in such circumstances, NIE economists would advocate keeping tax rates low (De Soto, 1989; Klitgaard, 1991).

The inefficiency of the tax system also affects the sustainable development of the hog and pork industry. Slaughter taxes collected locally are supposed to contribute to the local authorities' funds allocated to improve the market infrastructure (e.g. slaughterhouse waste management, plants conforming to international standards, wholesale markets). The legal institutional environment of the Vietnamese hog market is weak: the enforcement of property rights and contracts as well as the regulation of marketing are not efficient. Therefore traders resort to illegal institutions to govern their transactions and reduce the costs resulting from uncertainties.

## 4.2 MARKETING CONTRACTS AND INFORMAL ENFORCEMENT

### *4.2.1 RELATIONAL CONTRACTS*

In the hog subsector in Vietnam, actors have to cope with several kinds of uncertainties: uncertainty about the quality of the exchanged product and uncertainty about the behaviour of the partner in the transaction. For every transaction of the marketing process, these uncertainties lead to asymmetrical information exchanges between actors involved, therefore leading to greater likelihood of opportunistic behaviour. To reduce these uncertainties and to reduce the resulting costs of acquiring reliable information about the quality of the product or about market prices, the parties resort to specific bilateral agreements for each transaction (Williamson, 1985). Uncertainties can be reduced by formal and informal macro-institutions. Finally, the evolution of technology (production techniques as well as slaughtering techniques) can also reduce uncertainties.

In our case study, interviews with both traders involved in each transaction reveal that people coordinate their activities through verbal contracts (Le Goulven, 1996). Such agreements are identified as relational contracts (Goldberg, 1976), where the identity of the parties and the social milieu within which the contract is consummated are relevant.

Relational contracts between producers and collectors, as well as between collectors and retailers structure the exchange of property rights concerning the animals. The terms of the contracts are such that information costs and actors' risks are reduced (Le Goulven, 1996). But the terms have to be respected by both parties for the contract to have any meaning. This leads us to the question of enforcement of the contracts.

### *4.2.2 ENFORCEMENT THROUGH MUTUAL TRUST AND REPUTATION*

Along the marketing chain, contracts are enforced by mutual trust built up between the two parties in repeated transactions. Mutual trust is a calculated trust which is a «particular level of the subjective probability with which an agent assesses that another agent or group of agents will perform a particular action » (Williamson, 1993). This trust is exchanged in the same way as the product is exchanged. If party A cheats at $t^0$ (e.g. the producer does not tell the collector that his hog is sick), party B will either cheat at $t^1$ (e.g. the collector gives a price which is not the real market price) or break the contract as a result of loss of trust in the partner (the collector will never buy hogs from this producer again). In substance, the social content of personal relations « carries strong expectations of trust for abstention of opportunism » (Granovetter, 1985).

Moreover, the trust experienced by A towards B can be spread to other actors and then build up the reputation of B. From a specific group of people's point of view, every actor has a reputation capital which reflects the attitude adopted by this individual in all the relations he has experienced with the other members of the group. Groups convey and spread a reputation. Reputation is also used to enforce the verbal contracts made between actors during transactions. If a hog producer cheats in a transaction with a collector (e.g. by selling sick hogs or fattening up hogs just before the sale), he will lose his reputation of « good producer » or « good seller » amongst groups of collectors or producers respectively. For reputation to be able to enforce contracts effectively, the existence of groups is crucial. Groups must exert a pressure on members who do not respect internal rules, and punish cheaters.

In our case study, we included villages where actors (mainly producers) need to keep building their social positions: « One of the historical characteristics of Vietnam is that social organisation of rural areas is based on village collectivity » (Nguyen Trong, 1997).

Hog traders are also organised in networks. We identify collector networks and urban retailer networks in the hog marketing chain (Le Goulven, 1997). During the collectivisation (and it may also have been true before), actors involved in centralised and State-controlled hog trade or in the black market accumulated (i) specific knowledge of the hog trade and (ii) relational social capital. Actors acquired skills in buying hogs (identifying the lean hogs from the fat ones etc.), transporting them (without damaging their quality) and in identifying consumer preferences. Moreover, during this period, actors knew each other well and built strong trading and social relations (social capital) which constitute the cement of present trader networks. Indeed, we found that most of the traders of the present market were involved in the State-controlled marketing chain or in the black market before 1986 (Le Goulven, 1997).

The relations during the previous collective period built up or reinforced a path dependency (North, 1990) within which people are now organised to bring a product from the producers to the consumers. The legacy of the past plays a significant role in today's system, as is the case in other former communist countries (Chavance, 1994; Pouliquen, 1994).

Our example illustrates the social embeddedness of markets (Granovetter, 1995) and the importance of taking social structure into account when considering economic issues. East Asian countries have been referred to as « network » or « collectivist » societies by many economic sociologists. « Collectivist » systems as opposed to « individualist » ones (as found in western countries) « are more efficient in supporting intragroup agency relations and require less costly formal organisation (such as law courts) but it restricts efficient intergroup agency relations » (Greif, 1996). The following paragraph provides some insights into how a collectivist society can affect trade and the efficiency of the market.

### *4.2.3 LIMITS OF SOCIAL ENFORCEMENT AND EMERGENCE OF POWERFUL ACTORS*

Reputation and the threat of its loss are associated with groups or networks. « The relative efficiency of informal contract enforcement based on collective punishment thus depends on the extent to which these social and cultural factors happen to lead to segregation that is economically efficient » (Greif, 1996). In fact, a contract between two parties from different groups or networks (e.g. a collector from the rural area and a retailer from the urban area) cannot be enforced, because losing one's reputation amongst a group of retailers does not constitute a credible threat for a collector, or vice versa. Moreover, such intergroup relations require huge investments of time and credibility for the transaction to become effective in a trustful environment.

In the Vietnamese hog marketing chain, slaughtermen connect the supply from the countryside to the demand in the city and therefore belong to both associated networks. Their connections to both networks are inherited from acquaintances they had and relations they built up during the previous collective system or on the black market. As the distance (both social and physical) between rural collectors and urban retailers does not allow for social reinforcement, contracts are enforced by the slaughtermen's own trustworthiness and reputation. In addition, we have shown that they provide credit to both parties of the transaction in an environment of poor credit markets (Ronnas, 1992).

Slaughtermen derive the largest annual benefits from the hog trade (see Table 6.1). Such a position should « classically » attract more competitors to the profession. But 75% of the largest slaughtermen in Ha Noi were involved in the previous State system or in the black market and were the first to take advantage of their previous position when liberalisation was initiated. Later, newcomers could have joined the profession if entry barriers had not existed. The essence of the entry barriers to the hog market appears to be a mix of (i) inheritance of the former system, i.e. social capital built up between traders, and (ii) access to the capital market.

As contracts are enforced, trade is effective and urban stalls are supplied on a daily basis, the market and its structure can be said to perform well. Nevertheless, in similar contexts, economists and sociologists have shown that collectivism and enforcement through social sanctions have limited the flexibility in the face of changing situation (Greif, 1996; de Soto, 1989). Finally, in a context where the State does not provide effective law enforcement, which then becomes assumed by social rules, a Mafia is likely to emerge.

In 61% of the large slaughterhouses surveyed (which slaughter at least 30 hogs a day), tenants/managers hired someone to protect the hog stock from theft. This person is also in charge of using violence to make the parties respect the terms of the contract when reputation or trust is not sufficient. This primitive mechanism of physical sanctions also fosters reputation. A plausible hypothesis to explain the emergence of violent means of enforcement is the following: in the countryside where social links are strong, reputation, trust and social capital are enough to enforce trade contracts, whereas in the city these fundamentals may lose their intensity and their enforcing power. There the lack of both social and legal rules to structure trade gives way to the birth of privately organised violence surrounding selling and buying. In Hai Phong, the largest slaughterman works with a gang in charge of protecting property rights to animals and punishing cheaters (Le Goulven, to be published).

When formal or traditional institutions no longer exist or cannot effectively enforce contracts, informal ones like a Mafia organisation emerge. This phenomenon threatens democracy and the legitimacy of the State and thus the performance of markets.

## 5 CONCLUSION

This paper attempts to illustrate the need to enlarge the classical economic framework when analysing the performance of a market. This enlargement should reflect the fact that history matters (North, 1990) and that markets have a social dimension. It also highlights the fact that market performance can only be approximated if institutions and their own performances are internalised.

In Vietnam, the withdrawal of the State has stimulated agricultural production and helped to achieve food security. At first glance, the hog market appears close to a perfect market where there is no concentration, no large profits captured and a daily adjustment of prices. Nevertheless, the organisation of the market in the context of the absence (or inefficiency) of basic functions of the State, endows power to actors in a position to take advantage of such deficiencies. As a matter of fact, the environment reveals neither efficiency nor equity, but only advantage to actors endowed with decisive factors such as positions of authority and access to capital or information before liberalisation.

As a consequence, our case study illustrates that the lack of efficient institutions endangers the development and thus the performance of a market by not providing conditions for its long-term expansion. Our findings go against the direct and sometimes accepted relationship between the withdrawal of the State from the economy and an increase in economic performance. Laws are a determinant for development and the State is crucial for providing laws that will ensure property rights, contracts and extracontractual rules. Our findings support the propositions that (i) « government intervention has played a critical role in successful development efforts » (Stiglitz, 1989) and (ii) « the great differences in wealth of nations are mainly due to differences in the quality of their institutions and economic policies » (Olson, 1996).

## NOTES

[1] In Vietnam, retailing is an activity carried out by women only.

[2] In Figure 6.1, actors encircled are people who have, at least for a short time, property rights to livestock, carcasses or meat.

[3] The vernacular term used is « chi tro » which means showing, indicating in Vietnamese.

[4] « nguoi gom lon » or « nguoi bat lon » are the Vietnamese terms translated into collectors in this paper.

[5] « cai » is the vernacular term used to qualify slaughtermen and means conductor or works foremen.

[6] Rice contributes for more than 60% to the daily energetic supply of the Vietnamese ration (Lebailly, 1995).

[7] The average meat consumption in 1992 is 5.76 kg/head/year in rural areas and 11.4 in urban areas where 20% of the population live (data from the Vietnam Living Standards Survey). Pork stands for 80% of the meat consumed in Vietnam.

[8] The data has been collected in 15 slaughterhouses we interviewed in Ha Noi in 1997. The 15 plants surveyed stand for 37% of the carcasses marketed in the capital and include the biggest plants of the city .

[9] $H=\Sigma (qi/Q)^2$ , the index measures the sum of the quadrats of firms' market shares. It takes values between 0 and 1. A market of four identically large firms consequently gives a Herfindahl index of 0.25.

[10] In 1995, a delegate from the government reported to the media: « legal enforcement is a big challenge for us, but we don't know how to tackle it » (Schwarz, 1995).

[11] In 1994, the rate of vaccination coverage for the hog population was estimated at 30% by the Department of Veterinary Service.

[12] 120 tons of meat are required to cover Ha Noi citizens' everyday needs and 70% of it is supplied by inner city area slaughter houses, 60% of which are located in Hai Ba Trung district (Courrier du Vietnam, 1997).

[13] GNP was 170 US dollars per capita in 1993 according to the World Bank.

**REFERENCES**

ABBOTT, J. C. (1987) *L'amélioration de la commercialisation dans le monde en développement*. Rome : Food and Agricultural Organisation of the United Nations.

BOUTONNET, J.-P. & J.-P. SIMIER (1995) *Les viandes*. Paris : Economica.

CHAVANCE, B. (1994) *La fin des systèmes socialistes*. Paris : L'harmattan.

COURRIER DU VIETNAM (1996) 'Renforcement du contrôle des aliments.' In *Courrier du Vietnam*, pp.1-2.

COURRIER DU VIETNAM (1997) 'Alerte au contrôle vétérinaire à Ha Noi.' In *Courrier du Vietnam* pp.4-5.

DAO THE TUAN (1997) 'Les transformations rurales récentes au Vietnam.' In *Cahiers Agricultures*, vol 6, no. 5, pp.13-18.

DE SOTO, H. (1989) *The Other Path : The Invisible Revolution in the Third World*. New York : Harper and Row.

FFORDE, A. & S. DE VYLDER (1996) *From PLan to Market : The Economic Transition in Vietnam*. Boulder : Westview Press.

GENERAL STATISTICAL OFFICE (1996) *Impetus and Present Situation of Vietnam Society and Economy After Ten Years of Doi Moi*. Ha Noi : Statistical Publishing House.

GOLDBERG, V. P. (1976) 'Regulation and administrated contracts.' In *The Bell Journal of Economics*, vol 2, no. 7, pp.426-448.

GRANOVETTER, M. (1985) 'Economic Action and Social Structure: The Problem of Embeddedness.' In *American Journal of Sociology*, vol 91, no. 3, pp.481-510.

GRANOVETTER, M. (1995) 'La notion d'embeddedness.' In A. Jacob & H. Vérin (eds.) *L'inscription sociale du marché*. Paris : L'Harmattan, pp.11-46.

GREIF, A. (1996) *Contracting, Enforcement, and Efficiency : Economics Beyond the Law*. Annual World Bank Conference on Development Economics. Washington : The World Bank.

HA HUY KHOI (1996) *An overview of nutritional situtation in Vietnam*. Ha Noi : Nha xuat ban Y hoc.

HUBBARD, H. (1997) 'The 'New Institutional Economics' in Agricultural Development : Insights and Challenges.' In *Journal of Agricultural Economics*, vol 48, no. 2, pp.239-249.

KLITGAARD, R. (1991) *Adjusting to reality : beyond "State" Versus Market in Economic Development*. San Francisco : International Center for Economic Growth.

LE BA LICH (1996) *Livestock production of Vietnam in the passing year and development orientations toward 2000*. National Seminar on Animal Sciences and Development to year 2000, Ha Noi.

LE GOULVEN, K. (1996) *Les formes de coordination de la filière de viande porcine dans le delta du Fleuve Rouge*. Montpellier : ENSA.M.

LE GOULVEN, K. (1997) *Impact of liberalization on commodity chain organization : The case of hog/pork subsector in Northern Vietnam*. Mini-Conference of RC 40, Toronto : Rural Sociological Society.

LE GOULVEN, K., BOUTONNET J.-P. & J.-M. CODRON (to be published) 'Commercialisation d'un produit agricole dans un contexte économique "de transition". La filière de viande porcine de Nam Thanh à Hai Phong (Vietnam).' In *Revue d'Élevage et de Médecine vétérinaire des Pays tropicaux*.

LEBAILLY, P. & P. BUNNY (1995) 'Accroissement des exportations agricoles et sécurité alimentaire au Vietnam.' In *Économies et Sociétés*, vol 22, no. 3-4, pp.361-371.

MADESCLAIRE, Y. (1994) 'Au-delà des apparences: un autre regard sur le Vietnam des années 90.' In *Revue Tiers-Monde,* vol 35, no. 140, pp.891-906.

NGUYEN TRONG, N. T. (1997) *Les nouveaux paysans vietnamiens : retour et mobilité (1986-1996).* Paris : Institut d'Études du Développement Économique et Social.

NORTH, D. C. (1990) *Institutions, Institutional Change and Economic Performance.* New York : Cambridge University Press.

OLSON, M. (1996) 'Distinguished Lecture on Economics in Government.' In *Journal of Economic Perspectives,* vol 10, no. 2, pp.3-24.

PHUONG VAN (1996) 'Tax dodgers face hefty penalties under new drecree.' In *Vietnam Investment Review.* Ha Noi.

POULIQUEN, A. (1994) 'L'ordre social soviétique et l'économie administrée. Une grille utile pour l'analyse du post-communiste.' In *Revue d'études comparatives Est-Ouest,* vol 25, no. 2, pp.5-25.

RONNAS, P. (1992) *Employment Generation through Private Entrepreneurship in Vietnam.* Amsterdam : SIDA ILO.

SCHERER, F. M. & D. ROSS (1990) *Industrial market structure and economic performance.* Boston : Houghton Mifflin Company.

SCHWARZ, A. (1995) 'Nation Builders. In *Far Eastern Economic Review*, November 16, pp.22-23.

STIGLITZ, J. E. (1989) 'Markets, Markets Failures, and Development.' In *The American Economic Review*, vol. 79, no. 2, pp.197-203.

VAN POTTEN, A. J., J. M. E. JONKER, & al. (1996) *Food and Agribusiness in Vietnam.* Utrecht : Rabobank International.

VIETNAM, SOCIALIST REPUBLIC OF, STATE PLANNING COMMITEE AND GENERAL STATISTICAL OFFICE, (1994). *The Vietnam Living Standards Survey 1992-1993.* Ha Noi : Statistical Publishing House.

VU TUAN ANH (1994) *Vietnam's economic reform : results and problems.* Ha Noi : Social Science Publishing House.

WILLIAMSON, O. E. (1985) *The Economic Institutions of Capitalism*, The Free Press.

WILLIAMSON, O. E. (1993) 'Calculativeness, Trust and Economic Organisation.' In *Journal of Law and Economics*, vol 36, pp.453-486

# PART 3

## MARKET LIBERALIZATION IN AFRICA

# 7 THE EFFECTS OF LIBERALIZATION ON FOOD MARKETS IN AFRICA

Ousmane Badiane

## 1 INTRODUCTION

As recently as 15 years ago, the state dominated most aspects of agricultural production, marketing and trade in many African countries. Over time, this type of government control resulted in the suppression of producer prices and incentives, inefficiencies in agricultural marketing, stagnation of agricultural production, and an excessive budgetary burden that could no longer be sustained. With the adoption of market reform and stabilization programs, starting in the mid 1980's, most countries in Africa began the process of dismantling state control in favor of a more market-oriented economy. Despite extensive market reform efforts, the transition phase from a state-dominated economic environment to a market-based economy has not been easy. The few evaluations of domestic market liberalization efforts that exist indicate that the reforms have been partial and have not always produced the expected economic benefits.

The objectives of the paper is to review the impact of the liberalization programs on domestic agricultural markets in selected African countries. It covers the following countries: Benin, Ghana, Madagascar, Malawi, and Senegal. The countries have been selected in order to capture a range of institutional and structural characteristics of the economic environment where market reforms have been initiated. These characteristics include:

a) *geographic factors*: differences between Southern and West African environments, Sahelian, coastal or landlocked economies, and the implications in terms of potential for and incentives to develop local marketing systems;
b) *degree of private market development*: differences in the pre-reform level of development of marketing infrastructure and the degree of involvement of private traders;
c) *nature and extent of adopted reforms*: the different approaches to reforms (complete elimination of parastatals; reforming and scaling down parastatals but allowing for price competition; reforming parastatals and continuing price intervention; reforming only one subset of markets or some segments of a given marketing chains; etc.).

A common characteristic of all five countries is the importance of agriculture in the economy. The contribution of agriculture to overall GDP is the highest in Ghana at 47 percent and the lowest in Senegal at 24 percent. For all countries, agriculture is a major source of employment, and in most cases absorbs more than three-quarters of the labor force. All countries have an important agricultural export sector which generates most of the country's foreign exchange. Most countries also import some of their main cereal crop (maize for Malawi, Ghana and Benin; and rice for Madagascar and Senegal).

The paper is organized in four sections. The next section presents an overview of the history and evolution of agricultural market reforms in the five countries. It is followed by a section on the adjustment within the trader population and across local markets to policy changes. The analysis focuses here on three interrelated aspects of the market adjustment process: i) the investment response of private traders; ii) the effect of reforms on market prices; and iii) the impact on the cost of operating in local markets. The final section summarizes the study results and discusses their implications for policy making.

## 2 BRIEF OVERVIEW OF AGRICULTURAL MARKET REFORMS

In practically all five countries, parastatals that were created in the 60's and 70's have controlled the supply and distribution of most agricultural crops and inputs. Although, with the introduction of market reforms, private traders were allowed to enter these agricultural markets, parastatals and State Owned Enterprises (SOEs) (whether in their old structure or under a new reorganized shape) still dominate the input and output marketing sectors or regulate the entry and activities of the private sector in some countries. Technically, these previous marketing boards should nowadays be mainly responsible for managing buffer stocks and insuring food security, but in a few countries, such as Malawi, they are still major players in food crop marketing. The status of reforms in the five study countries at the time of the study is presented in Table 7.1.

In Benin, it was not until 1983 that the Office National des Céréales (ONC) was created to manage the marketing of cereal products. In 1990, it was still not able to control more than 5% of the maize market.

In Ghana, the Government launched in 1983 a structural adjustment program to liberalize the economy and restore growth. The agricultural sector became a prime target of the Economic Recovery Program (ERP) on both the input and output side. The ERP, resulted in the liberalization of the foreign exchange market, the phased removal of input subsidies, liberalization of agricultural input and output marketing, elimination of the Guaranteed Minimum Price (GMP), phased withdrawal of the government from direct production and distribution related activities, and the promotion of the private sector in agricultural marketing (Asante 1997). Food crop marketing in Ghana has traditionally been dominated by small independent traders, who currently handle about 95 percent of the marketable surplus.

The Malagasy government intervened from 1960 to 1975 in agricultural marketing through a system of pan-territorial and pan-seasonal input and output price subsidies and an extensive credit scheme created to facilitate access to fertilizers, pesticides and machinery. In that system, the private sector was in charge of distributing and marketing inputs and outputs and worked in tandem with the Government to ensure adequate food distribution throughout the country. Starting in 1976, the socialist government nationalized most large private enterprises including agricultural processing companies and agri-business firms. It also took complete control of agricultural input and output marketing. While input prices continued to be subsidized, the Government also took charge of the collection, marketing, processing, transportation, and distribution of rice. In 1982, the Government started to gradually disengage itself from input distribution and output marketing and the private sector was once again allowed to operate in agricultural markets. Large SOE's such as the rice production and marketing development schemes in Lac Alaotra and Maravoay were opened up to the private sector in 1986.

*Table 7.1 Status of market reforms in selected countries*

| | Benin | | Ghana | | Madagascar | | Malawi | | Senegal | |
|---|---|---|---|---|---|---|---|---|---|---|
| | Input | Output | Input | Output | Input | Output | Input | Output | Input | Output |
| Reform year | 1991 | 1990 | 1990 | 1983 | 1986 | 1986 | 1995 | 1987 | 1988 | 1994 |
| Product | Fertilizer | Maize | Fertilizer | Maize | Fertilizer Pesticides Seeds | Rice | Fertilizer Seeds | Maize | Fertilizer | Rice |
| Type of market intervention/policies in 1997 | | | | | | | | | | |
| Pricing policies | yes | no | no | no | no | no | no | yes[10] | no | no |
| Marketing Board | yes[1] | yes[3] | no | yes | yes | no | yes | yes | no | no |
| Movement restrictions | yes | yes[4] | no | yes[6] | no | yes | yes[9] | no | no | no |
| Licensing | yes | no | yes | yes[7] | yes | yes | yes | yes[11] | yes | no |
| Trade Bans | yes/no[2] | yes[5] | no | no | no | yes | no | yes[12] | no | no |

*[1] Reduced to control functions; [2] Import quota; [3] Security stocks and sales to civil servants; [4] By local trader associations; [5] Periodically; [6] By district assemblies and GPRTU; [7] Endorsement by Ministry of Food and Agriculture needed to import; [8] Not available; [9] Distributors have to procure fertilizers within a given district; [10] Floor and ceiling prices; [11] Registration with ministry of agriculture; [12] Exports; [13] Enacted by law but never implemented.*

Prior to 1981, smallholder agricultural production and marketing in Malawi were monopolized by the government. The parastatal ADMARC (Agricultural Development and Marketing Corporation), distributed subsidized inputs to and purchased output from smallholder farmers at guaranteed fixed prices. The country embarked on a series of macro-economic and structural adjustment programs in 1981. In 1986, its economy deteriorated further due to falling tobacco and tea export prices, severe droughts, and the cut-off of transport routes through Mozambique. In 1987, a new series of programs were initiated, freeing smallholder output markets for all crops except for cotton and tobacco. However, although private trading was allowed, producer prices were still fixed by the Government. In May 1993, a policy was announced to open up smallholder fertilizer markets (both imports and distribution) to the private sector. By April 1995, all input and output prices had been liberalized, and a maize price band established, with floor prices for smallholders and ceiling prices for consumers determined by ADMARC. Maize exports were prohibited unless national requirements are met (Ngongola et al., 1997).

Senegal, finally, did not embark on serious market reforms in its agricultural sector until after the devaluation of the country's currency in 1994, which forced the Government to accelerate the reform in other sectors of the economy to avert massive budgetary losses. Until June 1994, the Government of Senegal, through the SAED (Société d'Aménagement et d'Exploitation des Terres du Delta) managed production and marketing of paddy rice in the Valley. In 1994, SAED relinquished completely the buying and processing of paddy and the distribution of rice to local regions. In March 1995, all rice marketing margins were liberalized. The reform of the domestic rice market in Senegal was followed by the liberalization of the import rice sector in 1995. Previously, the parastatal the CPSP (Caisse de Péréquation et de Stabilisation des Prix) monopolized the purchases, distribution, transportation, storage, and sales of imported rice to major urban centers. Before the CPSP was completely dismantled in late 1995, the private sector was allowed to import rice into Senegal.

The most striking feature of the reforms described above is the tendency towards a partial approach to liberalizing domestic markets. Marketing reforms often do not go beyond the dismantling of public marketing boards. However, competition from public marketing agencies has, in many cases, been less constraining to efficient private sector participation than the various concomitant measures that were adopted to reinforce state monopolies. These frequently survive the elimination or restructuring of marketing parastatals. Worse, the elimination of public marketing boards is sometimes accompanied by a new set of measures that conflict with the objective of increased private sector involvement.

Partial liberalization also occurs when distorting policies and regulations in sectors that are directly related to agricultural marketing, such as rural transport and financing, are left outside of the reform process, jeopardizing the objective of efficient take over by private traders. A further example is when countries choose to exclude some output or input markets or certain segments of the distribution chains from the reform process, thereby narrowing the portfolio choices of private traders and raising the risk and/or reducing the profitability of investments in marketing services.

One of the principal reasons why partial liberalization of markets is the rule - rather than the exception - in many countries has to do with the political economy of reforms. Reforming governments are usually subjected to conflicting pressure from different social and economic groups, from within and outside of government. It is the response to that pressure that often leads to the exclusion of certain activities and the adoption of new measures to protect others from the effects of the reform process. When coupled with structural and institutional deficiencies, the partial approach to reforms puts domestic marketing systems in an extended and critical transition period, characterized by paralyzed parastatals and a private sector that is unwilling or unable to take over.

Table 7.1 illustrates the partial character of the reform process in the five study countries. With the exception of the input sector in Malawi and the rice sector in Senegal, the liberalization process has been launched in all countries by the end of the eighties, lead by Ghana which started in 1983. However, market intervention was quite prevalent as late as 1997. Pricing policies were still in effect in Malawi's maize sector and Benin's fertilizer sector. Export bans and other practices that are open to administrative abuse such as licensing are observable in almost all countries. Similarly, marketing boards were still active in at least one sector in all countries, except in Senegal. The same applies to restrictions to commodity movements, even though the latter were often the result, not of deliberate policy, but of the institutional void created by the state's withdrawal from certain parts of the sector.

## 3 POLICY CHANGES, TRADER RESPONSE, AND MARKET ADJUSTMENT

### 3.1 POLICY CHANGES AND INVESTMENT RESPONSE BY PRIVATE TRADERS

The ultimate objective of domestic market reforms is to transfer the marketing activities formerly carried out by parastatals to the emerging private trading sector. For this transition to successfully take place, private traders must be able to mobilize the resources that are necessary to fund marketing activities, including investment in buildings and storage facilities, vehicles, and other equipment. The speed at which they succeed in doing this will determine the speed of their response to the policy changes associated with the reform process. An overview of the value of assets and investments among the surveyed traders in the five countries is presented in Tables 7.2a and 7.2b. The tables distinguish two main types of assets: building and storage facilities, and equipment which mainly include transport vehicles and weighing and handling tools. In cases where reforms have been introduced recently and where the availability of data allows it, assets and investments are estimated for both the pre-reform and post-reform periods.

*Table 7.2a Assets and investment patterns of private traders in Benin and Senegal (Sample average in US$)*

| Asset category | Current value of assets[a] | | Value of assets acquired after 1994[b] | |
|---|---|---|---|---|
| | Benin | Senegal | Benin | Senegal |
| Buildings & storage | 52 (10) | 155 (13) | 16 (2) | 97 (5) |
| Equipment | 1,133 (10) | 474 (31) | 76 (2) | 42 (9) |
| Others | 9 (56) | 15 (7) | 2 (26) | 2 (3) |
| All assets | 1,194 (63) | 644 (38) | 94 (30) | 141 (15) |

*[a)]Figures in parentheses are shares of traders who own assets (%); [b)]Figures in parentheses are shares of traders who have acquired assets (%)*
*Source of basic data: IFPRI/LARES and IFPRI/ISRA Surveys, 1995/6*

*Table 7.2b Assets and investment patterns of private traders in Ghana, Madagascar, and Malawi (Sample average in US$)*

| Category of assets | Value of assets acquired after the introduction of reforms | | |
|---|---|---|---|
| | Ghana After 1983 | Madagascar After 1987 | Malawi After 1987 |
| Buildings & storage | 1,356 (32) | 523 (46) | 824 (50) |
| Equipment | 2,181 (76) | 1,018 (67) | 2,262 (87) |
| Others | 190 (30) | 19 (27) | 84 (53) |
| All assets | 3,727 (89) | 1,560 (81) | 3,170 (84) |

*Figures in parentheses represent share of traders who own assets.*
*Source of basic data: IFPRI / FOFIFA survey (Madagascar); IFPRI / Bunda College suvery (Malawi); and IFPRI / GIMPA survey (Ghana).*

The average value of assets owned by private traders varies greatly between the study countries. It ranges from about US$150 in Madagascar to more than US$ 3,500 in Ghana. It is the highest in Malawi and Ghana and lowest in Benin and Senegal. The average trader in the first two countries owns 3 to 5 times more assets than the counterparts in the latter countries. Two factors may explain the higher levels of assets in Ghana and Malawi. The sample in Malawi includes traders representing the well-endowed parastatal ADMARC, which still dominates maize marketing in Malawi. As for the case of Ghana, private maize traders have hardly been constrained by the public parastatals in the past. They are well established and have had time to expand and accumulate and invest resources in maize trading.

The tables also show the composition of assets and the distribution of ownership. More than 80 percent of traders in Ghana, Madagascar and Malawi own assets, compared to barely 63 percent in Benin or only 38 percent in Senegal. The very low number of traders who have invested in and accumulated trading assets in Senegal is most likely a reflection of their very recent entry into the marketing of domestically produced rice.[1] Equipment represent the largest category of assets, making up 60 to 90 percent of the average value of assets per trader. The low levels of investment in building and storage facilities indicate a commensurately low level of demand for storage space. Furthermore, the ownership of storage and other building facilities is less broadly distributed than that of equipment, as should be expected. Ownership of marketing equipment also seems to be fairly concentrated in Benin and to a lesser degree in Senegal. In these two countries, 95 and 75 percent of total assets are owned respectively by 10 and 30 percent of traders. In the case of Benin, the concentration of assets may be related to the prevalence of strong oligopoly in the main maize producing areas (See Soule 1996). As for Senegal, the concentration of assets may be explained by the recent nature of the private marketing system and the time lag needed to acquire these assets.

The extent to which traders react to market liberalization with new investments will depend on a series of factors that ultimately dictate the response of domestic markets to the reforms. Benin and Senegal are the only two countries where the availability of data allowed for an analysis of investment activities before and after the adoption of reforms. A major policy change that occurred in the two countries besides the liberalization of domestic markets is the devaluation of their common currency, the CFA Franc, by 50 percent in early 1994. Maize markets in Benin had already been liberalized for the four preceding years. The marketing of domestic rice was liberalized later that year in Senegal. The liberalization measures that were adopted by the Senegalese government include the liberalization of the price for paddy and the complete withdrawal of the state from the procurement and marketing of domestically produced rice. Private traders were still prohibited from competing with the state in importing rice directly. Once imported by the state, private traders then took control of the distribution of rice across the domestic market through the intermediation of licensed wholesalers. Rice imports were eventually liberalized in early 1996, just after we completed our survey of local traders.

The CFA devaluation was a significant policy change with regard to the investment behavior of local rice traders in Senegal, as it renders rice imports more expensive and raises demand for local rice. Table 7.2a shows that traders in Senegal acquired 20 percent of their assets within the two year-period that followed the liberalization and devaluation measures. The corresponding share of post-devaluation investments is only 10 percent for maize traders in Benin. Maize markets in Benin were, however, liberalized long before devaluation. Thus, traders here responded differently to the change in the exchange rate than their Senegalese counterparts. But when the entire period since the introduction of the reforms is considered, the share of investment by traders in Benin increases to around 25 percent.

There are significant differences in the investment response by private traders in the two countries, when individual investment categories are taken into consideration. In Senegal, two-thirds of investment in building and storage facilities were undertaken after the introduction of reforms, with hardly any additional investment in equipment (less than 10 percent). In contrast, only about one-third of investment in both asset categories were made after the introduction of reforms in Benin. Whereas private traders have long been excluded from the marketing of domestic rice in Senegal, parastatals have never been able to effectively dominate maize marketing in Benin. Consequently, the demand for additional investment that was triggered by the introduction of policy changes has been less important in Benin. Nevertheless, combined average investments per trader during the post reform period are 30 percent higher in Benin in terms of absolute values. The maize market is much larger than that of domestic rice in Senegal and traders in Benin have had six years to accumulate these assets against only two years for traders in Senegal.

Differences in investment patterns are also observed in terms of the share of traders who made investments after the introduction of reforms. In the case of Benin, 21 percent of traders made investments during the first 4 years after the reforms. The share rises to 30 percent during the 2 years following the CFA devaluation. For most of the traders who invested before devaluation (about 90 percent), the investments were made in categories other than building / storage facilities and equipment. The sums involved are also very modest, totaling about US$7 on average. The bulk of investment in this period was concentrated on equipment and was undertaken by only 2 percent of traders. In the case of Senegal, 15 percent of traders made investments after the introduction of reforms and the devaluation of the CFA in 1994. About 50 percent of traders who made investments did so in the categories of building / storage and equipment. The average sums involved here range between US$40 for equipment and US$100 for building and storage facilities.

Several factors may explain the investment behavior of private traders in the individual study countries. The following variables have emerged in the analysis as the key determinants of private trader investment in marketing activities: market location, area coverage, liberalization of marketing policies, currency devaluation (where it applies), profitability, education, and experience. The legacy of the public sector dominated pre-reform marketing systems is usually a considerable level of segmentation of domestic agricultural markets. In face of that segmentation, the supply and infrastructure (soft and hard) conditions in the immediate marketing area should become a key factor in the speed of the response by the emerging private sector to reform policy changes. Similarly, wider area coverage is linked to greater social capital, which in turn should provide a stimulus for higher investment levels.[2] Liberalization and devaluation measures, on the other hand, raise the incentives to invest in local marketing activities by reducing the obstacles to entry and boosting the competitiveness of local agriculture. The relationship between profitability and investment is straightforward and does not warrant further discussion. Education and experience are expected to raise the managerial ability of private traders and thereby the propensity to expand their business activities.

The empirical relationships between these variables and trader investment behavior in the individual study countries are presented in Table 7.3.

*Table 7.3 Determinants of investments by private traders after the introduction of market reforms (estimated regression coefficients and t-values in parentheses).*[3]

| Independent variables | Benin | Madagascar | Malawi | Senegal |
|---|---|---|---|---|
| | | estimates (t-value) | | |
| Intercept | 1.17 (1.81) | -1.03 (1.04) | -1.26 (2.03) | -1.61 (4.85) |
| Market location | | -1.01 (3.48) | 0.13 (0.79) | 0.55 (2.89) |
| Area coverage | -0.01 (2.88) | | 0.01 (1.81) | 0.01 (2.61) |
| Liberalization | 0.33 (2.07) | 0.35 (0.83) | | 0.36 (1.90) |
| Devaluation | 0.24 (1.78) | | | 0.22 (1.00) |
| Profitability | | 1.36 (1.67) | | 0.63 (2.35) |
| Education | | 0.10 (0.70) | 0.31 (1.99) | |
| Experience | 0.01 (0.92) | 0.04 (1.35) | 0.03 (1.66) | |
| No of observations | 219 | 189 | 154 | 217 |
| *'od* | *-124.84* | *-73.82* | *-89.55* | *-130.76* |

*Note: The reported equations are the ones with the best fit for each country. In all estimations, the dependent variable has the value of 1, if a trader has invested and 0 otherwise. Market location is given the value of 0, unless the trader operates in the following areas: La Alaotra (Madagascar), Central Region (Malawi), and Dagana (Senegal). Area coverage is the distance between the purchase and sales points. The liberalization variable has the value of 1, if the trader perceives the impact of liberalization to be positive. The devaluation variable is treated like the liberalization variable. The value for the profitability variable is set to 1 for the traders who have experienced high levels of profits in the preceding years. Additional variables that were included in the estimations for Madagascar and Malawi are, respectively, gender and specialization as traders. The estimated coefficients for these two variables are .77(2.98) for gender and .55(2.11) for specialization.*

The equations presented in the table represent the best fits for each of the country data set. The second row shows the results with respect to the market location variable, which is a dummy variable for the main production and marketing regions of Dagana (Senegal); Lac Alaotra (Madagascar); and Central Region (Malawi). The factor "market location" seems to have plaid a significantly positive role in stimulating trader investment response in Senegal. Its role is, in contrast, insignificant in Malawi and even negative in the case of Madagascar. The Central Region of Malawi is the main maize producing region. It also includes the main consumption center of Lilongwe. The thinness of the private maize marketing system, even in this part of Malawi, due to the still overwhelming presence of ADMARC, may explain the absence of any stimulating effect on trader investments. The fact that investment data for the local ADMARC stations were not available and therefore not included in the estimations also may explain the outcome.

In the case of Madagascar, the Lac Alaotra region is the main producing region and also seems the enjoy greater access to credit. However, markets in that region were the last to be liberalized, about 5 years after the rest of the country and just a few before the survey. The region is also fairly isolated due to poor infrastructure. The average distance covered by local traders is the shortest amongst the three regions included in the study. Finally, according to the respondent traders from this region, its marketing sector is characterized by a high level of competition. The effect on investment incentives resulting from these latter factors may have exceeded the stimulating influence induced by the abundance of marketed surplus and the better access to credit.

The second variable in the equations is that of area coverage by private traders, for which the distance between the principal sales and purchase points is used as proxy. Area coverage seems to have played a positive but limited role in trader investment response in Senegal and Malawi. In contrast, the contribution of area coverage to trader response appears negative albeit also modest in size in Benin. The difference may come from the fact that traders in Senegal and Malawi are still at the beginning of the entry phase compared to traders in Benin, where markets have been effectively liberalized for a much longer period of time. Even before liberalization, private traders have always plaid a significant role in maize marketing in Benin. It is most likely that by the time that the data on area coverage was being collected, expansion-minded traders in Benin had already reached the limit of the manageable size of their businesses. It is true that markets in Malawi have been liberalized since 1987. However, the system is still dominated by ADMARC, which handles more than 90 percent of the marketed surplus. Private traders who are still seeking to enter the system are therefore more inclined to invest and expand the geographic coverage of their activities. The outcome in Benin is supported by the evidence from other countries, which indicates that one of the greatest problems facing the emerging private marketing sector in reforming countries is as much one of entry as one of expansion. Private traders may be fast in entering the marketing system after the lifting of regulatory obstacles. But often after the entry phase, they face difficulties in acquiring the management skills and resources needed to significantly expand their business. It is therefore likely that traders in Benin, who have been operating for years, may have reached their expansion limit. This would be reflected in a weaker link between area coverage and the demand for investment by private traders.

The next two rows of the table show the investment response of private traders to the liberalization of domestic markets and the devaluation of the CFA Franc in Benin and Senegal. The proxies for the liberalization and devaluation variables are the perceived impact that traders associate with these policy changes, using a value of 1, if the impact is positive and zero otherwise. The coefficient for the liberalization variable is positive and has about the same value for the three countries for which it was included in the estimated equation. It is strongly significant in the cases of Benin and Senegal. Its insignificance in the case of Madagascar may be explained by several factors. Liberalization in Madagascar has been carried out only partially, with markets in the main producing and marketing region of Lac Alaotra being excluded for most of the study period. Even after liberalization was extended to that region, market segmentation and isolation due to poor infrastructure may have undermined the investment response by local traders. In fact the discussion in the previous section pointed out the negative influence of the Lac Alaotra location on trader investments.

The last policy change to be considered with respect to its impact on trader response in terms of investments in marketing activities is the devaluation of country currencies, which often has accompanied the liberalization programs. The CFA Franc, which is used in Benin and Senegal, had just been devalued before the inception of the case studies in these two countries, providing us with the opportunity to examine the adjustment of private traders with respect to their investment behavior. Again here the estimated coefficients are positive and of the same magnitude. However, it is significant only in the case of Benin. As mentioned earlier in the report, the Senegal study has focused only on the marketing system for domestic rice, which was controlled by public parastatals before the introduction of reforms and which has become quite segmented from the much larger market for imported rice, which has always had strong private sector participation. The domestic and imported types of rice hardly compete in consumption. Domestic rice is produced and primarily consumed in the relatively remote markets of the North and barely sold to the other main consumption regions of the country. Although devaluation has rendered imports more expensive, it coincided with the elimination of input subsidies, which made the already uncompetitive domestic rice production more costly.[4] Consequently, the effect of the CFA devaluation on the demand for local rice, and therefore on investments in domestic rice trading, has been quite limited. In contrast, the change in the CFA parity has raised the competitiveness of the maize sector in Benin, particularly in the export markets towards Nigeria and Burkina Faso. In the case of Burkina Faso, the gain in competitiveness is not with respect to local producers who share the same currency, but with respect to Ghanaian exports. Hence, it is understandable that traders in Benin responded more positively to the CFA devaluation in terms of higher investment in trading activities.

The case of the remaining variables in the equations is quite straightforward and should not warrant further discussion. For instance the fact that profitability has a strong impact on the investment behavior of private traders is normally expected. Looking at the entire results, however, it is interesting to note that, apart from the profitability variable, the liberalization variable shows the highest coefficient estimates. The results further indicate that the effective transition to a viable private marketing sector depends, beyond key sector-specific factors, on a variety of other factors that often are specific to a given country or even region. A factor that seems to be specific to the emerging marketing sector is the apparent inability of private traders, after the initial entry phase, to continuously and significantly expand their business activities to cover wider areas. It might therefore be necessary for reforming countries to encourage broad-based entry in local markets by private traders, and not only in the remote regions, as has been emphasized so far. The apparent constraints to business expansion means that a situation with few market entrants would not lead to oligopoly situations as often feared, but rather to a breakdown in the marketing system for certain regions.

Some of the factors that may be country or region-specific and that may interfere with the transition to a comprehensive private marketing system are market segmentation and thinness. The results above indicate that, even though traders may feel good about the reforms and may have adequate access to credit, they may still lack the incentives to invest in expanding marketing activities due to segmentation and thinness of the local markets. In other words, the local characteristics of the marketing system itself may end up being a obstacle to the transition process. It would seem, therefore, that reforming countries would raise the ability of the emerging private sector to satisfactorily fill the gap left by the dismantling of marketing parastatals by adopting measures to alleviate localized market segmentation and thinness.

## 3.2 THE EFFECTS OF MARKET REFORMS ON PRICES IN LOCAL MARKETS

One important aspect of marketing policy reforms is their impact on the level of domestic prices. The direction of price changes is difficult to predict, as it depends on the direct effect of pre-reform policies on price levels and on the response in the supply and demand mix within the reforming sectors. The price analysis carried out in this section examines the evolution of prices in local markets. In cases where the length of the available price series allows it, a comparison of price trends before and after the introduction of reforms has been carried out. Table 7.4 presents the average annual growth rates of real monthly prices in the individual countries.[5] Real prices in most markets appear to have declined over the respective study periods. It is most likely that the downward trend in prices is only partly the direct impact of the changes in the countries' marketing sectors. It is true that direct price controls in the pre-reform era may have prevented food prices to rise as fast as the general price levels in most of these countries. For the post-reform period, on the other hand, it is probable that prices have fallen in real terms, due to the strong (short term) inflation pressures that have been associated with overall economic reforms which most of the study countries have implemented along with the liberalization of agricultural markets. For instance, all of the study countries have had to repeatedly or at least significantly devalue their currencies.

*Table 7.4 Trends in real retail prices before and after reforms*[a]

| Benin | | | Malawi[b] | | | Ghana | |
|---|---|---|---|---|---|---|---|
| Regional Market | Average Price Growth rate | | Regional Market | Average Price Growth rate | | Regional Market | Average Price Growth rate |
| | 1985-1989 | 1990-1995 | | 1984-1987 | 1988-1991 | | 1987-1996 |
| Natitingou | 0.04 | -0.36 | Blantyre | 0.00 | 0.03 | Techiman | -0.68 |
| Azove | 0.07 | -0.46 | Lilongwe | -0.00 | 0.01 | Accra | -0.51 |
| Ouando | 0.02 | -0.29 | Mzuzu | -0.04 | -0.01 | Kumasi | -0.52 |
| Parakou | -0.01 | -0.50 | Zomba | -0.04 | -0.05 | Ejura | -0.61 |
| Bohicon | -0.05 | -0.52 | Karonza | -0.09 | -0.08 | Mankessim | -1.09 |
| Cotonou | -0.09 | -0.44 | Nkhotakota | -0.16 | 0.05 | Tamale | -0.60 |

*[a]Figures indicate average annual growth rates of monthly prices. [b]The post reform period stops in 1991, although the data was available up to 1995. The decision to shorten the post-reform period is due to the need to eliminate the effects of the serious repeated droughts that occurred during the first half of the nineties.*

*Source of basic data: Enquêtes, Laboratoire d'Analyse Régionale et d'Expertise Sociale (Benin); Ministry of Food and Agriculture (Ghana) and National Statistical Office and Ministry of Agriculture (Malawi.).*

In the case of Benin, although prices had been falling before the introduction of reforms in 1990, the decline is much stronger during the post-reform period. A key contributor to the decline in real prices was the devaluation of the CFA Franc in 1994, which resulted in unprecedented inflation rates in the subsequent couple of years. Similarly, prices in Ghana, which had introduced market reform programs in 1983 and subsequently proceeded to a series of devaluations of the Cedi, have fallen continuously throughout the 10 years for which data are available. The downward trend in maize prices in Ghana has, however, been observed since the inception of the reform process, as shown in Badiane and Shively (1998).

The next aspect of local prices that would be affected by the reform programs and that needs to be examined is their variability. One of the motivations behind market reforms is that the decentralized mode of operation of a private-sector based marketing system is more flexible and therefore responds better to changes in market conditions than under the usually centralized system of marketing boards. The expected improved response by private traders to changes in supply and demand conditions across local markets would lead to less instability as a more responsive flow of commodities would spread local price shocks to other markets more efficiently.

Table 7.5 illustrates the changes in price stability patterns that accompanied the liberalization of domestic agricultural markets in two of the study countries. The figures in the table confirm the sharp decrease in price variability in Benin, following the liberalization of markets. With the exception of Blantyre and Lilongwe, the two major cities, price instability has also declined across local markets in Malawi, albeit to a lesser extent. The price levels of price instability go up substantially, however, if the drought year between 1991 and 1995 are included in the analysis. The decline in price instability between the onset of reforms and the drought years can hardly be directly attributed to the entry of private traders into the marketing system, knowing their activities cover on average a distance of merely 15 km. It is, however, possible that the changes in ADMARC's operations due to the increasing competition with private traders following the liberalization of markets have had a stabilizing effect on local prices.[6] Equally important as the effects on price levels and instability patterns are the expected effects of market reforms on the costs associated with the movement of commodities over time and across space, the subject of the next section.

*Table 7.5 Index of instability of retail prices of maize in Benin and Malawi*

| Benin | | | Malawi | | |
|---|---|---|---|---|---|
| Regional Market | Average Price Growth rate | | Regional Market | Average Price Growth rate | |
| | 1985-89 | 1990-95 | | 1984-87 | 1988-91 |
| Natitingou | 28.94 | 16.34 | Blantyre | 14.89 | 20.29 |
| Azove | 27.76 | 14.89 | Lilongwe | 12.49 | 16.28 |
| Ouando | 29.99 | 13.97 | Mzuzu | 14.93 | 12.35 |
| Parakou | 29.91 | 14.17 | Zomba | 27.03 | 22.77 |
| Bohicon | 28.73 | 12.19 | Karonza | 15.18 | 8.59 |
| Cotonou | 24.94 | 8.98 | Nkhotakota | 16.58 | 15.14 |

*Figures are trend-corrected coefficients of variations.*
*Source of basic data: Enquêtes, Laboratoire d'Analyse Régionale et d'Expertise Sociale (Benin); and National Statistical Office and Ministry of Agriculture (Malawi.).*

## 3.3 THE EFFECTS OF MARKET LIBERALIZATION ON MARKETING COSTS AND PRICE MARGINS

Marketing costs in local markets in the study countries and their individual components are presented in Tables 7.6a and 7.6b. The average costs per ton of marketed product during the period of the survey lie around the range of 20-30 US cents per ton, except in Malawi, where they are less than 10 cents. The bulk of marketing costs being accounted for by transport, prices become more comparable, when the differences in average distances covered by private traders are taken into consideration. The average marketing costs per ton/km range from 17 US cents in Benin to almost 60 cents in Malawi. While unit costs in the other three countries are about similar, they are nearly at least two-thirds lower in Benin, where private traders cover the longest average distance. The comparison of the costs in Benin and Senegal may suggest the existence of economies of scale related to area coverage. Private traders in the Senegal sample have marketed one-third less quantities and have traveled on average about one-third of the distance covered by traders in Benin. Their unit costs per ton/km is, however, about three times as high as that of traders in Benin. It is true that fuel and vehicle prices are all lower in Benin, but they can not explain the large difference in marketing costs, especially given the much better infrastructure in Senegal.

*Table 7.6a Marketing costs in the study countries*

| Country | Product | Average marketing costs[a] | | Average marketed quantities[a] (tons) | Average distance traveled (km) | Unit marketing costs | |
|---|---|---|---|---|---|---|---|
| | | domestic currency | US$ | | | US$ per ton | US$ per ton/km |
| Benin | Maize | 1,275,000 | 2,493 | 131 | 113 | 19.03 | 0.17 |
| Madagascar | Rice | 8,857,000 | 2,046 | 72 | 55 | 28.42 | 0.52 |
| Malawi | Maize | 22,500 | 1,471 | 161[b] | 16 | 9.14 | 0.59 |
| Senegal | Rice | 487,000 | 930 | 44 | 47 | 21.15 | 0.45 |

*[a]Quantity and cost refer to the survey periods in the respective countries; [b]Excluding ADMARC. When ADMARC traders are included, the average marketed quantity increases to 520 tons.*

*Table 7.6b Structure of average marketing costs per trader*

| | Benin (maize) | | Madagascar (rice) | | Malawi (maize) | | Senegal (rice) | |
|---|---|---|---|---|---|---|---|---|
| | US$ | % | US$ | % | US$ | % | US$ | % |
| Transportation | 1,484.48 | 60 | 845.68 | 41 | 471.48 | 32 | 350.94 | 38 |
| Wages & salaries | 357.32 | 14 | 139.07 | 7 | 313.90 | 21 | 102.60 | 11 |
| Communication & electricity | 5.54 | 0 | - | - | 66.19 | 4 | 27.36 | 3 |
| Storage rent | 139.17 | 6 | * | * | 101.96 | 7 | 3.17 | 0 |
| Other rent | 111.89 | 5 | 45.32 | 2 | 8.22 | 1 | 44.46 | 5 |
| Taxes | 78.75 | 3 | 57.72 | 3 | 112.47 | 8 | 57.45 | 6 |
| Bagging & handling | 310.85 | 12 | 100.15 | 5 | 323.63 | 22 | 167.85 | 18 |
| Other costs | 5.23 | 0 | 858.30 | 42 | 73.30 | 5 | 176.59 | 19 |
| Total | 2,493.23 | 100 | 2,046.25 | 100 | 1,471.15 | 100 | 930.41 | 100 |

** Included in other rent.*
*Source of basic data: IFPRI survey*

The effect of scale economies also appears with respect to cost levels in Malawi. The average quantities marketed by traders in Malawi are comparable to that of their counterparts in Benin. However, they cover on average about one-tenth of the distance covered by traders in Benin, and average costs in Malawi are four times higher than in Benin. Similarly, traders in Madagascar market about 50 percent less quantities and for 50 percent of the distance covered by traders in Benin. But, their average unit costs are three times higher than in the former country. It should also be added that both Malawi and Madagascar have poor infrastructure compared to Benin.

The importance of infrastructure in determining average costs of marketing in the study countries can be observed from Table 7.7. The table shows the average distances traveled by traders, the average quantities transported, and the time spent per trip in the respective countries. It takes traders in Malawi more than 25 percent of the time it takes traders in Benin to transport 70 percent less maize on distances that are about 90 percent shorter. Similarly, it takes traders in Malawi the same amount of time to transport similar quantities on distances that are 3 times shorter than their counterparts in Senegal. Traders in Madagascar cover about 50 percent of the distance traveled by traders in Benin, but need the same amount of time for similar quantities. The better quality of infrastructure in Senegal appears in the relatively shorter duration of trip for the distance traveled. The disparities in the performance of the transport sector in the study countries is reflected in the unit cost of transport presented in the last column of the table. They are 3 and 14 times higher in Madagascar and Malawi, respectively, compared to Benin. The cost difference between Senegal and Benin is again a reflection of lower fuel and vehicle costs in the latter country, due to the proximity to the Nigerian market.

*Table 7.7 Transport cost in country marketing sectors*

| Country | Average distance (km) | Average duration (hrs) | Average volume transported (tons) | Cost per shipment in local currency | Transport cost per ton/km | |
|---|---|---|---|---|---|---|
| | | | | | local currency | US$ |
| Benin (maize) | 113 | 3.90 | 34 | 23,000 | 5.9 | 0.012 |
| Madagascar (rice) | 55 | 3.03 | 30 | 208,000 | 127.1 | 0.029 |
| Malawi (maize) | 16 | 1.12 | 10 | 329 | 2.1 | 0.136 |
| Senegal (rice) | 47 | 1.05 | 15 | 6,750 | 9.5 | 0.018 |

*Source of basic data: IFPRI/LARES survey (Benin); IFPRI / FOFIFA survey (Madagascar); IFPRI / Bunda College Survey (Malawi); IFPRI /ISRA survey (Senegal).*

To the extent that the changes in marketing costs are transmitted to prices, these changes should be reflected in the spatial margins between local prices. Spatial spreads between local markets in the study countries are presented in Tables 7.8a to 7.8e. For the sake of comparison between countries, the absolute spreads are also presented as a proportion of the lower price between each of the considered pairs of markets. The relative price spreads between local markets range from 20 to 30 percent in Benin and Ghana. In Malawi, absolute price spreads are within the 10 to 20 percent price range. The drought period of 1992 to 1995 has not been included in the analysis of Malawi prices, due to the bias it would bring to level and pattern of price spreads. In Madagascar, price spreads between local markets are around the 10 percent mark. In the case of Senegal, spatial price spreads exhibit sharp differences between markets for local rice and that for imported rice. Price spreads for local rice range between 10 and 30 percent, whereas spreads for imported rice is less than 1 percent between all market pairs.

*Table 7.8a Spatial spreads between real retail prices for maize in Benin before and after reforms*

| Regional market pairs | Approximate distance between markets (km) | Average spatial Margins[a] | |
|---|---|---|---|
| | | 1985-89 | 1990-95 |
| Natitingou-Cotonou | 480 | 33 | 29 |
| Azove-Cotonou | 108 | 32 | 19 |
| Ouando-Cotonou | 312 | 8 | 9 |
| Parakou-Cotonou | 360 | 63 | 19 |
| Bohicon-Cotonou | 96 | 17 | 14 |

[a] *Ratio of spatial spread to the lower price between the two markets in percent.*
*Source of basic data: Enquêtes, Laboratoire d'Analyse Régionale et d'Expertise Sociale.*

*Table 7.8b Spatial spreads between real retail prices for maize in Malawi before and after reforms*

| Regional market pairs | Approximate distance between markets (km) | Average spatial Margins[a] | |
|---|---|---|---|
| | | 1984-87 | 1988-91 |
| Blantyre-Lilongwe | 295 | 11 | 11 |
| Mzuzu-Lilongwe | 315 | 14 | 12 |
| Zomba-Lilongwe | 250 | 15 | 18 |
| Karonga-Lilongwe | 485 | 15 | 21 |
| Nkhotakota-Lilongwe | 145 | 24 | 21 |

[a] *Ratio of price spread to the lower price between the two markets in percent.*
*Source of basic data: the National Statistical Office and Ministry of Agriculture, Malawi.*

*Table 7.8c Spatial spreads between real retail prices for maize, Ghana, 1987-1996*

| Regional market pairs | Approximate distance between markets (km) | Average spatial margin[a] |
|---|---|---|
| Techiman-Accra | 256 | 37 |
| Kumasi-Accra | 176 | 5 |
| Ejura-Accra | 240 | 26 |
| Mankessim-Accra | 72 | 25 |
| Tamale-Accra | 365 | 28 |

[a] *Ratio of price spread to the lower price between the two markets in percent.*
*Source of basic data: Ministry of Food and Agriculture, Ghana.*

*Table 7.8d Spatial spreads between real retail prices for rice, Madagascar, 1996*

| Regional market pairs | Approximate distance between markets (km) | Average spatial margin[a] |
|---|---|---|
| Andravoahangy-Anosibe | 1 | 0.9 |
| Anjoma-Anosibe | 285 | 3.3 |
| Atalata-Anosibe | 285 | 2.4 |
| Ambatondrazaka-Anosibe | 175 | 11.0 |
| Anjoma-Andravoahangy | 285 | 4.3 |
| Atalata-Andravoahangy | 285 | 1.2 |
| Atalata-Anjoma | 3 | 5.6 |
| Ambatondrazaka-Andravoahangy | 150 | 12.1 |
| Ambatondrazaka-Anjoma | 450 | 7.5 |
| Ambatondrazaka-Atalata | 450 | 13.4 |

[a] *Ratio of the spread to the lower price between the two markets in percent.*
*Source of basic data: IFPRI-FOFIFA 1996 survey.*

*Table 7.8e Spatial retail price spread for rice, Senegal, 1996*

| Regional market pairs | Approximate distance between markets (km) | Average spatial margin[c] |
|---|---|---|
| A. Imported rice[a] | | |
| Dakar-St. Louis | 150 | 0.4 |
| Kaolack-St. Louis | 300 | 0.8 |
| Dakar-Kaolack | 200 | 0.4 |
| B. Local rice[b] | | |
| Dagana-Ndioum | 175 | 28 |
| Podor-Ndioum | 75 | 11 |
| Richard Toll-Ndioum | 130 | 22 |

*[a]For February 10 - May 7, 1996. [b]For February 12 - May 28, 1996. [c]Ratio of the spread to the lower price between the two markets in percent.*
*Source of basic data: IFPRI-ISRA 1996 survey.*

For the three maize countries, spatial spreads seem much higher in Ghana and Benin compared to Malawi. The lower price spreads in Malawi are probably more the outcome of ADMARC's ongoing price band policy than the result of greater arbitrage activities and thus stronger price interdependence across local maize markets in Malawi. It is clear that ADMARC's domination of the maize distribution system and its strategy to keep prices within a given range imposes a boundary on potential price margins between local markets. It is more difficult, on the other hand, to explain the considerably lower spatial price spreads in Madagascar. Malagasy traders have the second highest transport costs and unit costs of operations and the second smallest area coverage. Even though the period for which local price data were collected covers only six months, the margin between local prices should be expected to be much wider than shown by these numbers.[7]

Even much lower are the spatial price spreads observed for imported rice in Senegal. There is also a wide discrepancy between the price spreads for imported rice and that for local rice, the latter being rather comparable to the spreads observed for maize in Benin and Ghana. Imported rice arrives through the port of Dakar from where it is channeled to local markets throughout the country. Before liberalization, it was sold at prices that were set by the government and which were based on the procurement from public warehouses in Dakar and a lump sum transport fee between Dakar and the major regions of the country. Private traders have operated for decades under that system, which did not allow for significant spatial margins. Furthermore, during most of the 12 months that have preceded the collection of local prices for the present study, the associations of private traders have been involved in a series of dispute with the government about the liberalization of rice import activities and its consequences for local price stability. It is therefore likely that private traders have come to consider the low price margins not only as normal but even as desirable. Moreover, exactly one year before the liberalization of rice imports, the national currency had been devalued by 50 percent, raising local prices for imported rice to unprecedented levels. Consequently, there may not have been enough room, economically, to raise prices further, independently of the aforementioned psychological and tactical factors.

From the point of view of the reforms and the expected objectives, the changes in spatial price spreads that have taken place after the liberalization of domestic markets turn out to be more interesting than their relative levels. Results reported in Badiane et al. (1997) indicate that spatial price spreads have declined considerably across local markets, particularly in Benin and Ghana. In contrast no significant change can be discerned with respect to price spreads in Malawi. The difference in the evolution of price spreads between Benin and Malawi is reflected in the figures in Tables 7.8a and 7.8b. The second table shows that maize price spreads in Malawi have hardly changed, both in absolute as well as in relative terms. The unchanged price spreads are well in line with the continued domination of the marketing system by ADMARC and its drive to stabilize local prices. In contrast, absolute price spreads in Benin have declined across all market pairs. More importantly, price spreads have fallen also in relative terms, except for the pairs Ouando-Cotonou and Malanville-Cotonou. The fall in relative spatial price spreads among the remaining market pairs ranges from 12 to 70 percent. Although comparable data were not available for rice markets in Senegal and Madagascar, the data from Benin, Ghana, and Malawi indicate that effective market liberalization has been associated with lower spatial price spreads across local markets.

In addition to lowering spatial margins, market liberalization is also expected to lower the cost of carrying supplies over time and thereby reduce the temporal spread of prices in local markets. Greater movement of commodities between markets also contribute to lowering temporal price gaps as periodic price pressures are spread across a larger number of markets. Results reported in Badiane et al. (1997) indicate that, even after exclusion of the drought years after 1991, temporal price spreads in Malawi, if not constant, exhibit an upward trend. The only major changes in price spreads have taken place in the Karonga market in the North, where they have declined by about 60 percent, and in the Blantyre market in the South, where they have more than doubled. In contrast, price spreads in Benin have dropped by up to 80 percent after the liberalization of markets. Similarly, the study by Badiane and Shively (1998) shows that price spreads have also declined significantly in Ghana after the introduction of reforms.

## 4 SUMMARY AND IMPLICATIONS FOR POLICY

The results of the preceding analysis indicate that local markets in the study countries have adjusted to the changes in marketing policies, albeit at greatly varying degrees. In most cases, changes have occurred as expected and are observable quite early in the reform process. Local markets have adjusted in terms of price seasonality. Price instability has also decreased quite significantly across local markets. Market reforms have also resulted in a marked reduction in the cost of intermediation. The decrease in unit costs of operations are reflected in lower spatial and temporal price spreads in most markets.[8]

## 4.1 THE NATURE AND EXTENT OF MARKETING POLICY REFORMS

The review of the policy environment in the five selected countries suggest that in most cases market liberalization has been partial and the transition to a private-sector based marketing system is not yet complete. The government still intervenes in many aspects of agricultural marketing, and in areas where the government has withdrawn its activities, the private sector has not always been able to effectively replace the role of the state due to various constraints which are discussed below. Further progress in market reform will require not only further liberalization, but a more concerted effort to go beyond the withdrawal of the public sector from agricultural marketing. While previous reform efforts led to government savings and a reduction in the budgetary burden of the state, progress in market reform will require both more costly investment and a partnership between the government and the private sector to promote the development of markets and the institutions that support them.

The analysis of output markets in the five African countries included in the study suggest that the extent of government involvement in agricultural output marketing varies from country to country. In Ghana and Benin, for example, marketing boards for cereals still exist, but their share of the market is very minimal and the private sector has taken over most of the marketing activities of cereals. The marketing boards in these countries have reduced their activities to holding buffer stocks or strategic reserves and insuring food security in periods of shortage. In contrast, in Malawi, although a lot of informal trading does occur, the marketing board, ADMARC still absorbs a lion share of the official maize market. Therefore, private sector traders find themselves competing with a giant that dominates in terms of distribution networks, storage and transportation infrastructure and financial facilities. In Senegal, on the other hand, the trade marketing board for rice has been completely dismantled and the country has witnessed far reaching sector reforms in just a couple of years. The rice market in Senegal (whether imported or domestic) is now operating fully and successfully in the hands of the private sector, without any need for government intervention.

## 4.2 RESPONSE BY PRIVATE TRADERS AND ADJUSTMENT IN THE MARKETING SECTOR

Although liberalization has led to the emergence of a fairly competitive private trading sector, in a few countries, trading in certain markets is monopolized or restricted through local trader associations or an informal network of traders which create barriers to entry in the market. In Benin, for example, an established network of traders regulates the entry and the activities of new traders in their markets. In Ghana, transporters' unions regulate the transport industry and have a noticeable influence on transportation charges for maize. The government should make sure that pockets of monopoly or oligopoly power do not develop in certain areas and that access to markets follows a set of clear and transparent rules which do not impede free entry and exit. A common feature in all countries, is that although a lot of private traders have entered agricultural markets, expansion of area coverage and quantity traded has been very limited. Traders are not able to go much beyond entry into large-scale trading either in terms of geographical or volume expansion. Traders are often limited by unavailability of credit, limited storage and transportation facilities, and various restrictions by existing organizations or parastatals.

Private traders have responded to the changes in marketing policies through higher investment in trading activities. There are, however, quite important differences across countries in terms of the value, composition, and the distribution of asset ownership among private traders. As expected from the limited storage activities in most countries, the average values of accumulated investments in buildings and storage facilities are quite low. The ownership of this category of assets is also less broadly distributed than that of equipment. Private traders appear to be responding to policy changes with higher investments. The quantitative analysis of the relationship between investment by private traders on the one hand, and on the other, several factors that, next to the direct changes in policy, are expected to influence investments, such as market location, area coverage, profitability, and managerial skills, confirms the strong response of traders to the liberalization of marketing policies. Among the above variables, only the profitability variable appears to have a higher impact on the decision of traders to invest in marketing activities than the change in policy. The comparison of the results across countries and regional markets within individual countries reveals that the extent to which private traders respond to market reforms with higher investment is affected by factors such as the quality of infrastructure, the level of integration across local markets, and the thinness of these markets. Trader investment response also seems to depend on whether the private sector is predominantly in the entry or expansion phase. The response tend to be higher at the beginning (entry phase) of the reform process than at later stages, when most expansion-minded traders tend to be at the limit of the financially and technically manageable size of their businesses.

More importantly, the analysis of the determinants of trader investment behavior indicates that the effective transition to a viable private marketing sector depends, beyond key sector specific factors, on a variety of other factors that often are specific to a given country or even region. A factor that seems to be specific to the emerging marketing sector is the apparent inability of private traders, after the initial entry phase, to continuously and significantly expand their business activities to cover wider areas. It might therefore be necessary for reforming countries to encourage broad-based entry into local markets by private traders, not only in the remote regions, as has been emphasized so far. The apparent constraints to business expansion means that a situation with few market entrants would not lead to oligopoly situations as often feared, but rather to a breakdown in the marketing system for certain regions.

Some of the factors that may be country or region specific and that may interfere with the transition to a comprehensive private marketing system are market segmentation and thinness. The results above indicate that, even though traders may feel good about the reforms and may have adequate access to credit, they may still lack the incentives to invest in expanding marketing activities, due to segmentation and thinness of the local markets. In other words, the local characteristics of the marketing system itself may end up being a obstacle to the transition process. It would seem, therefore, that reforming countries would raise the ability of the emerging private sector to satisfactorily fill the gap left by the dismantling of marketing parastatals by adopting measures to alleviate localized market segmentation and thinness.

Local markets in the study countries have adjusted to the changes in marketing policies, albeit at greatly varying degrees. In most cases, changes have occurred as expected and are observable quite early in the reform process. One finding that seems to cut across all study countries is the continuous decline of real local prices over the study period. It is unlikely that the downward trend in prices is the direct impact of the changes in the countries' marketing sectors, at least in the post reform period. It is true that direct price control may have prevented food prices to rise as fast as the general price levels in most of these countries. For the post-reform period, on the other hand, it is probable that prices have fallen in real terms, due to the strong inflation pressures that have been associated with overall economic reforms, which most of the study countries have implemented along with the liberalization of agricultural markets.

Price instability has also decreased quite significantly across local markets after the introduction of reforms. The only exception appears to be Malawi, where price instability in local markets signals an upward trend after the introduction of reforms. If the drought period is excluded, however, one observes a decline in market price variability after the introduction of reforms.

No data were available to compare the level of marketing costs in the study countries before and after the introduction of reforms. The approach chosen is therefore to use the changes in spatial and temporal price spreads as indicators of changes in the performance of local marketing systems in moving goods across space and over time. Although an analysis of price spreads before and after reforms could not be carried out for Madagascar and Senegal, the analysis of price data from the other three countries indicates that effective market liberalization has been associated with lower spatial price spreads across local markets. Both absolute and relative price margins between local markets have declined considerably in the two countries, Benin and Ghana, where effective reforms have been carried out.

Furthermore, market liberalization is expected to lower the cost of carrying supplies over time and thereby reduce the temporal spread of prices in local markets. Greater movement of commodities between markets also contribute to lowering temporal price gaps, as periodic price pressures are spread across a larger number of markets. Here again, the data show a significant decline in temporal price spreads, particularly in Benin and to a lesser extent in Ghana.

The preceding discussion indicates that there is clear evidence of adjustment, in the expected direction, of local markets and private traders to the changes in policies. Market reforms, where effective, have been associated with desired changes in terms of price instability patterns. Similarly, reforms have lead to lower unit costs of marketing and a reduction of price margins between markets and seasons. There is, however, considerable variation in terms of adjustment and changes in individual countries. Generally, the response in terms of improved market performance seems to be more positive in countries where i) there has been a sizable private trading sector before the reforms; ii) market reforms have been introduced earlier; and ii) partial liberalization in terms of continued presence of marketing parastatals or continued intervention in certain market segments or regional markets has been avoided.

The absence of partial liberalization as defined above improves market performance in the post-reform period for two reasons. The first one is that such an environment present private traders with greater opportunity and reinforce their confidence in the change in policy direction. The second is that the survival of parts of the pre-reform marketing parastatals is more often than not a sign of insufficient commitment on the side of policy makers to the reform process. The process in such cases does not only lack consistency and credibility, it also carries some of the same obstacles from the pre-reform period and often breeds new types of restriction to accompany the adopted policy changes. The examples of partial liberalization in the study countries include the continued dominance of ADMARC in maize marketing in Malawi, the exclusion of the main production and marketing region of the Lac Alaotra for years from the reform process in Madagascar, and the control of rice import activities as well as the attempt to dictate rice prices in Senegal long after the liberalization of domestic markets. In spite of the changes described above and which point to improving market conditions following the adoption of effective marketing reforms, the study data also show falling real prices throughout the sample.

## NOTES

[1] Although they were banned from importing rice directly, private traders have always been allowed to buy imported rice from the marketing board and distribute it at fixed prices to all regions of the country. The marketing of domestic rice, on the other hand, has been under tight state control until June 1994.

[2] On area coverage and social capital see Fafchamps and Minten, 1998a and 1998b.

[3] Ghana is not included in the estimation due to unavailability of trader investment data.

[4] See Benz et al., 1995; Fusillier et al., 1995; and Benz, 1996

[5] Data on local prices were not available for Madagascar and Senegal, besides the daily prices gathered during the respective 6 month-surveys carried out in the two countries. The analysis of the long term price trends is therefore restricted to the other 3 countries.

[6] See Ngongola et al., 1997.

[7] It is most likely that the prices in this case apply to different categories of rice and therefore do not reflect the actual margins. The Madagascar survey identified two types of rice: a high quality and a low quality rice. The quality standard did, however, vary across markets area, such that a low quality rice in a market x could correspond to a high quality rice in another market y. Comparing the price for the latter variety with that of the high quality rice in market x would yield a much lower margin than if it were compared, as it should, with the price of the lower quality rice in x.

[8] Another aspect of market liberalization that was examined is the effect on integration in the cereal sector of the study countries. The data did not allow for an effective comparison of the situation before and after the introduction of reforms in order to document eventual changes that have been associated with the latter. Instead, the analysis had focused on a comparison of the

behavior of local markets in the study countries during the last few years. The case of Benin stood out one more time, where markets seemed to be much better integrated than in the comparator countries. The higher degree of price interdependence in Benin and Ghana, which also have liberalized their markets more completely, suggest that the reforms have resulted in improved integration amongst local markets.

**REFERENCES**

ASANTE, E.O. 1997. The Maize market in Ghana. Report submitted to the International Food Policy Research Institute (IFPRI). Accra, Ghana: Ghana Institute of Management and Public Administration.

BADIANE, O. AND G. SHIVELY. 1998. Spatial integration, transport costs, and the response of local prices to policy changes in Ghana. *Journal of Development Economics*.

BADIANE, O., F. GOLETTI, M. KHERALLAH, P. BERRY, K. GOVINDAN, P. GRUHN, AND M. MENDOZA. 1997. Agricultural Input and Output: Marketing Reforms in African Countries. Final report submitted to Bundesministerium für wirtschaftliche Zusammenarbeit und Entwicklung. Washington, D.C.: International Food Policy Research Institute.

BENZ, H. 1996. Riz local et riz importé en Afrique: Les déterminants de la compétitivité. Thèse de Doctorat, Ecole des Hautes Etudes en Sciences Sociales, Paris, France.

BENZ, H., J.L. FUSILLIER, S. HASSANE, L. LIAGRE, P. MENDEZ DEL VILLAR, AND J.M. SOURISSEAU. 1995. *Les premiers effets de la dévaluation sur les filières riz irrigué au Sahel Synthèse*. Research Report 25. Paris, France: Centre de Coopération Internationale en Recherche Agronomique pour le Développement.

FAFCHAMPS, M. AND B. MINTEN. 1998a. Returns to Social Captial Among Traders. Discussion Paper No. 23. Markets and Structural Studies Division. Washington, D.C.: International Food Policy Research Institute.

FAFCHAMPS, M. AND B. MINTEN. 1998b. Relationships and Traders in Madagascar. Discussion Paper No. 24. Markets and Structural Studies Division. Washington, D.C.: International Food Policy Research Institute.

FUSILLIER, J.L., L. LIAGRA AND P.Y. LE GAL. 1995. *Les premiers effets de la dévaluatin sur les filières riz irrigué au Sahel: Le cas du Sénégal*. Paris, France: Centre de Coopération Internationale en Recherche Agronomique pour le Développement.

NGONGOLA, D., R.N. KACHULE, AND P.H. KABAMBE. 1997. The Maize, fertilizer, and maize seed markets in Malawi. Report submitted to the International Food Policy Research Institute (IFPRI). Lilongwe, Malawi: University of Malawi, Bunda College of Agriculture.

SOULE, B. G. 1996. Le Fonctionnement du marche des intrants agricoles et du mais dans le contexte des reformes economiques en Afrique: Le cas du Benin. Report submitted to the International Food Policy Research Institute ((IFPRI). Cotnou, Benin. Laboratoire d'Analyse Regionale et d'Expertise Sociale (LARES).

# 8 PRICE SETTING POWER AMONG WHOLESALERS AND RETAILERS: MAIZE IN BENIN AND SORGHUM IN BURKINA FASO

W. Erno Kuiper, Clemens Lutz and Aad van Tilburg

## 1 INTRODUCTION

Local market places, scattered throughout Benin and Burkina Faso, play a pivotal role in the functioning of cereal markets in both countries. Many transactions take place in spot markets; buyers and sellers meet in the market place where the cereals for sale are displayed on the market day. The grains are transferred from producers to final consumers through conventional marketing channels. In each market place a retail and a wholesale segment can be distinguished. In the large towns, local retailers generally buy on the wholesale segment of the market, while retailers in rural centres can choose to obtain their stocks either from the wholesale segment or directly at the farm gate. The prices operating in these markets reflect market supply and demand conditions. This study focuses on the vertical arbitrage process for two staples: maize in the south of Benin and white sorghum in Burkina Faso.

In earlier studies on spatial price arbitrage in the wholesale segment of the maize market in Benin, we concluded that all wholesale markets played a role in this process (see, for example, Kuiper et al., 1999). None of the price series of any of the wholesale market places were found to be dominant: all price series were interdependent. The arbitrage process corresponded to a network with a number of interdependent wholesale markets; there were no autarkic markets and transportation costs did not show a variable trend. A study of Bassolet (forthcoming) (see also Bassolet & Lutz, 1999) on sorghum and millet in Burkina Faso, led to similar conclusions. These studies, however, focused on arbitrage between market places and neglected the arbitrage process within a market place. In the present study we concentrate on this omission, questioning the relationship between prices on wholesale and retail market segments in the same spot market. We focus on the question whether there is any evidence for wholesale market dominance vis-à-vis the retailers.

In the literature on industrial organisation retail prices are often assumed to be determined by wholesale market conditions (see, for example, Tirole, 1988, Chapter 4; and Martin, 1993, Chapter 12). Likewise, in the marketing literature on the functioning of food markets in tropical countries, the vertical price leadership of wholesalers is often assumed. This assumption is sometimes based on popular complaints from retailers and market organisations. Looking at the Benin maize market, Lutz (1994) found that retail and wholesale price series in the same market place cohere, which implies that retail margins are stationary. This result suggests that retailers are indeed passive decision makers, following wholesale prices without taking local supply and demand conditions into account. However, a survey among the retailers in the towns indicated that only a minority of them buy from the same wholesaler; most buy from the broker who proposes the lowest price (Lutz, 1994). Bassolet (forthcoming) found similar results for Burkina Faso. Moreover, in rural areas retailers can choose to buy either from wholesalers or at the farm gate. Buying directly from farmers gives retailers some freedom to set prices. This implies that intermediaries in both market segments may have some market power and be able to influence price formation.

In order to verify whether the price formation process is driven by retailers, wholesalers, or both, we consider the long-run equilibrium (i.e. cointegrating) relationship between the wholesale and retail prices. We test whether or not wholesale prices and retail prices respond to deviations from the equilibrium price (i.e. the equilibrium error). Most studies on vertical price relationships published to date in marketing and industrial organisation (e.g. Gerstner & Hess, 1991; Lee and Staelin, 1997 and the references they cite) have used comparative statics to study channel behaviour; the long-run relationships derived have not been empirically tested. Our study differs in that its main focus is empirical analysis.

The basic assumption we make is that the common stochastic trend observed in the cointegrated wholesale-retail price series is generated by local supply and demand conditions (seasonal price trend). Three models then become interesting for the study of price adjustment: *Model* 1 in which both retail and wholesale prices respond to the equilibrium error; *Model* 2 in which only the retail prices respond to the equilibrium error; and *Model* 3 in which only the wholesale prices respond to the equilibrium error. In *Model* 1 wholesalers have sufficient power vis-à-vis the retailers to behave as vertical price leaders, although retailers can still maximise their profits, conditional on the wholesale price being set by the wholesalers. This model applies if both retail and wholesale traders exercise some market power, for example, if alternative market opportunities exist for both actors. By contrast, in *Model* 2 the retailers do not allow wholesalers to influence short-run retail price deviations and leave them with only the option of setting wholesale prices on the basis of the wholesalers' unit costs (i.e. local farm gate price plus a margin to enable the wholesaler to survive). Market power for retailers may be the result of a temporarily abundant supply in the wholesale segment. Lastly, in *Model* 3, only wholesalers are able to set their prices in the sense of Stackelberg leadership and to respond to price deviations from the equilibrium. This situation corresponds to a market where retailers are entangled in local arbitrage whereas wholesalers ship to other market places.

The chapter is structured as follows. Section 2 formulates the long-run model and derives its testable implications on the short-run price system. Section 3 presents the empirical results. Section 4 is the conclusion.

# 2 METHOD

## 2.1 THEORETICAL FRAMEWORK

We assume that wholesalers and retailers in each market can be represented by the behaviour of a representative agent. Hence, a two-member channel is considered and channel members' long-run supply decision behaviour is modelled. A single upstream firm, the wholesaler, buys an intermediate good at the local farm gate price $p_f$, distributes it at a constant (i.e. independent of the quantity of the good) unit cost $c_w$, and sells it to a single downstream firm, the retailer, at a wholesale price $p_w$ on the wholesale market. The retailer faces constant unit retailing cost, $c_r$, and resells the product to the consumer at a price $p_r$ on the retail market. It is assumed that the wholesaler and retailer do not have losses with respect to the intermediate good, implying that the quantity purchased by the wholesaler, denoted by $q$, is equal to the quantity finally consumed.

Let the consumer demand function be of the constant elasticity type (see Von Ungern-Sternberg, 1994):

[1] $$\ln(q) = \delta\ln(p_r) + d \qquad (\delta < -1)$$

where $\delta$ is the price elasticity of demand and $d$ represents exogenous shifts in the demand curve.

First, we consider the Stackelberg model in which the wholesaler is the vertical price leader, i.e. the retailer maximises profit conditional on the wholesale price that has to be paid to the wholesaler, and the wholesaler then determines $q$ or, similarly, $p_w$, by maximising profit while taking the conditional profit-maximising behaviour of the retailer into account. The conditional profit-maximisation problem of the retailer can be represented by:

[2a] $$\max_{q} (p_r - c_r - p_w)q$$

or equivalently,

[2b] $$\max_{p_r} (p_r - c_r - p_w)q$$

subject to [1]. The first-order condition for this problem is:

[3a] $$p_r + (dp_r/dq)q - c_r - p_w = 0$$

or equivalently,

[3b] $$q + (dq/dp_r)(p_r - c_r - p_w) = 0.$$

From each of both [3a] and [3b] it can be derived that:

[4] $$p_w = \delta^{-1}(1 + \delta)p_r - c_r.$$

The wholesaler maximises individual profit while taking the conditional profit-maximising behaviour of the retailer into account, so that

[5a] $$\max_{q}\,(p_w - c_w - p_f)q$$

or equivalently,

[5b] $$\max_{p_w}\,(p_w - c_w - p_f)q$$

is subjected to [4] and has the following first-order condition:

[6a] $$p_w + (dp_w/dq)q - c_w - p_f = 0$$

or equivalently,

[6b] $$q + (dq/dp_w)(p_w - c_w - p_f) = 0.$$

Using [4], we can derive from each of both [6a] and [6b] that the total unit cost of the wholesaler, $c_w + p_f$, can be represented by a linear combination of the wholesale and retail prices:

[7] $$p_w + \delta^{-2}(1+\delta)p_r = c_w + p_f.$$

We shall return to this relationship in the next subsection when considering the econometric implications of our theoretical framework. Further, each of both [6a] and [6b] can be solved for $p_r$ by using [4], giving:

[8] $$p_r = [\delta/(1+\delta)]^2[c_r + c_w + p_f]\ .$$

If the retailer does not take $p_w$ as given as was assumed in [3], then it can be derived that:

[9] $$p_r = \delta(1+\delta)^{-1}(c_r + c_w + p_f)$$

which is the price that results when the retailer and the wholesaler determine $q$ by maximising total channel profits as if they were an integrated industry.

In order to solve for $p_w$, we substitute [8] into [4] to get:

[10] $$p_w = (1+\delta)^{-1}[\delta(c_w + p_f) - c_r]\ .$$

It is interesting to observe that if prices are set according to [8] and [10], then the wholesaler has enough power vis-à-vis the retailer to behave as a vertical price leader in choosing $p_w$. However, if the retailer dominates, then we may have a situation as modelled by [9], in which case the retailer maximises profit and forces the wholesaler to set prices based on total unit costs alone:

[11] $$p_w = c_w + p_f.$$

Until now two models have been considered: the model made up by [8] and [10] (or similarly, [2] and [5]), *Model* 1, according to which the wholesaler is able to manipulate the retail price by $dp_w/dq$ in [6a] (or, similarly, by $dq/dp_w$ in [6b]), and the model formed by [9] and [11], *Model* 2, implying that the retailer dominates. In addition, a third model, *Model* 3, is obtained if we assume that the retailer is able to buy directly from the farmer and set $p_r$ on the basis on $p_f$ as follows:

[12] $$p_r = c_w + p_f$$

where $c_w$ is added to cover the costs that would otherwise be made by the wholesaler. Competition among retailers tends to be perfect and as a consequence, a retailer is not able to charge a price that is different from the one determined by [12]. Nevertheless, in spite of the retailer's ability to purchase directly from the farmer, the retailer can also buy from the wholesaler while having $p_r$ still determined by [12]. Although the wholesaler is not able to influence the local retail market because $p_r$ is fixed by [12], the wholesaler can still determine $p_w$ by [10] if involved in market arbitrage, so that $p_w$ can be based on the reaction function of retailers in other local markets where the retailers are unable to buy directly from the farmer. The testable implications of the three models being considered will be discussed in the next subsection.

## 2.2 ECONOMETRIC CONSIDERATIONS

Many economic time series, such as $p_{rt}$ and $p_{wt}$ ($t = 0, 1, \ldots, T$), do not fluctuate around a constant or linear trend in a seemingly random way, but their first differences, $\Delta p_{rt} = p_{rt} - p_{r,t-1}$ and $\Delta p_{wt} = p_{wt} - p_{w,t-1}$, do (Granger & Newbold, 1986). Consequently, the variables in levels, $p_{rt}$ and $p_{wt}$, are assumed to be nonstationary, while in first differences they will be stationary. In time series analysis this is expressed by saying that $p_{rt}$ and $p_{wt}$ are integrated of order one, denoted $p_{rt} \sim I(1)$ and $p_{wt} \sim I(1)$, and $\Delta p_{rt}$ and $\Delta p_{wt}$ are integrated of order zero, denoted $\Delta p_{rt} \sim I(0)$ and $\Delta p_{wt} \sim I(0)$.

The nonstationarity is caused by a 'stochastic trend' (Banerjee et al., 1993, 153), which can be interpreted as the driving force of the variable. If two variables are driven by the same stochastic trend, then a linear combination of the two will be stationary, which is expressed by saying that the two variables are 'cointegrated' (Engle & Granger, 1987) or, equivalently, have a 'common stochastic trend' (Stock & Watson, 1988).

At first sight, a stochastic trend in the prices $p_{rt}$ and $p_{wt}$ can originate from four variables: $d_t$, $c_{rt}$, $c_{wt}$ and $p_{ft}$. However, according to [8]-[12] the prices depend only on the cost variables $c_{rt}$, $c_{wt}$ and $p_{ft}$, and not on $d_t$; $q_t$ fully captures $d_t$ in [1] after $p_{rt}$ is set by the pricing decisions of the channel members. The justification for this assumption is that the theoretical framework is static, and hence describes long-run (i.e. strategic) behaviour. Because long-run shifts in demand are gradual, they can easily be anticipated by long-run supply decisions. Further, we may assume that $c_{rt}$ and $c_{wt}$ do not contain a stochastic trend of importance when compared with the stochastic trend generated by the prices of the raw product as represented by the farm gate price $p_{ft}$. Consequently, we assume that $c_{rt}$ and $c_{wt}$ are stationary whereas $p_{ft}$ introduces the stochastic trend in the price system, expressing local supply and demand conditions and seasonal factors.

To illustrate the relationship between the concept of a stochastic trend and the concept of cointegration, let us consider the retail price $p_{rt}$ and the wholesale price $p_{wt}$ in a vector autoregression of order $k$, denoted VAR($k$), as follows:

[13] $$\mathbf{X}_t = \sum_{i=1}^{k} \Pi_i \mathbf{X}_{t-i} + \Phi \mathbf{D}_t + \varepsilon_t$$

where $\mathbf{X}_t = [p_{rt}, p_{wt}]' \sim I(1)$, the $\mathbf{\Pi}_i$ ($i = 1,\ldots,k$) are (2×2) parameter matrices, $\mathbf{\Phi}$ is a ($2\times m$) parameter matrix, $\mathbf{D}_t$ is an ($m\times 1$) vector with first element equal to one for the intercepts and the other elements are, for example, centred seasonal dummies which sum to zero over a full year, $\boldsymbol{\varepsilon}_t = [\varepsilon_{rt}, \varepsilon_{wt}]'$ are disturbances that follow a two-dimensional Gaussian white noise process, and the values of $X_{-k+1}, \ldots, X_0$ are fixed. The VAR in [13] can be rewritten to obtain:

$$[14] \qquad \Delta\mathbf{X}_t = \mathbf{\Pi}\mathbf{X}_{t-1} + \sum_{j=1}^{k-1} \mathbf{\Gamma}_j\Delta\mathbf{X}_{t-j} + \mathbf{\Phi}\mathbf{D}_t + \boldsymbol{\varepsilon}_t$$

where $\Delta\mathbf{X}_t = \mathbf{X}_t - \mathbf{X}_{t-1}$, $\mathbf{\Pi} = \sum_{i=1}^{k} \mathbf{\Pi}_i - \mathbf{I}$ and $\mathbf{\Gamma}_j = \sum_{i=j+1}^{k} \mathbf{\Pi}_i$. Notice that there can never be a relationship between a variable with a stochastic trend and a variable without a stochastic trend. So, if $\Delta\mathbf{X}_t \sim I(0)$ and $\mathbf{X}_t \sim I(1)$ (and hence, $\mathbf{X}_{t-1} \sim I(1)$), then $\mathbf{\Pi}$ will be a zero matrix except when a linear combination of the variables in $\mathbf{X}_t$ is stationary, i.e. when $p_{rt}$ and $p_{wt}$ are cointegrated (or when one of the prices is stationary so that we should also test for the absence of each individual price in the cointegrating relation to justify our assumption that both prices are $I(1)$). Because this linear combination is unique, the rank of $\mathbf{\Pi}$ will be equal to one, i.e. rank($\mathbf{\Pi}$) = 1. Hence, rank($\mathbf{\Pi}$) = 0 if there is no cointegration and rank($\mathbf{\Pi}$) = 2 if $\mathbf{X}_t \sim I(0)$. The Johansen procedure (e.g. Johansen & Juselius, 1990; and Johansen, 1995) estimates [14]; to test for cointegration, trace statistics are used to determine the rank of $\mathbf{\Pi}$, and asymptotic $t$ statistics are used to test for the absence of each individual price in the long-run equilibrium, in order to check whether both price series are $I(1)$.

Clearly, the result of interest will be rank($\mathbf{\Pi}$) = 1. In this case $\mathbf{\Pi}$ can be decomposed into $\mathbf{\Pi} = \boldsymbol{\alpha\beta}'$, where $\boldsymbol{\alpha} = [\alpha_r, \alpha_w]'$ is the adjustment vector and $\boldsymbol{\beta} = [\beta_r, \beta_w]'$ is the cointegrating vector, so that [14] becomes a vector error-correction model (VECM):

$$[15] \qquad \Delta\mathbf{X}_t = \boldsymbol{\alpha\beta}'\mathbf{X}_{t-1} + \sum_{j=1}^{k-1} \mathbf{\Gamma}_j\Delta\mathbf{X}_{t-j} + \mathbf{\Phi}\mathbf{D}_t + \boldsymbol{\varepsilon}_t$$

where $\boldsymbol{\beta}'\mathbf{X}_t \sim I(0)$ and represents the deviations from the long-run equilibrium, i.e. cointegrating, relationship between $p_{rt}$ and $p_{wt}$, and the changes in at least one of the prices, $\Delta\mathbf{X}_t$, respond to these deviations from the previous period, $\boldsymbol{\beta}'\mathbf{X}_{t-1}$, through the adjustment parameters $\boldsymbol{\alpha}$ in such a way that the equilibrium errors $\boldsymbol{\beta}'\mathbf{X}_t$, $\boldsymbol{\beta}'\mathbf{X}_{t+1}$, ... converge to zero, ceteris paribus (given that $\boldsymbol{\beta}'\mathbf{X}_t$ is zero-mean reverting).

Premultiplying [15] by $\boldsymbol{\beta}'$ and rearranging, gives:

$$[16] \qquad \boldsymbol{\beta}'\mathbf{X}_t = (1 + \boldsymbol{\beta}'\boldsymbol{\alpha})\boldsymbol{\beta}'\mathbf{X}_{t-1} + \sum_{j=1}^{k-1} \boldsymbol{\beta}'\mathbf{\Gamma}_j\Delta\mathbf{X}_{t-j} + \boldsymbol{\beta}'\mathbf{\Phi}\mathbf{D}_t + \boldsymbol{\beta}'\boldsymbol{\varepsilon}_t .$$

Because $\Delta\mathbf{X}_{t-j}$ ($j = 1,\ldots,k-1$) and $\boldsymbol{\varepsilon}_t$ are stationary, the condition $|1 + \boldsymbol{\beta}'\boldsymbol{\alpha}| < 1$, or equivalently, $-2 < \boldsymbol{\beta}'\boldsymbol{\alpha} < 0$, allows $\boldsymbol{\beta}'\mathbf{X}_t$ to be stationary as well. If we return to our theoretical framework in the previous subsection, then given the assumption that $c_{rt}$ is stationary, [4] is the linear combination of $p_{rt}$ and $p_{wt}$ in *Model* 1 that represents $\boldsymbol{\beta}'\mathbf{X}_t$: $c_{rt} = \delta^{-1}(1+\delta)p_{rt} - p_{wt} = \boldsymbol{\beta}'\mathbf{X}_t$. In *Model* 2 the cointegrating relationship is found by substituting [11] for $c_{wt} + p_{ft}$ in [9], giving: $c_{rt} = \delta^{-1}(1+\delta)p_{rt} - p_{wt}$. Lastly, in *Model* 3, substituting [12] for $c_{wt} + p_{ft}$ in [10] yields the following long-run relationship between $p_{rt}$ and $p_{wt}$: $c_{rt} = \delta p_{rt} - (1 + \delta)p_{wt}$. Notice that each time the equilibrium error is given by $c_{rt}$.

The long-run equilibrium implies a common stochastic trend in the prices $p_{rt}$ and $p_{wt}$. To study the econometric implications of this trend, let $\boldsymbol{\alpha}_{\perp} = [\alpha_{r\perp}, \alpha_{w\perp}]'$ be a (2×1) vector of parameters of full rank such that $\boldsymbol{\alpha}'\boldsymbol{\alpha}_{\perp} = 0$. Premultiplying [15] by $\boldsymbol{\alpha}_{\perp}'$ shows that $\boldsymbol{\alpha}_{\perp}'\Delta\mathbf{X}_t$ does not respond to the equilibrium errors, while its solution, $\boldsymbol{\alpha}_{\perp}'\mathbf{X}_t$, is driven by the stochastic trend, $\boldsymbol{\alpha}_{\perp}'\sum_{i=0}^{t}\varepsilon_i$ . Further define $\boldsymbol{\beta}_{\perp} = [\beta_{r\perp}, \beta_{w\perp}]'$ as a (2×1) vector of full rank such that $\boldsymbol{\beta}'\boldsymbol{\beta}_{\perp} = 0$. Using the relation

$$\boldsymbol{\beta}_{\perp}(\boldsymbol{\alpha}_{\perp}'\boldsymbol{\beta}_{\perp})^{-1}\boldsymbol{\alpha}_{\perp}' + \boldsymbol{\alpha}(\boldsymbol{\beta}'\boldsymbol{\alpha})^{-1}\boldsymbol{\beta}' = \mathbf{I}$$

it can be seen that:

$$\text{[17]} \qquad \mathbf{X}_t = \boldsymbol{\beta}_{\perp}(\boldsymbol{\alpha}_{\perp}'\boldsymbol{\beta}_{\perp})^{-1}\boldsymbol{\alpha}_{\perp}'\mathbf{X}_t + \boldsymbol{\alpha}(\boldsymbol{\beta}'\boldsymbol{\alpha})^{-1}\boldsymbol{\beta}'\mathbf{X}_t\,.$$

Note that $\boldsymbol{\beta}$ does not contain zero elements, neither does $\boldsymbol{\beta}_{\perp}$. Moreover, $\boldsymbol{\alpha}_{\perp}'\boldsymbol{\beta}_{\perp}$ is a scalar unequal to zero. Consequently, the stochastic trend introduced by $\boldsymbol{\alpha}_{\perp}'\mathbf{X}_t$ is the *common* stochastic trend in the prices $p_{rt}$ and $p_{wt}$. This is why Gonzalo & Granger (1995) define $\boldsymbol{\alpha}_{\perp}'\mathbf{X}_t$ as the common factor. Note that if $\alpha_{r\perp} = 0$ ($\alpha_{w\perp} = 0$), then $\alpha_w = 0$ ($\alpha_r = 0$). The implication is that $p_{wt}$ ($p_{rt}$) captures the common factor, whereas $p_{rt}$ ($p_{wt}$) does all the correction to eliminate any deviation from long-run equilibrium, see also [17]. In turn, this is equivalent to saying that there is long-run causality running from $p_{wt}$ to $p_{rt}$ ($p_{rt}$ to $p_{wt}$). See Hall & Milne (1994), Granger & Lin (1995) and Gonzalo & Granger (1995) for the concept of long-run causality and see, for example, Tiffin & Dawson (1996) and Dawson & Tiffin (1998) for empirical applications.

Using the econometric concepts introduced above, we can now derive the testable implications that discriminate between our three strategic channel pricing models: *Model* 1, given by [8] and [10] and implying price leadership of the wholesaler; *Model* 2, formed by [9] and [11] and implying that the retailer dominates since the wholesaler is only allowed to set its price on the basis of the farm gate price; and *Model* 3, composed of [10] and [12] to capture the fact that the retailer buys directly from the farmer while the wholesaler is involved in market arbitrage. Given that $p_{ft}$ introduces the stochastic trend, while $c_{rt}$ represents the equilibrium error, it follows that $c_{wt} + p_{ft}$ forms the common factor. In *Model* 1 the common factor is captured by a linear combination of both prices, $p_{rt}$ and $p_{wt}$, see [7]. Consequently, both $p_{rt}$ and $p_{wt}$ will display error correction. In *Model* 2 $p_{wt}$ is not error correcting, see [11], but $p_{rt}$ is: [9] includes $c_{rt}$. Lastly, in *Model* 3, $p_{rt}$ does not show error-correcting behaviour, see [12], but $p_{wt}$ does, see [10].

## 3 EMPIRICAL ANALYSIS

We will now analyse the relationship between the wholesale price and the retail price for a selection of three local maize markets in the south of Benin and three local white sorghum markets in Burkina Faso: one national center for interregional arbitrage in each country (Cotonou in Benin, and Pouytenga in Burkina Faso) and two rural centres (Azové and Dassa in Benin, and Djibasso and Solenzo in Burkina Faso). In both countries maize is marketed through a private commercial system; the state does not intervene as an active buyer or seller. There are wholesalers who are involved in spatial arbitrage. A retail and a wholesale segment can be distinguished in each market place. In Cotonou, local retailers buy on the wholesale segment of the market, while in Pouytenga and the rural centres not only wholesalers but also the local surpluses from farms meet retailers' demand.

Cotonou is the largest town in and capital of Benin and its market Dantokpa can be considered as the most important market place in the country. Dantokpa is a centre for retail trade and wholesale trade: it is here that the brokers sell their products to retailers. Dantokpa is open every day, but every fourth day the market is more busy as it is the traditional market day. A number of wholesalers in Cotonou also sell maize on the market of Bohicon, the other urban market in the south of Benin. Consequently, they have opportunities for arbitrage. Moreover, most of the retailers in Cotonou buy from brokers instead of provisioning themselves in nearby villages, because it is difficult to combine collection in villages with selling services, as the Dantokpa market is open every day. Nevertheless, from a survey conducted amongst retailers in Cotonou it appeared that only a minority of them buy from the same wholesaler. In fact, they buy from the broker who offers the lowest price (Lutz, 1994). Furthermore, the urban area of Cotonou is always a maize-deficit area with relatively high prices. Consequently, the maize available at the market place has to be sold to local retailers. This implies that wholesalers are constrained in their search for clients. To summarise, in Cotonou the wholesalers have some opportunities for arbitrage, but the monopolistic competition among the retailers is based on buying from the broker who offers the lowest price. Consequently, we may expect *Model* 1 to hold for Cotonou although the results may also tend to be in favour of *Model* 2.

Important surpluses of maize are traded from Azové and Dassa to Cotonou and other market places in the south of Benin. Consequently, the wholesalers in these rural markets are involved in market arbitrage, and hence anticipate supply and demand conditions in the different, but spatially price-integrated, wholesale markets. Although retailers in Dassa are able to buy directly from the farmers, the pure competition arising from their large number forces them to set the retail price on the basis of the farm price according to [12]. In general, more than 50 and sometimes even more than 100 retailers are operational in these market places. Therefore, our empirical results with respect to Dassa are expected to comply with *Model* 3. In Azové the situation in the retail segment is not much different from Dassa, although an important group of the retailers is not permanently active on the market; this group consists of farmers, students, etc. Consequently, the retailers who are permanently active may have some price setting power arising from monopolistic competition. Therefore, the empirical results for Azové are expected to resemble *Model* 1, although they may also tend to be in favour with *Model* 3.

A similar situation as that for Dassa in Benin applies to Pouytenga in Burkina Faso. This market place in the middle of Burkina Faso is an international trading centre where wholesalers are involved in market arbitrage while fierce competition among retailers only allows for the common factor retail price [12]. In the other two market places that we consider in Burkina Faso, Djibasso and Solenzo, wholesalers are involved in spatial market arbitrage as well. However, in these two rural markets the number of retailers is rather small so that their existence is based on monopolistic competition. Consequently, we hypothesise that *Model* 1 describes the wholesale-retail price relationship in Djibasso and Solenzo.

Daily wholesale and retail prices are available for the three markets in Benin where the market days are held in a four-day cycle: for 190 market days at Cotonou (4 September 1987 to 29 September 1989), 184 market days at Azové (28 September 1987 to 29 September 1989), 174 market days at Dassa (3 November 1987 to 25 September 1989). Weekly data are available for the three markets in Burkina Faso: 152 observations for Pouytenga (92.01 - 94.48) and 151 observations for Djibasso and Solenzo (92.01 - 94.47).[1]

For each market we considered the retail and wholesale prices and estimated bivariate VARs of order $k$ = 1,..., 10.[2] To determine the appropriate order, the Schwarz criterion (SC) was computed. The estimate for $k$, denoted $k$*, was chosen so that SC was minimized. The results are presented in Table 8.1 and show that a VAR(1) is selected for all markets, except for Pouytenga where a VAR(2) must be considered. Note that the estimates of $k$ are much smaller than 10, suggesting that it is unnecessary to conclude that it is more fruitful to increase the information set rather than to automatically increase the lag length.

*Table 8.1 VAR order determination*

| Prices in | SC at $k$ = | | | | | | | | | | |
|---|---|---|---|---|---|---|---|---|---|---|---|
| | 0 | 1 | 2 | 3 | 4 | 5 | 6 | 7 | 8 | 9 | 10 |
| Cotonou | 9.51 | 6.06* | 6.09 | 6.17 | 6.26 | 6.30 | 6.40 | 6.49 | 6.59 | 6.70 | 6.81 |
| Azové | 9.03 | 5.90* | 5.97 | 6.03 | 6.09 | 6.17 | 6.28 | 6.37 | 6.43 | 6.52 | 6.62 |
| Dassa | 9.55 | 6.37* | 6.41 | 6.48 | 6.59 | 6.68 | 6.77 | 6.88 | 6.95 | 7.04 | 7.14 |
| Pouytenga | 5.79 | 4.97 | 4.75* | 4.91 | 5.12 | 5.42 | 5.57 | 5.26 | 5.38 | 5.68 | 5.52 |
| Djibasso | 6.63 | 5.11* | 5.31 | 5.66 | 5.93 | 6.10 | 6.40 | 6.72 | 7.14 | 7.12 | 7.11 |
| Solenzo | 7.80 | 6.01* | 6.36 | 6.69 | 6.92 | 7.11 | 7.15 | 7.39 | 7.52 | 7.76 | 7.95 |

* *indicates the lowest value; SC is Schwarz criterion*

Next, the Johansen procedure to test for cointegration was applied. We found the wholesale price and the retail price to be cointegrated for each market; this supports the assumption that $c_{rt}$ can be considered to be stationary. The results are presented in Table 8.2 and are based on [14] and [15]. The intercepts in [14] and [15] can be restricted so that they are only part of the cointegrating space (i.e. only the error-correction term must be corrected for a non-zero mean) if it appears that they are not significant when the error-correction term is de-meaned. Moreover, if the mean of the error-correction term is not significant either, then [14] and [15] do not include intercepts at all when computing and evaluating the trace statistic (in Table 8.2 indicated as 'absent in VECM'). Comparing the trace statistics with their critical values shows that for each market $r = 0$ (i.e. no cointegration) must be rejected while $r \leq 1$ (i.e. cointegration) cannot be rejected. The Johansen procedure is invariant to the choice of the variable selected for normalisation (Hamilton, 1994). In our presentation we chose the retail price to be the left-hand variable of the cointegrating relationship ($e_{rt}$ is the equilibrium error, see Table 8.2). For each market place we found the parameter of the wholesale price to be highly significant (using asymptotic $t$ values). If we took the wholesale price as the left-hand variable and estimated the parameter of the retail price, we found all parameters to be significant as well. Consequently, both prices must be $I(1)$ and their relationship is a real cointegrating relationship.

*Table 8.2 Testing for cointegrationa*

| Prices in | Intercept | $r \leq$ | Trace | Critical Values[a] 5% | 1% | Cointegrating Relationship (standard error in parentheses) |
|---|---|---|---|---|---|---|
| Cotonou | absent in | 0 | 50.34 | 12.53 | 16.31 | $p_{rt} = 1.11\, p_{wt} + e_{rt}$ |
| | VECM | 1 | 0.13 | 3.84 | 6.51 | (0.01) |
| Azové | restricted in | 0 | 87.28 | 19.96 | 24.60 | $p_{rt} = 2.51 + 1.05\, p_{wt} + e_{rt}$ |
| | coint. space | 1 | 1.43 | 9.24 | 12.97 | (1.23) (0.02) |
| Dassa | restricted in | 0 | 48.89 | 19.96 | 24.60 | $p_{rt} = 5.67 + 1.07\, p_{wt} + e_{rt}$ |
| | coint. space | 1 | 3.08 | 9.24 | 12.97 | (2.54) (0.04) |
| Pouytenga | absent in | 0 | 42.24 | 12.53 | 16.31 | $p_{rt} = 1.06\, p_{wt} + e_{rt}$ |
| | VECM | 1 | 0.00 | 3.84 | 6.51 | (0.01) |
| Djibasso | absent in | 0 | 49.61 | 12.53 | 16.31 | $p_{rt} = 1.19\, p_{wt} + e_{rt}$ |
| | VECM | 1 | 0.02 | 3.84 | 6.51 | (0.01) |
| Solenzo | absent in | 0 | 62.30 | 12.53 | 16.31 | $p_{rt} = 1.13\, p_{wt} + e_{rt}$ |
| | VECM | 1 | 0.23 | 3.84 | 6.51 | (0.01) |

[a] *Critical values obtained from Osterwald-Lenum (1992, Table 1* for Azové and Dassa and Table 0 for the other markets)*

From the cointegrating relationships in Table 8.2 we can also deduce that the parameter of the wholesale price is significantly greater than one in all markets. Retailers' long-run margin behaviour is characterised by charging a percentage mark-up. In Azové and Dassa, however, retailers also add an absolute component to their margin (about 5 percent of the average wholesale price level in Azové and 10% in Dassa). A possible reason for this is that retailers in these market places are not always easy to distinguish from the other actors, as they may also function partially as petty wholesalers (Lutz, 1994, 202, 208). Hence, they may integrate part of the wholesale distribution function, such as transportation, the cost of which is related more to the quantity traded than to the price, which is the economic rationale for imposing an absolute margin.

Based on the long-run parameter estimates presented in Table 8.2 (these are the estimates of the $\beta$ parameters in [15], i.e. we replaced $\beta_r p_{r,t-1} + \beta_w p_{w,t-1}$ by $e_{r,t-1}$), we estimated the short-run parameters $\alpha$ and $\Gamma_j$ (this time, without restricting the intercept). The $\alpha$ parameters, which can be interpreted as adjustment parameters, are of particular interest, because they were used to test our models. The estimates of the adjustment parameters are presented in Table 8.3 and appear to be in favour of our hypotheses. *Model* 1 applies to Cotonou: retailers and wholesalers have some price setting power, because both $p_{rt}$ and $p_{wt}$ are error correcting (each of both $\alpha_r$ and $\alpha_w$ are significant and the parameter estimates of $\beta$ and $\alpha$ obey the stationarity condition put forward by [16]). Nevertheless, the estimate of $\alpha_w$ is rather small and not significant at the one percent significance level of the one-sided $t$ test. This result shows the tendency to *Model* 2. The results for Azové also comply with *Model* 1, although now the weak significance of the estimate of $\alpha_r$ shows a tendency to *Model* 3. For Dassa and Pouytenga the results fully comply with *Model* 3, as $p_{wt}$ is error correcting ($\alpha_w$ is significant and lies within 0 and 2) and $p_{rt}$ is not ($\alpha_r$ is insignificant), indicating that there is a direct link between the retail price and the farm gate price, while the wholesalers are able to be involved in market arbitrage, leaving them some leeway to influence wholesale prices. Finally, the results for Djibasso and Solenzo favour *Model* 1, because both $p_{wt}$ and $p_{rt}$ are error-correcting.

Finally, note that a higher percentage mark-up could indicate stronger competition among the retailers. If we take this into account and compare the estimated percentage mark-ups in Table 8.2 with the magnitude and significance of the estimated $\alpha_r$ parameters in Table 8.3 for each country, it appears that higher mark-ups go hand in hand with retail price-setting behaviour getting closer to or significantly complying with *Model* 3.

*Table 8.3 Testing for long-run causality*

| Prices in | Effective sample size | Estimate $\alpha_r$ | $t$ value | Estimate $\alpha_w$ | $t$ value |
|---|---|---|---|---|---|
| Cotonou | 189 | -0.35* | -5.48 | 0.11* | 2.31 |
| Azové | 183 | -0.19 | -2.28 | 0.52* | 5.95 |
| Dassa | 164 | -0.05 | -0.59 | 0.30* | 4.12 |
| Pouytenga | 107 | -0.24 | -1.31 | 0.48* | 2.61 |
| Djibasso | 71 | -0.35 | -3.05 | 0.54* | 5.02 |
| Solenzo | 116 | -0.22* | -2.07 | 0.54* | 5.47 |

** indicates significantly different from zero (one-sided test at the 0.05 level)*

## 4 CONCLUSIONS

Recall that we proposed a method for empirically testing whether or not wholesalers have some price setting power vis-à-vis retailers. The method was applied to three models that were considered as possible candidates for describing the vertical price relationships in the marketing channels of local maize markets in Benin and white sorghum markets in Burkina Faso. A salient feature of our method is that the common stochastic trend and the long-run equilibrium error must be assigned to the variables in each decision structure being considered. Doing this for the application in this study, we found that the exclusion restrictions on the error-correction structure led to testable implications discriminating between the three models.

As far as our limited evidence goes, we conclude that wholesalers behave as vertical price leaders in the sense of Stackelberg leadership, in particular if they are involved in market arbitrage. Nevertheless, in the maize-deficit urban area of Cotonou, where local wholesalers can only sell to local retailers and are not involved in market arbitrage, retailers show a clear tendency to dominate the marketing channel and behave as if they had integrated the local wholesalers. In the other markets wholesalers dominate as they have alternative arbitrage opportunities, giving them some freedom to influence prices. Retail prices follow the common factor when the number of local retailers is large and perfect competition prevails (Dassa and Pouytenga) or are error-correcting when the number of retailers is small and maintained by monopolistic competition (Azové, Djibasso and Solenzo).

Our empirical results indicate that relations between wholesalers and retailers vary between market places. In contrast to common understanding, retailers can play a crucial role in the price formation process. Consequently, we conclude that in addition to spatial arbitrage, local market conditions are decisive for the distribution of market power among retailers and wholesalers, as well.

## NOTES

[1] Data are available from authors on request.

[2] All computations were performed in EViews, Version 2.0 with the correct Schwarz criterion.

**REFERENCES**


BANERJEE, A., J.J. DOLADO, J.W. GALBRAITH & D.F. HENDRY (1993) *Co-Integration, Error-Correction, and the Econometric Analysis of Non-Stationary Data*. Oxford: Oxford University Press.

BASSOLET, B. (forcoming) *Organisation et Fonctionnement des Marchés Céréaliers sous Libéralisation. L'Expérience de Burkina Faso*. Groningen: University of Groningen.

BASSOLET, B. & C. LUTZ (1999) 'Information Service and Integration of Cereal Markets in Burkina Faso'. In *Journal of African Economies*, vol. 8, pp.31-51.

DAWSON, P.J. & R. TIFFIN (1998) 'Estimating the Demand for Calories in India'. In *American Journal of Agricultural Economics*, vol. 80, pp.474-481.

ENGLE, R.F. & C.W.J. GRANGER (1987) 'Cointegration and Error Correction: Representation Estimation, and Testing'. In *Econometrica*, vol. 55, pp.251-276.

GERSTNER, E. & J.D. HESS (1991) 'A Theory of Channel Price Promotions'. In *American Economic Review*, vol. 81, pp.872-886.

GONZALO, J. & C. GRANGER (1995) 'Estimation of Common Long-Memory Components in Cointegrated Systems'. In *Journal of Business & Economic Statistics*, vol. 13, pp.27-35.

GRANGER, C.W.J. & J.-L. LIN (1995) 'Causality in the Long Run'. In *Econometric Theory*, vol. 11, pp.530-536.

GRANGER, C.W.J. & P. NEWBOLD (1986) *Forecasting Economic Time Series*. Orlando: Academic Press.

HALL, S.G. & A. MILNE (1994) 'The Relevance of *P*-Star Analysis to UK Monetary Policy'. In *The Economic Journal*, vol. 104, pp.597-604.

HAMILTON, J.D. (1994) *Time Series Analysis*. Princeton: Princeton University Press.

JOHANSEN, S. (1995) *Likelihood-Based Inference in Cointegrated Vector Autoregressive Models*. Oxford: Oxford University Press.

JOHANSEN, S. & K. JUSELIUS (1990) 'Maximum Likelihood Estimation and Inference on Cointegration: With Applications to the Demand for Money'. In *Oxford Bulletin of Economics and Statistics*, vol. 52, pp.169-210.

KUIPER, W.E., C. LUTZ & A. VAN TILBURG (1999) 'Testing for the Law of One Price and Identifying Price-Leading Markets: An Application to Corn Markets in Benin'. In *Journal of Regional Science*, vol. 39, no. 4, pp.713-738.

LEE, E. & R. STAELIN (1997) 'Vertical Strategic Interaction: Implications for Channel Pricing Strategy'. In *Marketing Science*, vol. 16, pp.185-207.

LUTZ, C. (1994) *The Functioning of the Maize Market in Benin: Spatial and Temporal Arbitrage on the Market of a Staple Food Crop*. Amsterdam: University of Amsterdam.

LUTZ, C., A VAN TILBURG & B. VAN DER KAMP (1995) 'The Process of Short- and Long-Term Price Integration in the Benin Maize Market'. In *European Review of Agricultural Economics*, vol. 22, pp.191-211.

MARTIN, S. (1993) *Advanced Industrial Economics*. Oxford: Blackwell.

OSTERWALD-LENUM, M. (1992) 'A Note with Quantiles of the Asymptotic Distribution of the Maximum Likelihood Cointegration Rank Test Statistics'. In *Oxford Bulletin of Economics and Statistics*, vol. 54, pp.461-471.

STOCK, J.H. & M.W. WATSON (1988) 'Testing for Common Trends'. In *Journal of the American Statistical Association*, vol. 83, pp.1097-1107.

TIFFIN, R. & P.J. DAWSON (1996) 'Average Earnings, Minimum Wages and Granger-Causality in Agriculture in England and Wales'. In *Oxford Bulletin of Economics and Statistics*, vol. 58, pp.435-442.

TIROLE, J. (1988) *The Theory of Industrial Organization*. Cambridge (Mass.): The MIT Press.

VON UNGERN-STERNBERG, T. (1994) 'Percentage retail mark-ups', Cahier no 9412, DEEP, University of Lausanne.

# 9 COMPARING LIBERALISATION IN AGRICULTURAL INPUT AND DRAUGHT ANIMAL MARKETS IN BENIN

Houinsou Dedehouanou and Paul Quarles van Ufford

## 1 INTRODUCTION

One of the main development problems in Sub-Saharan African countries in general and the Republic of Benin in particular stems from the conflicting nature of the role of the state (Krueger, 1990; Thorbecke & Morrisson, 1989).[1] In the years following the independence of most African countries, the public sector was thought to play a crucial role in the national economic development of the newly emerging states. However, two decades of experience with government-led economic development drew many developing countries into a deep economic crisis, which affected all sectors of the economy. The root-cause of this was attributed to government failure rather than market failure (Srinivasan, 1985). Thus, from the 1980's onwards, international financial institutions such as the International Monetary Fund (IMF) and the World Bank persuaded many countries to start a Structural Adjustment Programme (SAP). Benin was no exception to this and has adopted various SAPs since 1989. The programme's objectives for the agricultural sector were to increase the efficiency of the principal services supplied to the producer by making them more flexible, by facilitating the redistribution and training of personnel and by taking steps to encourage the private sector to become more involved. Accordingly, a reduction and restructuring of state expenditure in this sector could be achieved. Measures included the liberalisation of the market for agricultural inputs in 1992 in order to stimulate private sector involvement through the awarding of licenses to a selected number of private enterprises. However, the liberalisation process was accompanied by sharp price increases on the one hand, since the usual government subsidies on agricultural inputs lapsed, and by additional cost increases for farmers on the other, as a result of transport costs they incurred because the private firms did not sell at the farm gate but only from major district towns.

These developments provided the context for two research projects carried out in Benin, which examined the role of the state in the agricultural input sector in Mono province and the functioning of cattle markets (of which draught animals represent a significant segment) in Borgou province, respectively.[2] It appeared that both studies contained some interesting clues to help analyse the problems that occur with the liberalisation of the agricultural input sector in general, and the role of the public sector therein, in particular.

The aim of the present chapter is to compare the functioning of the agricultural input market and the draught animal market in Benin and to draw lessons with regard to the position of the state in a liberalised agricultural input market. The observed dynamics between the supply of and the demand for draught animals will be highlighted in order to inspire the on-going reforms in the agricultural input sector in Benin. While from a theoretical, non-institutional perspective such reforms are not necessarily mistaken, analysis is lacking with regard to the actual implementation of policies and the perverse economic dynamics they may create. Through its focus on institutional environments and arrangements, the new institutional economics approach can provide the appropriate framework for such an analysis. It is suggested in this paper that the liberalisation reforms underway are of limited scope. First, they tend to be characterised by limitations in the economic and political assumptions on which they are based. Second, they seem to suffer from a lack of knowledge about the extent to which socio-cultural and spatial factors influence the demand for agricultural inputs. Consequently, it can be readily seen that achieving reform targets in such a limited context may not be a positive step forward.

As stated above, the two studies were undertaken in two different regions: the Mono region in South Benin and the Borgou region in North Benin. In both regions, the liberalisation of the agricultural input market is underway. The situation with respect to the draught animal market is, however, different. In particular, the adoption rate of the use of animal traction is substantially higher in the Borgou region than in the Mono region. In the latter, unfavourable climatic conditions and socio-economic characteristics of farmers hamper the development of livestock keeping. In the former, a good livestock keeping environment coupled with the steep growth in cotton production have particularly favoured the adoption of animal traction, notably from the 1980's onwards. In the same region, the role of the state has been progressively reduced because farmers have become increasingly involved in livestock transactions themselves. In the Mono region on the contrary, the state is still a major actor on the draught animal market, providing credit and subsidies.

These general characteristics of the two regions and the two markets provide the starting point for this chapter. After presenting the conceptual framework in section 2, the case studies of the agricultural input market and the draught animal market will be presented in sections 3 and 4, respectively. Finally, sections 5 and 6 will synthesise the analyses and derive some implications for the role of the state in the agricultural input sector in Benin.

## 2 CONCEPTUAL FRAMEWORK

According to the proponents of neo-classical economic theory, three arguments could justify government intervention: "*the public goods argument, the equity or 'merit good' argument and the market failure argument*" (Batley, 1994, p.494). First, some goods and services do not lend themselves to individual charging because of collective use. Therefore, the state as the collective interest holder is the best provider. Second, other goods and services are to be provided to all community members regardless of their ability and willingness to pay the market price. Finally, market failures thought to result from 'structural rigidities' due to a lack of responsiveness to price signals constitute the main justification for government intervention in the economic sector. In fact, the principles guiding market functioning were said to not satisfactorily reconcile problems of equity and efficiency.[3] Henceforth, government should take a leading role in the allocation of investment, control the functioning of the economic mechanisms, and otherwise intervene to compensate for market failures.

Recently, the macroeconomic environment in most of Sub-Saharan Africa has changed with the emergence of democracy and the decentralisation process. This has been compounded with the implementation of Structural Adjustment Programmes (SAPs). Although the SAPs seemed to alleviate the budget deficit and reduce strains on public finance through liberalisation, privatisation and devaluation, the programme has proven to bear high social costs.[4] Of course, the provision of 'public' goods and services could be considered with regard to their efficiency as well as their equity. Efficiency, in the sense of achieving development goals with minimum per unit cost, depends upon the internal characteristics of government institutions pertaining to the delivery of goods and services. Equity relates to the extent to which vulnerable groups, such as the large majority of the rural poor, are incorporated and taken care of. This is characteristic of the conflicts (centralisation versus decentralisation) regarding the distribution dimension of development as mediated through government institutions. What matters most is the lack of evidence that government sponsored institutions are internally organised in such a way that they can satisfactorily carry out efficient or equitable development activities. This leads us to the issue of the nature of government intervention.

The theoretical approach to government intervention calls for the analysis of the reasons for intervention. According to Batley (1994), these are:

> "the control of monopoly abuse, the supply of information and subsidy to consumers, the management of the supply of 'public goods', etc. For different services in differing national contexts there will be different reasons for government to be involved, and these imply diverse levels and forms of involvement" (ibid., 1994, p.495).

The wide range of intervention possibilities between the public, private and participatory sectors becomes clearer as soon as two components of supply are considered: provision, i.e. the act of ensuring that a given good or service is available, and production which is the act of physically generating, delivering and maintaining the good or service.[5] In other words, liberalisation is not an all or nothing choice for the public sector. We suggest in this chapter that the public sector could take up a specific role, not directly connected to the production or the delivery of agricultural inputs, but linked to the provision of services for instance.

The key features of the goods or services appear to be paramount to the scope and the scale of government intervention (see Thorbecke, Chapter 2 of this book). Analysing goods and services and designing institutions constituted a breakthrough in the study of the role of the public, private and participatory sectors in the development process. Of greater significance is the analysis of institutions themselves. The New Institutional Economics approach is therefore the relevant framework for making such a goal operational. In this respect, Williamson (1990) has put forward the concepts of institutional environment and institutional arrangement. The institutional environment represents the rules of the game in the above-mentioned sectors, the so-called provision component of the development service supply. It should be noted here that these rules are not exclusively defined by the government but also by communities at the local level. For instance, farmers' groups and self-help organisations define their proper rules of functioning. The institutional arrangement, though, refers to aspects of production. In Benin, for instance, the government used to play a dominant role in the production and delivery of agricultural inputs before the liberalisation measures. Following liberalisation, the involvement of the private and participatory sectors increased.

The present chapter will include both an analysis of the main features of the goods and services related to the draught animal and agricultural input supply markets as well as an analysis of relevant institutions.

## 3 THE AGRICULTURAL INPUT MARKET

Agriculture supplied approximately 70% of household income and contributed to about 40% of the GNP in 1991. Throughout the 1980's, it was the most dynamic sector of the economy, having grown by 63% between 1982 and 1991, compared with 21% for the overall GDP (at 1985 prices). Cotton constitutes the main cash crop and provides essential foreign exchange earnings for the government and revenues for farmers. The government has assigned particular importance to matters associated with agriculture and rural development, for they can reduce poverty in rural areas and contribute to the goals of the social dimension of the SAPs (Benin, 1990).

The agricultural input sector (mainly fertilisers, pesticides and insecticides) used to be the domain of the organisation in charge of extension (CARDER) until the early 1990's. The extension institutions were reported inefficient and in need of performance improvement. To reorganise these institutions, the World Bank imposed an adaptation of the Training and Visit (T&V) system introduced in Benin in 1985. The Bank emphasised the need to transfer some of the activities formerly carried out by CARDER, such as agricultural input delivery, credit and the primary collection of agricultural products, to the private sector, especially to farmers' organisations. This reorganisation is part of the liberalisation process in the agricultural sector. For instance, farmers' organisations called *'Groupements Villageois'* are currently taking over agricultural input delivery, credit distribution and primary collection of cotton at district, commune and village levels. Such a transfer of vital activities in cotton production is likely to empower farmers. As such, they can influence extension activities and agricultural production as well. There is still uncertainty about the pace and direction of economic development that will follow this drastic shift in the role performed traditionally by the public sector.

*Characterisation of demand for and supply of agricultural inputs*

In this section, two periods are considered with regard to demand for and supply of agricultural inputs. The first period is characterised by excess of supply and high transaction costs incurred by the delivery system at the village level (high transport & information costs, credit reimbursement risks and losses due to theft and to carry-over from year to year). The supply system was heavily organised, while the structure of demand seemed to be governed mainly by financial considerations like access to credit. This problem was partly solved by tying the credit to participation in a '*Groupement Villageois (GV, farmers' organisations)*'. During the second period, following the implementation of the liberalisation programme for the agricultural sector in 1992, the needs are recorded only once, one year before effective supply occurs. Before the reform, however, the village extension officer used to keep a buffer stock, which allowed for demand expressed at a later stage. Indeed, farmers and certainly small rain-fed farmers hardly plan their production ahead, and if they do, gradual changes are incorporated as soon as the natural and managerial conditions permit. Thus, the decision to purchase agricultural inputs rests on the day-to-day evolution of farm works.

The liberalisation of the agricultural input sector is a gradual process. It may be worth recalling Berg's warning about the uncertainty surrounding the operation of the private sector (World Bank, 1981). Shortly after the liberalisation programme was implemented in Mono province, only one private enterprise had contracted with the formal marketing board (SONAPRA) for the delivery of agricultural inputs, for the first two years of implementation, and in the only district characterised by a high input consumption. It forced SONAPRA to establish its own branches in the other four districts. To date, the situation has evolved with the establishment of new private agricultural input sales points in the districts that had previously been served only by SONAPRA.

An important reason given in the literature for the reluctance of the private sector to carry agricultural input supply is the low level of demand (Thompson, 1991). Economies of scale dictate the attitude of the private sector where transport costs, stock handling and capital mobilisation are concerned. This is largely substantiated by country studies in Lele et al. (1989). In addition to the substantial rise in agricultural input prices, several other demand as well as supply side factors can be identified which discouraged farmers, bringing about the prevailing low levels of demand.[6] The elaboration of these factors equally sheds light on the apparently reluctant attitude of private sector firms.

With regard to demand side factors, the problem of contracting demand from the large majority of individual and dispersed smallholders could be considered as a threat to private sector involvement. Whereas demand for agricultural inputs is still expressed at the local level, the private firms supply inputs at the district level and the coordination of demand therefore requires cooperative behaviour among farmers. At first sight, this appears contradictory because the existing farmers' organisations, the GVs, apparently serve as coordinating mechanisms between the locally fragmented demand and the wholesale supply of inputs. However, the participation in GVs is still low, notably in the Mono region where the rate does not exceed 32% (Dedehouanou, 1994; CARDER Mono, 1996). Participation rates in the Borgou region are higher, approximately 50% to 60%. This is largely due to the fact that participation in a GV is limited to cotton producers, of whom there are high numbers in the Borgou region traditionally. Notwithstanding this, the GVs only partially perform a demand coordinating function in both regions.

High prices have equally led to a decrease in demand. The phasing out of subsidies on agricultural inputs as well as the devaluation of the CFA franc in January 1994 had an impact on farmers' purchasing power. As a result, the financial risk of investing in agricultural inputs has increased significantly. In turn, this has contributed to the shrinking of demand and the reluctance of the private sector to engage in the agricultural input market. However, care must be taken not to blame the Structural Adjustment Programme (SAP) for all factors impinging on the demand. More general socio-economic factors, too, affect the demand for agricultural inputs in Benin, such as farmers' risk aversion, the relative price relationship between input and output as well as the productivity response of inputs given local climatic conditions. Also of great significance is the customary property rights structure and its influence on resource control and access within the wider rural development context.

Finally, the insufficient appropriation of market functioning and product knowledge by farmers further hinders the development of the agricultural input market. The issue of truth in labeling and trustworthiness imposes itself mainly on the parallel channel of the agricultural input market. In Benin, the parallel channel of the agricultural input market is represented by (fraudulent) imports of fertilisers and pesticides from Nigeria. Typically, these imported products are of dubious quality. They used to be imported from Nigeria on a substantial scale. Although the liberalisation programme in this country has to some extent reduced these flows, pesticides in particular are still imported. Most farmers are sceptical of the agricultural inputs from the informal channel, but they are often coerced to resort to it because of the ease of access compared to the formal channel. The damaging role of the parallel channel, in terms of low quality products, frustrates their belief in the effects of modern inputs on agricultural production.

With regard to the supply side factors that clarify the reluctance of the private sector, six constraints can be distinguished. First of all, the agriculture-input market in Benin has long been a public sector monopoly. More importantly, agriculture generally holds a low profile in the portfolio of private investors in developing countries. Consequently, involving private firms in agriculture is still at its infancy. Second, the spatial dispersal of the small family farms makes any individual delivery system prohibitive from the point of view of the private sector. In addition, the poor infrastructure contributes to the high costs that would be incurred by a system of individual delivery. This explains why the post-liberalisation supply outlets are located only at district level. Third, the thinness of the agricultural input market makes the supply unattractive for the licensed private enterprises. The share of the agricultural input market controlled by the parallel imports from neighbouring countries remained high after the liberalisation reforms. LARES (1996) reported the import of inputs from Nigeria on a substantial scale in the past. Although the inward-looking input policy in this country has to some extent reduced these flows, the paper provides evidence that pesticides in particular are still imported, following trends in the exchange rate between the Nigerian currency and the CFA franc. Fourth, the required timeliness of supply makes the activity very risky since the failure to provide the input at the correct time of the growing season implies a loss of sales. The inflexible recording of demand for inputs mentioned earlier may also lead to the inadequate timing of supply by private enterprises. Fifth, farmers buy their inputs on credit. During the period before the liberalisation reforms, the government-sponsored organisations used to fulfil the functions of both input supplier and credit provider. Today, though, farmers need collateral in order to obtain a credit. This not only requires some form of 'group cohesion' but specific arrangements between the private sector and the agricultural credit institutions as well. The best solution to this dilemma, as strongly advocated in the literature, is to establish contract farming (Grosh, 1994; Porter & Phillips-Howard, 1997) or contract-like schemes. This suggests a role for the input supplier (the private firms) on the output market, which at the moment is still held by SONAPRA.

Finally, the non-transparency in the licensing of private enterprises plays its part in curbing private sector willingness to boost the formal channel of the agricultural input market. In addition to the risks mentioned above, the possibility of failing to get the license renewed represents a threat for private firms (Bidaux et al., 1997).[7]

*Actors' view of private sector involvement*

Following the analysis of the impact of the liberalisation policy on the structure of demand and supply for agricultural inputs, the present section introduces the point of view of the main actors of concern; farmers, private entrepreneurs and civil servants. Their perception will give better insight into the problems that have accompanied the liberalisation process and the way in which these could be overcome.

The respondents in general and farmers in particular, attribute an important role to the spatial and temporal dimensions of the coordination mechanisms. They indicate that there would be a steady increase in the demand for agricultural inputs if supply were location specific. It should be recalled that the coordinating role played by the GV is mainly a technical one, the quantity of agricultural inputs requested being a mere function of the forecast cotton production. This explains why most of the respondents (including all actors) regard social values, which may dictate a retreat from the process or a curbing of demand, as marginal in the structuring of estimated demand for agricultural inputs.[8] Nonetheless, the role of social values in shaping institutional arrangements is worth analysing. In this respect, the crucial role of traditional grass roots organisations has to be underlined. Based on mutual confidence and group solidarity these organisations evolve around themes such as credit and farm work. In the villages where such groups are found, the chances for a successful establishment of a private firm are greater. For these indigenous institutions to play a greater role in the agricultural input sector, the challenge is to be represented at the district level. Otherwise, the private firms should seek representation at the village level.

Each group of actors has its own understanding of the changes and the motives behind the functioning of the system. The farmers are all aware of their losses after the withdrawal of subsidy and the retrenchment of the state sponsored organisations. According to them, the situation before liberalisation was much better, though they implicitly recognise the bureaucratic and inflexible character of the former organisational structure. However, the burden of transport costs and risks initially incurred by government sponsored organisations is now to be borne by them. In the opinion of farmers, this could explain their reluctance to purchase agricultural inputs through the formal channel.

According to extension service reports, which reflect the official government stance, the transfer of roles from the public sector to both the private sector and the GV has been successfully achieved (CARDER Mono, 1996). Whereas, during the agricultural campaign of 1991-92, 1.56% of the marketing activities had been transferred to the GVs, a 100% score was obtained during the 1995-96 campaign (ibid., 12). The official reports represent only part of the story. The informal side is represented by dissatisfied extension officers who consider themselves as losers, because they lost their authoritative power. Besides, they have lost material advantages related to the services formerly rendered in the agricultural input marketing sector. The contradiction between the formal and informal position in the public sector is not novel. The reluctance of extension officers to forgo access to privileges leads to neglect of their remaining tasks such as agricultural extension on the field. The private firms, finally, are enjoying the profits of externalising most of the transaction costs: transport costs are taken care of by farmers' cooperatives, information costs are supposed to be assumed by extension services, storage and payment risks are suppressed by simply responding to the expressed demand, etc. With respect to credit for the purchase of inputs, the private firms acknowledge that indigenous credit associations could be a partner when represented at the district level.

Among all three groups of relevant actors, only the private firms perceive the success of the reforms because of new prospects in the sector. The other two groups of actors, farmers (taken individually and in GVs) and extension officers, are more cautious about the outcomes, and not sure about their post-liberalisation roles.

## 4 THE DRAUGHT ANIMAL MARKET

The introduction of animal traction in much of Sub-Saharan Africa is relatively recent and it became common practice only in the second half of the century. High adoption rates were realised in areas with cash crop production and well-developed crop marketing systems (Starkey, 1988). In Benin, animal traction has always been linked to the cotton production sector for which the state sponsored organisations used to play a leading role in the agricultural input, marketing and credit services. These conditions favoured the adoption of animal traction, particularly in the savannah zones of the country (Bruentrup, 1996).[9] Demand for draught animals in Benin has steadily risen over the years: in 1993, an estimated 37,000 working pairs were used in agriculture (ibid., 1994). It should be noted, though, that a pair of animals generally serves for more than one year before it is replaced. To start with, a farmer usually purchases a pair of immature bulls (1-3 years old), which are trained before working on his plots. Trained animals are used for up to 7 cropping seasons before the farmer sells them.

The draught animal market in Benin can be seen as an excellent example of a private sector market. There is no government intervention at either the provision or the production level. Demand and supply converge at formal and informal markets where private sellers and buyers meet directly. The absence of uniform quality and measurement standards for the product (cattle) makes direct contact and price negotiation per unit necessary. Although the cotton marketing board in Benin (SONAPRA) provides agricultural inputs on credit, such schemes (loans) are virtually absent with regard to the purchase of draught animals. Today, the financing of animal traction is largely based either on the farmers' revenues from cotton production or on traditional sources of capital. Animals are generally purchased for cash at cattle markets which are located in livestock production zones (such as northern Benin) throughout West Africa

*Characterisation of demand for and supply of draught animals*

The market for draught animals should be seen as part of a larger whole, namely the cattle market. At the cattle market, demand is expressed not only by farmers but also by traders (who are responsible for 75% of all purchases), butchers and pastoralists. As each actor demonstrates a preference for one or more specific types of cattle, this causes a certain degree of market segmentation. Following the preferences for cattle to be used for animal traction, farmers almost exclusively operate at the market segments for mature and immature bulls, notably the latter.[10]

Farmers' demand for draught animals has a typically seasonal character and therefore peaks from a few months after the last harvest to the beginning of the new agricultural season.[11] This period typically coincides with the payment of cotton revenues. During these periods of seasonal demand peaks, which result in temporarily higher prices, it was observed that farmers were often willing to pay more than traders, particularly for high quality bulls, both mature and immature ones.[12] This seems to relate to the fact that, from a farmer's point of view, the investment in draught animals is an attractive one. Given the purchase price of a young bull plus additional investment and maintenance costs, a farmer does generally not incur a loss when he sells the animal as an adult bull, a few years later. On the contrary, the value of the animal appreciates rather than depreciates.[13] Besides, the investment risk is also partly reduced since the state guarantees a fixed price for agricultural output, as is the case with cotton. On the other hand, the risk factor should not be totally excluded given the probability of animal mortality.

In general, the cattle market is characterised by a relatively high level of supply compared to demand. Traditionally, pastoralists have been the principal suppliers. The main reason for members of pastoral groups to sell their cattle is not so much demand oriented, but rather 'need driven': the sale of cattle is the main source of cash income with which necessary expenses can be met.[14] In Benin, pastoralists account for 65% of total supply at the main cattle markets. A slightly different supply picture emerges when the market segment in which farmers operate is considered. Of all animals sold to farmers, pastoralists supply 45% whereas traders and farmers account for a noticeable 21% and 34% of supply respectively.[15] The percentage of supply that traders account for is somewhat higher for draught animals than the percentage of supply they carry for the remaining market segments. This reflects the attractiveness of the draught animal market. Experienced traders who know the preference of farmers with respect to draught animals, visit distant markets or pastoralist camps to purchase the appropriate animals. These are then sold to those farmers who do not have the time, the contacts or the experience to make such a move. Finally, even though supply generally exceeds demand at the cattle market, stiff competition occurs in the draught animal segment notably during the above-specified periods.

The fact that 'farmers supply other farmers' (in 34% of all cases) marks a recent strategy that has been observed among them. It concerns the purchase of mature draught animals (adult bulls) which have already been used for traction purposes. The new owner will employ the animals, which can be put to use immediately, for an additional year or two. This phenomenon reveals the existence of dynamic processes within a particular market segment and shows how farmers have become acquainted with market functioning (see below). Instead of using traders as intermediaries, they now directly supply other farmers. This is partly explained by the fact that farmers are more likely than anyone else to be aware of the prospects of using mature draught animals which have already been used for traction purposes. Furthermore, they are likely to be a credible actor in the market segment for draught animals, and a transaction between farmers possibly benefits from mutual confidence.

Despite a growing acquaintance with the draught animal market, the position of farmers, either as sellers or as buyers, can in some cases still be qualified as vulnerable. Although market access is clearly not subject to noticeable restrictions, this vulnerability is mainly due to a certain obscurity in market functioning as well as to farmers' ignorance of existing price levels. Various intermediary persons (brokers and traders) capitalise on this, a situation which was observed at all cattle markets. It probably contributed to the increased importance of direct supply by farmers. Nevertheless, we cannot speak of a situation in which farmers are confronted with groups of buyers or sellers who are able to set prices. As has already been stressed, demand and supply are diversified in the market segment for draught animals and we observed considerable, seasonally confined competition, notably for high quality bulls. Stretching the argument, we can argue that competition has quite an individual character: except for a relatively small group of rich merchants, other traders do not collude in any way, and neither do farmers, butchers and pastoralists.

*Actors' view of the functioning of the draught animal market*

After the introduction of animal traction by the French around 1950, its adoption rate steadily increased from the mid-seventies onwards (Quarles van Ufford, 1999). To date, the main argument for farmers not to use animal power for agricultural purposes is the lack of capital and credit facilities (ibid.). Although the government sponsored extension service has always promoted the use of animal traction, state involvement in the northern parts of the country has been greatly reduced. Formerly, the state used to provide draught animals on credit and would organise training facilities. Today, this is only the case in the South (the Mono region in particular), with most farmers in the North taking care of purchase, training and selling themselves. Thus, animal traction in northern Benin has been subject to a progressively diminishing role of the state, becoming a predominantly private activity today. In the light of current developments, the same situation can be perceived in the agricultural input sector.

As was already mentioned, the attitude of farmers, towards the cattle market in general and the draught animals segment in particular, has altered considerably over the years. It used to be common practice for a farmer, who was in need of draught animals, to purchase them from nearby livestock keepers. Arrangements such as the exchange of food grains for cattle were common practice. To dispose of draught animals, farmers used to invite a local trader. From around the mid-1980's, though, farmers have become increasingly involved in marketing activities themselves. This gradual change in attitude was one of the consequences of the creation of new cattle markets in zones where the use of animal traction was widespread such as the Borgou region (Quarles van Ufford, 1999). As a result, farmers turned to the market for the purchase and sale of their draught animals. To them, markets represented a larger choice of animals as well as better sales prices, since there now was some competition among traders for their animals. New opportunities occurred. The above-mentioned case concerning the purchase, through other farmers, of mature draught animals, has certainly been facilitated by the emergence of new cattle markets.

Notwithstanding the fact that a large group of farmers continued to employ the intermediary services of traders or middlemen, others became increasingly involved in market activities. The increased knowledge of market functioning mechanisms brought about a diversification in the ways these farmers acted on the market. The vicinity of cattle markets, for example, encouraged them to act as petty cattle traders, using cotton revenues as their working capital. At present, some farmers even cooperate to organise joint transport of draught animals for sale at consumer markets in southern Benin. In general, it seems fair to say that the involvement of farmers in the cattle market expanded as their consciousness of the profitability of marketing activities increased. At the same time, farmers continue to invest in cattle herds and livestock keeping has become an integral part of their livelihood. All together, the above-mentioned developments resulted in an enhanced familiarity with and confidence in commercial practices on the cattle market in general and on the draught animal market in particular. In other words, farmers have increasingly appropriated the market.

## 5 COMPARISON OF THE TWO MARKETS

In general, both markets have experienced a high degree of state involvement in the recent past. Equally similar is the image of a state which is receding from the market. The reasons, though, are different. Whereas the liberalisation process in the agricultural input sector is part of the Structural Adjustment Programme, the state stepped out of the draught animal market because its role had become superfluous. This was mainly the case in the northern Borgou region where cotton farmers, encouraged by the creation of new cattle markets and pushed by the ever increasing use of animal traction, engaged *themselves*, i.e. without the intervention of traders or middlemen, in both the purchase and sale of draught animals, using their revenues from cotton production. In addition, it was argued that the typical structure of demand and supply on the draught animal market contributed to the active participation of farmers in the market. The specificity of demand for and supply of draught animals can be summarised as follows. First, supply is of local origin. The larger part involves so-called need driven sales by local pastoralists. Second, we noticed that demand for draught animals is stimulated by the fact that the purchase of cattle is an attractive investment. The value of the animal generally appreciates rather than depreciates, although the risk of mortality should not be obscured. Third, demand appears not to be constrained by high transaction costs or by the absence of credit facilities. Thus, we concluded that it was the farmers themselves who provided the conditions for a successful withdrawal of the state through their continuous appropriation of market functioning and product knowledge.

A contrasting situation pertains in the market for agricultural inputs. Here, the apparent reticence of private sector firms to engage in agricultural input supply is partly explained by the low level of demand. Demand for agricultural inputs depends on input and output prices, credit availability, costs related to the purchase of the product and farmers' strategies (e.g. risk avoidance). The partial removal of subsidies on inputs, according to SAP policies, has further contracted demand. However, special emphasis was laid on the explanatory value of non-price elements such as the absence of a properly functioning demand coordinating mechanism. The private firms generally supply inputs at the district level and it was observed that the '*Groupement Villageois*' do not channel demand, which originates at the level of the farm. Although farmers have shown innovative behaviour through the establishment of indigenous credit associations, the spatial fragmentation of demand remains precarious, especially since these organisations generally lack representation at the district level. The existing situation (high prices and concentrated supply points) is prone to further development of the parallel channel of the agricultural input market, i.e. the import of fertilisers and pesticides from Nigeria, which have been in existence for a long time already. Moreover, the imported products commonly are of rather dubious quality and this strengthens our argument below for improving farmers' ability to assess the quality of inputs.

An additional reason for private sector hesitation is the lack of product differentiation. Differentiation is useful for market expansion, while its absence could explain the retrenchment of the private sector. The fact that certain types of inputs are more risk prone than others implies a responsibility of private firms for environmental concerns in general and farmers' health in particular. For instance, the approval of certain pesticides by the national regulatory office includes assumptions about its field management (Tripp & Gisselquist, 1996, p.3-4). In the concluding paragraph, we will address the implications of this situation for government policy.

## 6 IMPLICATIONS FOR GOVERNMENT POLICY

Two future directions of government policy for the agricultural input sector are envisaged. The first involves a further enhancement of market liberalisation and the second concerns the provision of an appropriate institutional environment.

Intensification of the market liberalisation process would require a facilitating role for the state which should ameliorate the conditions for market entry, such as access to working capital for traders, improved infrastructure and knowledge, in order to overcome the 'thinness' of the market. In this respect, further `opening' the market could lead to improved coordination of demand. Moreover, the situation in the cattle market has shown that a legitimised and high participation rate of local entrepreneurs, a role which could be played by the farmers themselves, would assist in bringing together supply and demand. Also, involving indigenous institutions would diminish the reliance on former official structures such as the GV. The second policy direction is crucial in the light of the present article. It revolves around issues such as truth in labeling and farmers' knowledge of markets and products. In the draught animal market, most farmers appeared to be well informed about most product characteristics which depend on the breed, the strength and the well-being of the animal. Over the years, farmers were able to accumulate skills with regard to practices in livestock keeping and market functioning. This has undoubtedly contributed to the dynamism and participation rates that were observed on this market.

The agricultural input sector, on the contrary, is characterised by a standardisation of the products, and consequently only price-quantity relationships matter. This can be seen as a major weakness. The nature of the product and the difficulty for farmers in assessing quality and reliability are obstacles to a smooth market process (trust & confidence). Moreover, farmers do not have a wide range of choices whereby they could test the performance of each input (notably pesticides, fertilisers and seeds) under specific conditions: it is difficult to differentiate between products without technical expertise. Quality and reliability problems become crucial issues in a black market situation.

The free play of market relations is unlikely to guarantee the application of norms and standards, which means that conditions need to be enforced. For instance, raising the awareness about truth in labeling is unlikely to be assured through the market. Even in the draught animal market the problem of incorrect value labeling sometimes imposes itself, although the product knowledge of farmers is significantly higher than in the agricultural input market. Furthermore, a cost-effective use of agricultural inputs involves adequate research and dissemination of information, which the private sector will not endorse.

There is probably room for government intervention, and for its extension service in particular, to deal with these market imperfections. In this regard, intervention in the institutional environment means verifying the quality and reliability of agricultural inputs (assuring truth in labeling) as well as increasing farmers' knowledge about these aspects. It is unlikely that the private sector will effectively provide very much of this information, because the results are location specific and often the benefits cannot be captured by private firms. As the dynamic processes observed on the draught animal market have revealed, there is a need for building local capacity with respect to market and product knowledge. Enhancing farmers' market and product knowledge, and thus furthering their appropriation of the market for agricultural inputs, is vital for the expansion of the latter. In certain cases, building local capacity to assume some responsibilities for monitoring the sale and use of inputs may be more efficient than relying on a large centralised regulatory agency.

## NOTES

[1] Development refers to social and economic well-being of people and accommodates Sears' view: "*the extent to which poverty and unemployment decline compared with an initial stage*" (1977, 3).

[2] Data with respect to the first survey, which was financed by a WOTRO (Netherlands Institute for Research in Tropical Areas) grant, were collected among individual farmers and key informants in Mono province during 1996. The second survey, financed by both the European Union (STD-3 program) and the

University of Amsterdam, was carried out at four main cattle markets in Borgou province between December 1995 and November 1996.

[3] See Krueger (1990) and Bardhan (1990), for a detailed discussion of these arguments.

[4] Commander (1989) and Cornia et al. (1987) notice a departure from equity considerations that guided early government development policies.

[5] According to Batley (1994), provision entails roles such as establishing policies or plans and monitoring their implementation, monitoring standards of service and the availability of information, financing, advising, enabling, coordinating, regulating and licensing private producers as well as maintaining public ownership and contracting producers. Production, on the other hand, involves the construction of plants, agricultural production and the maintenance and management of the delivery of goods and services.

[6] In a case study on Guinea, Sahn & Sarris (1994) discovered that even a decrease in prices did not lead to an increase in demand for agricultural inputs. They concluded that more attention should be paid to non-price and related institutional factors in explaining the low level of demand (1994, 295).

[7] Some observers claim that instead of the application of professional criteria for the selection of private firms to be allowed entry on the input market, it is political, rent-sharing considerations which have determined access for some of these firms.

[8] The notion of social values refers, among other things, to the tenure status of the farmer (i.e. fear of appropriation by the land owner after good harvests), the intra-household power relations (i.e. the ability of individual household members to purchase inputs through the GV) or intra-community solidarity (the case of village saving groups, for instance).

[9] In Mono province, the use of animal traction developed at a later stage and at a slower rate. This is primarily due to its less important cotton production and unsuitability for livestock keeping. Reflecting this situation, state involvement in the draught animal market in the Mono region is still prominent.

[10] The following table shows the degree of segmentation of the cattle market in the Borgou region, Benin. It is based on a sample of 2,241 market transactions carried out from December 1995 to November 1996, at the four main cattle markets in the region. Note that traders, who mainly operate on the mature bull and cow segments of the cattle market, carried out nearly three quarters of all transactions.

*Table 9.1 Breakdown of demand at four cattle markets in the Borgou region according to actor and type of cattle, December 1995 - November 1996 (n=2241)*

| | Trader | Farmer | Pastoralist | Butcher |
|---|---|---|---|---|
| mature bull | 32% | 39% | 5% | 29% |
| immature bull | 20% | 55% | 28% | 3% |
| cow | 33% | 1% | 2% | 62% |
| heifer | 15% | 5% | 65% | 6% |
| | 100% | 100% | 100% | 100% |

*Source: Quarles van Ufford, 1999*

[11] In fact, the term *demand* only reflects the revealed demand, i.e. the number of animals purchased.

[12] Of course, farmers and traders each consider different qualities of the animal for sale. Whereas a farmer is more likely to look at the breed, the physical qualities and the docility of an animal, a trader tends to be more interested in its weight versus price ratio since he will, in the end, sell it to a butcher.

[13] The profitability of the investment can be calculated in greater detail if factors like (hired) labor and (increased) agricultural output are taken into account. Bruentrup (1996, 92-101), presents such an analysis. Still, we witness a rather peculiar phenomenon when, speaking in economic terms, the residual value exceeds investment cost.

[14] For a summary of the discussion on the marketing of cattle by pastoral groups in Benin, see Bierschenk & Forster (1991) and Quarles van Ufford & Djedjebi (1997).

[15] To a certain extent a specialization in supply exists with regard to the type of animal sold. As could be expected, farmers play a larger part in the supply of adult bulls whereas pastoralists, from a livestock keeping point of view, prefer to release young bulls.

**REFERENCES**

BARDHAN, P. (1990) 'Symposium on the State and Economic Development.' In *Journal of Economic Perspectives*, vol. 4, no. 3, pp.3-7.

BATLEY, R. (1994) 'The Consolidation of Adjustment: Implications for Public Administration.' In *Public Administration and Development*, vol. 14, pp.489-505.

BENIN (1990) 'Country Presentation'. Second United Nations Conference on the Least Developed Countries.

BIDAUX, A., G. RAYMOND & B. G. SOULE (1997) 'Evaluation du système d'approvisionnement et de distribution des intrants agricoles du Bénin, rapport principal'. Cotonou, MDR - Rep. du Bénin.

BIERSCHENK, T. & R. FORSTER (1991) 'Rational herdsman: Economic strategies of the agro-pastoral Fulani of northern Benin.' In *Applied Geography and Development*, vol. 38, pp.110-125.

BRUENTRUP, M. (1994) 'La production du coton et l'environnement: quelques considérations pour les zones de savane dans le Nord du Bénin.' In P. Ton & L. De Haan (eds.) *A la recherche de l'agriculture durable au Bénin*, Amsterdam: Université d'Amsterdam, pp.94-104.

BRUENTRUP, M. (1996) 'Conditions for the introduction of animal traction in Sub-Saharan Africa; lessons from North Benin for the introduction of complex innovations.' Paper presented on an international symposium on Food Security and Innovations, Hohenheim: University of Hohenheim.

CARDER Mono (1996) 'Rapport Annuel'.

COMMANDER, S. (1989) 'Prices, markets and rigidities: African agriculture, 1980-88.' In *Structural Adjustment & Agriculture: Theory & practice in Africa & Latin America*, London: ODI.

CORNIA, G. A., R. JOLLY & F. STEWART (1987) *Adjustment with a Human face. Volume 1: Protecting the Vulnerable and Promoting Growth*. Oxford: Clarendon Press.

DEDEHOUANOU, H. (1994) 'Régionalisation du Développement Rural: étude de cas des organisations chargées du développement rural intégré en République du Bénin.' In *Série d'Economie et de Sociologie Rurales de la FSA/UNB*, no. 4.

GROSH, B. (1994) 'Contract Farming in Africa: an application of the New Institutional Economics'. In *Journal of African Economics*, vol. 3, no. 2, pp.231-261.

KRUEGER, A. O. (1990) 'Government Failures in Development.' In *Journal of Economic Perspectives*, vol. 4, no. 3, pp.9-23.

LARES (1996) 'Les échanges d'engrais'. In *Bulletin régional de suivi des échanges transfrontaliers, Bénin - Niger - Nigéria - Tchad - Cameroun*, fiche thématique no. 4, pp.1-4.

LELE, U., R. E. CHRISTIANSEN & K. KARIRESAN (1989) 'Issues in fertiliser Policy in Africa: lessons from development programs and Adjustment lending, 1970-87.' MADIA Discussion Paper no. 5, Washington: World Bank.

MÖSCHEL, W. (1993) 'The New Institutional Economics meets Law and Economics: Comment.' In *Journal of Institutional and Theoretical Economics*, vol. 141, no. 1, pp.88-91.

PORTER, G., & K. PHILLIPS-HOWARD (1997) 'Comparing contracts: an evaluation of contract farming schemes in Africa'. In *World Development*, vol. 25, no. 2, pp.227-238.

QUARLES VAN UFFORD, P. (1999) *Trade and Traders: The making of the cattle market in Benin*. Amsterdam: Thela Thesis.

QUARLES VAN UFFORD, P. & T. DJEDJEBI (1997) 'Le carrefour de la commercialisation. L'éleveur face au marché à bétail: le cas du Bénin septentrional'. Paper presented at the conference on 'prospects of pastoralism in West Africa', Giessen: Justus-Liebig University.

SAHN, D. E. & A. SARRIS (1994) 'The Evolution of States, Markets, and Civil Institutions in Rural Africa.' In *Journal of Modern African Studies*, vol. 32, no. 2, pp.279-303.

SCOTT, K. E. (1993) 'The New Institutional Economics Meets Law and Economics: Comment.' In *Journal of Institutional and Theoretical Economics*, vol. 141, no. 1, pp.92-95.

SEARS, D. (1977) 'The New Meaning of Development.' In *International Development Review*, vol. 3, pp.2-7.

SRINIVASAN, T. N. (1985) 'Neoclassical Political Economy, the State and Economic Development.' In *Asian Development Review*, vol. 3, no. 2, pp.38-58.

STARKEY, P. (1988) 'The Introduction, intensification and diversification of the use of animal power in West-African farming systems: implications at the farm level.' In Starkey & Ndiamé (eds.) *Animal Power in Farming Systems*.

THOMPSON, A. M. (1991) 'Institutional changes in agricultural product and input markets and their impact on agricultural performance.' FAO Economic and Social Development Paper no. 98, Rome: Food and Agriculture Organization.

THORBECKE, E. & C. MORRISSON (1989) 'Institutions, Policies and Agricultural Performance: a comparative analysis.' In *World Development*, vol. 17, no. 9, pp.1485-1498.

TRIPP, R. & D. GISSELQUIST (1996) 'A Fresh Look at Agricultural Input Regulation.' Natural Resource Perspectives paper no. 8, London: ODI.

WILLIAMSON, O. E. (1990) 'A Comparison of alternative Approaches to Economic Organization.' In *Journal of Institutional and Theoretical Economics*, vol. 146, pp.61-71.

WORLD BANK (1981) *Accelerated Development in Sub-Saharan Africa: an agenda for action.* Washington: The World Bank.

# 10 THE PUZZLE OF THE ABSENT RURAL FORMAL FINANCIAL INSTITUTIONS

J.G.M. (Hans) Hoogeveen[1]

## 1 INTRODUCTION

Households relying on rainfall dominated agriculture face enormous fluctuations in income due to price and primarily production instability. Such income variability does not have to be of concern to a utility maximising farmer who is interested in consumption stability, unless income shocks get transferred into consumption. In that case a bad harvest may even pose a threat to survival. Financial markets are the obvious means to disconnect the dangerous liaison between income and consumption. For instance if they allow households to secure a stable income by purchasing income insurance policies or if they provide the possibility to borrow money during times of income shortfall and to repay in times of high income.

Empirical studies on the presence of formal financial institutions and the degree of income risk in Africa have uncovered two facts that together constitute a puzzle. First there is the observation that in most African rural areas private formal financial institutions are absent. The reasons advanced to explain this phenomenon are the high cost of monitoring to prevent moral hazard problems and adverse selection, the difficulty to re-insure covariate risks (for insurances) or the risk of massive default after the manifestation of a covariate risk (in the case of credit). Second there is the fact that income variability is high. Dercon (1992a) for instance reports a coefficient of variation of crop income of 67 percent in the Sahellian zone and of 52 percent in the Sudanian zone. The puzzle that is enclosed in these two facts is that not being able to smooth consumption yields a reduction in welfare to the order of 25 to 45 percent of average income.[2] If formal financial institutions could capture this welfare loss by offering a service that allows households to smooth their consumption, then they have a huge incentive to do so despite the above mentioned problems. The question then is whether it can be explained, given the large potential benefits, that financial services aimed at consumption smoothing remain absent.

In this paper a partly reconciliation of these seemingly contradictory findings is sought by exploring how rural households reduce the consumptive consequences of income variability through the accumulation of liquid assets in times of plenty and their sales during income shortfalls. If such a buffer stock strategy is successful in reducing consumption variability, the demand for financial services in the rural areas might be substantially reduced.[3]

The welfare gain that may be expected from the introduction of formal financial institutions, in an environment where households apply a self insurance strategy, is the topic of this chapter. To explore this, an approach developed by Deaton (1989) is followed. For the situation with an exogenous income process and a safe asset, he shows how an optimal consumption rule can be determined numerically for liquidity constrained households. Dercon (1992b) extended his model by allowing the safe asset to have more real life characteristics such as a value that is covariant with the exogenous income process. Here Deaton's work is extended in another direction by introducing a more realistic income process, namely one with a lognormal distribution and large standard deviation. Having determined the consumption rule a Monte Carlo simulation is employed to determine the variance in consumption that results after following the consumption smoothing rule.

The paper is organised as follows. In Section 2 a brief empirical background is sketched of the presence of income risk and the use buffer stocks by rural households in Zimbabwe. To this end information is used from a 7 year panel dataset for 400 rural households. Section 3 contains the derivation of the expression that serves as basis for the determination of the consumption rule. Section 4 presents the simulated consumption rule and presents the variance in consumption that results after optimal smoothing. In Section 5 the expected benefits from the introduction of financial markets are determined. A discussion of the results follows in Section 6.

## 2 INCOME RISK AND BUFFER STOCKS

The 400 rural households on which this section is based belong to the group of more than 70,000 families that have been resettled on land acquired after independence by the Zimbabwean government from large scale farmers. All households were resettled in the early 1980's into tightly clustered villages located in three of Zimbabwe's resettlement schemes (Mpfurudzi, Mutanda and Sengezi). These schemes were chosen so as to ensure representation of each of the three major agro-ecological zones in the country, which are suited for cropping. Farmers located in Mpfurudzi live in the area most favourable to farming, those in Mutanda have to deal with the least favourable conditions. Those resettled in Sengezi live in an intermediate area. On resettlement, families were allocated a 1 acre (0.4 ha) residential plot, 12 acres (5 ha) of arable land for individual production and the right to use a variable amount of grazing land on a communal basis. An important feature of the resettlement schemes is that families are subject to a wide range of rules and regulations that define what they can and can not do. Till 1992 crop production was the only legal source of income for the head of household. Till date, this has an important bearing on the way income is earned. Crop production is the most important source of income and in a normal agricultural season off farm income and remittances contribute less than 10 to 15 percent to total household income. This contrasts sharply with Reardon's (1997) estimate that 30 to 50 percent of rural income is earned in the non farm sector.

Another important feature of farming in the resettlement schemes is that crop production is almost exclusively rain fed making rainfall the dominant factor in production. Rainfall patterns are characterised by unpredictable variability both within and between years. Midseason droughts are frequent occurrences which can be particularly disastrous for drought vulnerable crops like maize (Scoones, 1996), the most important crop grown (Table 10.1). During the period under consideration, 1990-91 to 1996-97, farmers experienced two droughts, of which the one in 1992 resulted in near uniform crop failure. The 1995 drought was less severe, but hit at a crucial moment in crop development so that maize yields were still only about 35 percent of the period's average. Only farmers in one of the three survey sites (Sengezi) were affected less. Their maize yields were about 70 percent of the period's average. How dominant rainfall is for production is illustrated by the close correlation (a correlation coefficient of 0.89) between average rainfall and the average yield of the most important crop grown, maize. For the other two most widely cultivated, cotton and groundnuts this is 0.80 and 0.91 respectively.

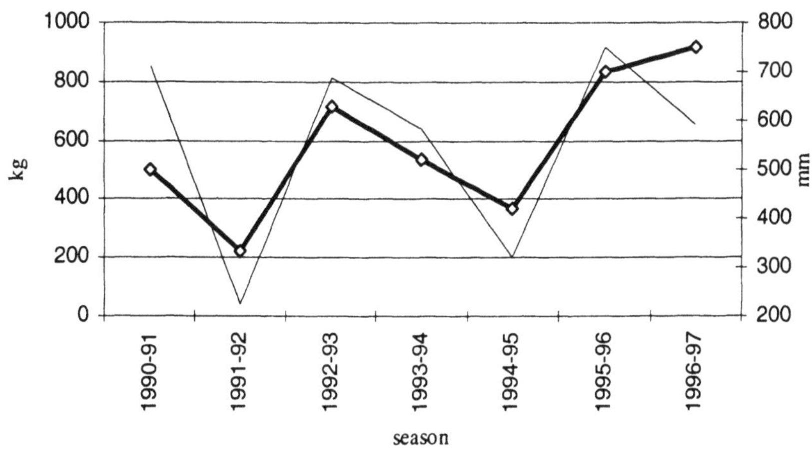


*Figure 10.1 Average rainfall (bold line) and average maize yield per acre (thin line)*
*Sources: Department of Meteorological Service (FEWS); Kinsey survey*

The yield variability with which individual farmers have to deal is large. The inter-year coefficient of variation of yield per acre for the most important crops lies around 90 percent (Table 10.1).

*Table 10.1 Acreage and coefficient of variation (1993-1996)*[4]

| | *Percent of acreage allocated* | *Coefficient of variation yield per acre* |
|---|---|---|
| Maize | 59% | 90% |
| Cotton | 23% | 75% |
| Groundnuts | 9% | 93% |

*Source: Kinsey survey*

With the incomes from the individual crops being so variable farmers will seek other possibilities to smooth their income. Diversification is a strategy that might generate some relief. Antonio, Shakespeare's merchant of Venice practices it successfully[5], but for the resettlement farmers diversification is less of an option as the correlation coefficients for the yield of their most important crops lie close to 0.60[6] and because the options for off farm employment are limited and mainly restricted to agriculture.

*Table 10.2 Coefficient of variation of total household income in Z$ 1990 (1992-1997)*[1]

| | *Coefficient of variation* |
|---|---|
| Household income | 71% |
| Household income (excluding 1992 drought) | 60% |
| Village income | 44% |

[1]*Total income comprises value added of crop income, female income (mainly from gardening), income from livestock produce, business income, income from aid, off farm income and remittances.*
*Source: Kinsey survey*

As Table 10.2 indicates, even if diversification into different crops and sources of income is taken into account households still have to deal with highly variable income with a coefficient of variation of 70 percent. Such variability is hard to imagine. The series 991, 9426, 4435, 1922, 6272, 13345 has a mean of 6065 and coefficient of variation of 77 percent. If these figures represent the past weeks prices of a share at the stock market then few people would be inclined to invest in it, unless for diversification purposes. But resettlement farmers have to cope with this sort of variability from year to year for their prime source of income as the sequence represents the total yearly real income realised by a randomly picked household between 1992 and 1997.

In view of such high and unavoidable income variability one might expect coping mechanisms aimed at smoothing consumption to be important. Financial services could fill the gap, but formal crop insurance is not available to smallholder farmers in Zimbabwe. Formal credit is, but lending is mostly seasonal and related to production. The Cotton Company of Zimbabwe and the Cotton Marketing Board provide inputs on credit. In 1996 45 percent of the farmers that grew cotton made use of this facility. Loans for activities like livestock fattening, breeding and stocking can be obtained from the Agricultural Finance Corporation (AFC), who also gives loans for the production of crops like tobacco. Little to no credit available from these institutions can be used for consumptive purposes so that a household that wants to delink consumption from income using credit has to make use of informal arrangements.[7] But informal credit for consumption smoothing purposes is only available to deal with idiosyncratic risks and not with the main source of covariate risk, lack of rain (Kinsey, Burger and Gunning, 1998). Another possibility is to rely on informal insurance such as income pooling at the village level. As Table 10.2 indicates, there is scope for such insurance within the village, but it is not clear to which extent such possibilities are explored. The table shows the minimum income variability that can be attained through complete income pooling within the village; 34 percent. In the literature that tests whether full insurance takes place at the village level such is generally not found (Townsend, 1994; Ravallion & Chaudhuri, 1997), so that even with a considerable degree of informal insurance, household level income variability remains large.

This leaves households with a strong incentive to accumulate stocks of liquid assets. Storing food is one possibility. Joseph recommended this to the Pharaoh of Egypt somewhere around 1800 BC, when he prophesied that Egypt would enjoy 7 years of prosperity followed by 7 years of famine. The sampled families still follow Joseph's advice. In bad agricultural seasons like 1992 or 1995 the amounts stored were insufficient to last most households till the next harvest, but in the seasons following a drought, storage is in excess of a year's requirement.

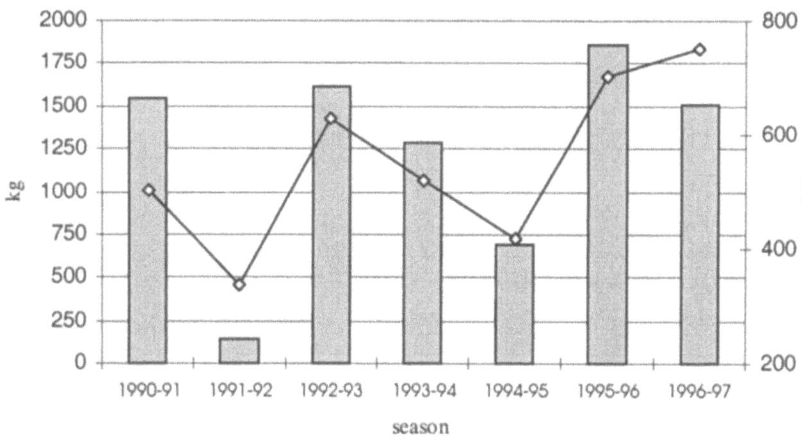


*Figure 10.2 Maize retentions (columns) and average rainfall (line)*
*Source: Kinsey survey*

Livestock is another potential buffer stock. Evidence that livestock is used to this end is presented in Fafchamps, Udry and Czukas (1998) who find that in West-Africa livestock is accumulated when there is a windfall income gain while disinvestment takes place in years with adverse weather shocks. Rosenzweig and Binswanger (1993) report for India that bullock sales increase significantly when weather outcomes are poor and incomes low, and that purchases of bullocks increase when rainfall is ample and incomes above average. A similar pattern can be found in the Zimbabwe data. Figure 10.3 presents household livestock sales expressed as trained oxen equivalents. These have been calculated by comparing for each of the years the median prices of the different types of livestock held and then expressing them in trained oxen equivalents. The average of these trained oxen fractions was then used to express all livestock except chicken, pigeons and rabbits as trained oxen equivalents.[8] The figure confirms that during the bad agricultural seasons of 1992 and 1995 livestock sales were highest, while they were low in better seasons.

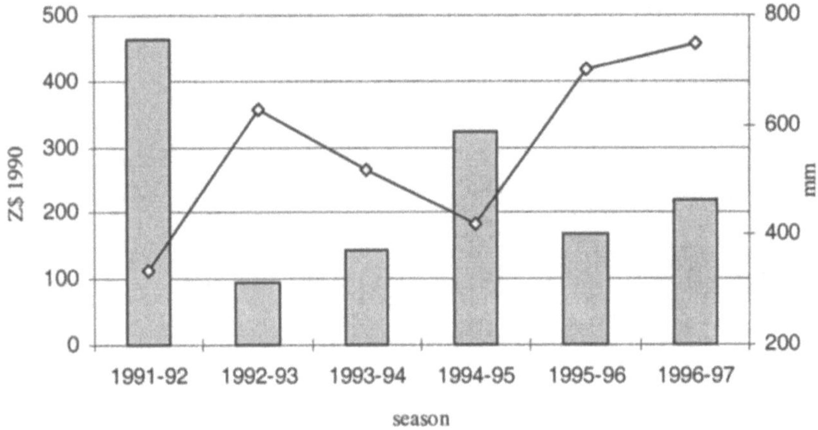


*Figure 10.3 Livestock sales (columns) and average rainfall (line)*
*Source: Kinsey survey*

Money balances also serve as buffer stock. Two third of the households owned a savings account in 1997, despite the fact that no banks are located in the resettlement schemes, forcing farmers to incur considerable transport costs, both in time and in money spent. Both during the 1992 and the 1995 drought, cash and savings were used to purchase food. This is confirmed below where it is shown that both in 1992 and 1995 the balances in the savings accounts fell following the dry agricultural conditions, to be restored in the subsequent years (Figure 10.4).

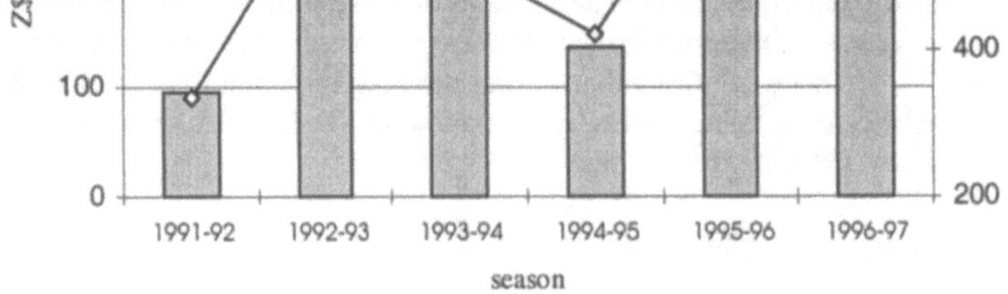


*Figure 10.4 Household money balances in Z$ 1990 (columns) and average rainfall (line)*
*Source: Kinsey survey*

## 3 THEORY

Now that it has been established that farmers have to deal with highly fluctuating income, that financial markets are absent and that farmers use buffer stocks to insulate consumption from income shocks, the question is to which extent farmers are successfully doing so. One way of approach would be to measure consumption directly. But as will be clarified in Section four, this requires exceptionally accurate expenditure data that are not available. Therefore, a different approach is followed which is based on Deaton's (1989) work.

Let households know the distribution of the income process. This is a strong condition but it can be defended on the grounds that in many rural areas crop yields are stationary and their variances well understood by farmers. Households live infinitely. This too is a restriction, but one that can be justified on the ground that households, or families, in developing countries are usually quite long lived.

With income stochastic, utility is maximised with respect to the expected value of lifetime utility:

[1] $$U_t = E\sum_{t=1}^{\infty}\left(\frac{1}{1+\delta}\right)^t u(c_t)$$

subject to the sequence of constraints:

[2] $$a_{t+1} = (1+r)a_t + y_t - c_t$$

where E is a mathematical conditional expectations operator –a probability weighted average of possible outcomes. $U_t$ is the utility function of consumption, $c_t$, $r$ is the rate of return on the stock of assets, $a_t$ , $\delta$ the rate of time preference, $y_t$ exogenous but random income. The rate of time preference and the rate of return are constant over time. The rate of return can be negative or positive. It is likely to be negative if assets are kept in the form of food stores that are affected by storage losses or as cash balances that reduce in value over time due to inflation. A positive rate of return can be expected for livestock whose value increases through breeding and weight increase.

As long as the utility function is strictly concave, households have an incentive to smooth consumption because a given increase in consumption above a certain value provides the household with less extra utility than the loss in utility the household would experience if it had to deal with a similar reduction of income below this value. Since the focus is on the consequences of income risk for consumption it is plausible that households have an interest in the expected value of future income and in its variability. Such a precautionary motive can be included by ensuring that marginal utility is convex.

As a first step, assume full financial markets. Households can now borrow as much as they wish at rates identical to the lending rate. Assets can be positive or negative in which case they represent borrowings. Ponzi schemes, in which the household has access to infinite resources by borrowing money and repaying the interest and the principal by borrowing even more money are not allowed, so that:

[3] $$\lim_{T\to\infty}(1+r)^{-T} a_{t+T+1} \geq 0$$

must hold.

One of the approaches to arrive at a consumption function in such an intertemporal setting is through dynamic programming. Applying this to the decision represented by the equations (1) and (2) allows to derive the stochastic Euler equation[9]:

[4] $$u'(c_t) = \frac{1+r}{1+\delta} E_t\left(u'(c_{t+1})\right)$$

This equation is central to the modern view on consumption, namely that households attempt to keep the marginal utility of consumption constant over time. It states that, in discounted utility terms rational forward looking households will not want marginal consumption to be worth more in one period than another (adjusted for a time factor) and shows that in the optimum the agent is indifferent between consuming her last dollar in the current period, or saving it and consuming it next period. In the presence of full financial markets, households would be able to optimally smooth consumption and have complete certainty about the level of consumption at all dates. If the household faces a bad income draw and is out of assets, it borrows money to repay later. If the household's cash on hands (income plus accumulated assets) exceeds the requirements for optimal consumption, it puts money in the bank to prepare for less fortunate circumstances.

As a second step, let formal and informal financial markets be absent. A trivial but important consequence of this is that households can no longer have negative asset holdings. Hence to equation [2] a borrowing constraint has to be added stating that household assets always have to be larger or equal to zero: $a_t \geq 0$.

The asset constraint ensures that consumption can, at maximum, be current income plus the value of the stock of assets. Now the household can be confronted with two situations:

1. in which the asset constraint makes itself felt. The household has consumed all its income and assets and would like to borrow but cannot. With all assets consumed, the marginal utility of an extra unit of current consumption is higher than the expected marginal utility to be derived from saving this extra unit until the next period.
2. in which the asset constraint does not limit household decisions and where current income plus assets are sufficient for the stochastic Euler equation to hold. The marginal utility of current consumption is equated to expected future marginal utility and the household would not wish to borrow money even if it were able to do so.

These two cases can be put together in a single expression, in which the second part is the stochastic Euler equation presented as equation [4]:

[5] $$u'(c_t) = \max\left\{u'(y_t + a_t); \mathrm{E}_t\left[\frac{(1+r)}{(1+\delta)} * u'(c_{t+1})\right]\right\}$$

With borrowing constraints a smooth profile of future consumption is no longer assured. Instead consumption becomes variable though, due to the accumulated savings it will be less erratic than income. In some years however, when the household is out of assets and income is low, the household faces a situation in which there is very little to consume, namely the low income only. If zero assets are carried forward for many periods then consumption changes will be equal to the erratic income changes. In periods in which income is sufficiently high however, the agent accumulates buffer stocks to be used in periods in which income is low.

To obtain an expression for consumption, expression [5] is inverted. This is allowed because $u'(c_t)$ is a monotonically decreasing function. Inversion gives:

$$[6] \qquad c_t = \min\left\{(y_t + a_t); u'^{-1}\, \mathrm{E}_t\left[\frac{(1+r)}{(1+\delta)} * u'(c_{t+1})\right]\right\}$$

Expression [6] shows that the level of consumption is solely determined by the current level of cash on hands (assets plus income). This allows to write the consumption function in a very general form as: $c_t = f(a_t + y_t)$. It is interesting to note that this equation for consumption is very different from the expression found in the permanent income theory, which states that consumption is constant over time. Here is stated that consumption is stochastic ($y$ is a random variable) and depends on cash on hand.

Substituting the expression for consumption for the current and the next period into the previous expression, yields:

$$[7] \qquad f(y_t + a_t) = \min\left\{(y_t + a_t); u'^{-1}\, \mathrm{E}_t\left[\frac{(1+r)}{(1+\delta)} * u' f(y_{t+1} + a_{t+1})\right]\right\}$$

Finally replace $a_{t+1}$ by the sequence from which it originates:

$a_{t+1} = (1+r)(y_t + a_t - f(y_t + a_t))$ to obtain equation [8]:

$$[8] \qquad f(y_t + a_t) = \min\left\{(y_t + a_t); u'^{-1}\, \mathrm{E}_t\left[\frac{(1+r)}{(1+\delta)} * u' f(y_{t+1} + (y_t + a_t - f(y_t + a_t))\right]\right\}$$

The latter equation is not an optimal solution to the problem in terms of exogenous variables as the choice variable $c_t = f(a_t + y_t)$ is still present on the right side of the equation. What it is, is the projection of a function onto itself. Deaton (1989) who derived this equation, argues that an explicit functional form for the consumption rule is unlikely to be found and uses a numerical approach to find the shape of the consumption function for particular values of the parameters of the model. To do so, the expectation sign is replaced by an integral and an initial guess is made for $f(a_{t+} y_t)$. The latter is substituted in the right hand side of equation [8], after which the integral is solved to yield an initial guess for $f_0(a+y)$. This first round allows to determine a new function, $f_1(a+y)$ which can subsequently be put back in the right hand side of equation [8]. This process is repeated until it converges.

## 4 DETERMINING OPTIMAL CONSUMPTION VARIABILITY

To find the optimal consumption rule numerically, and in view of the income density presented in Figure 10.5, an income distribution is chosen that is skewed to the right.

There are evolutionary reasons to expect a priori that the income distribution from crop production is skewed to the right. Farmers have been dealing with risk for thousands of years. If, in the beginning, agricultural income had a symmetric distribution, then households were facing an equal probability of low and high outcomes. Given the shape of their utility function in which the utility loss from outcomes below a given point is larger than the benefit associated with an outcome above that point, farmers would have had a strong incentive to alter the shape of their income distribution to a form where low incomes are less probable. Through the selection of plants and development of cultivation methods they might have been able to do so with a distribution that is skewed to the right as the result. And as crop income is skewed to the right, so is non agricultural wage income (which cannot be negative) and so are profits from entrepreneurial activities (because negative profits cannot be sustained for a prolonged period), so that the total income distribution is also skewed to the right.

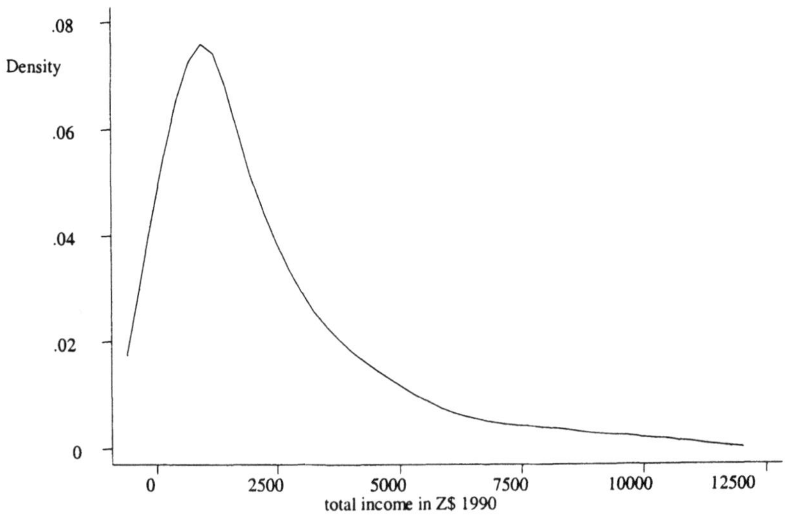


*Figure 10.5 Kernel density estimate of total income in Z$ 1990 (1993-1998)*
*Source: Kinsey survey*

Following the empirical results, it is assumed in the simulations that the income process follows a lognormal distribution with mean 100. Two cases are considered: one in which income has a standard deviation of 50 and one where the standard deviation is 80.[10] 50 represents a situation in which there is considerable income pooling within the village, 80 where this is largely absent and the household is primarily dependent on crop income.

Apart from a different income distribution, the simulation model uses the same assumptions and parameters as Deaton (1989) did. The utility function chosen is the iso-elastic one: $u(c) = \frac{c^{1-(1/\sigma)}}{1-(1/\sigma)}$, where $\sigma$ is the intertemporal elasticity of substitution. For marginal utility to be convex it is required that $\sigma < 1$. Presented is the consumption function for $\sigma = 1/2$, implying a rate of relative risk aversion of 2. Having taken a lognormal distribution, negative income draws are of no concern, so that the distribution does not need to be truncated. The rate of return to asset holdings is set at 5 percent and the rate of time preference at 10 percent.[11]

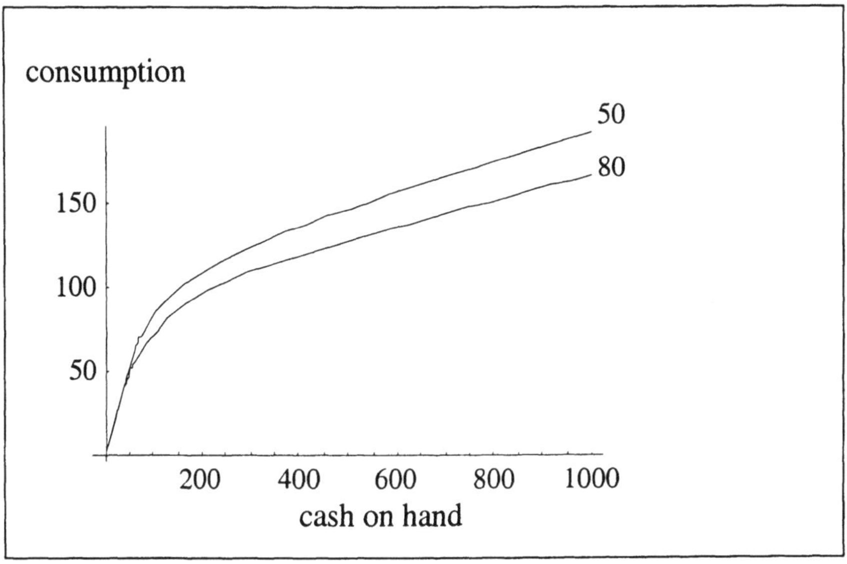


*Figure 10.6 Simulated consumption rules with mean income of 100 and standard deviations of 50 and 80 respectively*

Figure 10.6 presents the derived consumption rules. The functions are increasing in cash on hands to reflect that in the absence of borrowing, perfect smoothing is feasible nor optimal. The consumption rule for the case with higher income variability lies below the other, to reflect that at higher levels of income risk households are more inclined to save for bad days. Both functions are kinked at about half of mean income. At values of cash on hand below the kink, households cannot cushion the effects of low income and their only option is to equate consumption to cash on hand.

The shape of the functions is rather flat: for the higher consumption rule an increase in cash on hands from 500 to 750 induces an increase in consumption of 24 units only from 147 to 171. This has consequences for researchers interested in measuring empirically the extend of consumption variability. Imagine a household that owns cash on hand worth 500. Its consumption rule tells it to consume 147 units and to keep the remainder as assets. Suppose that in the next period an exceptional income is realised of say three times average income. Now the household has cash on hand of 671 ((500-147)*1.05+300). Following the consumption rule the household increases consumption by about 17 units. But to measure such a tiny response a degree of accuracy in measurement is required that will be difficult to attain in practice. Measuring the difference in assets creates much less problems, if only because recall on an increase in the stock of assets from 353 to 507 is a lot easier than on a 12 percent increase in consumption.[12]

This is further illustrated in Figure 10.7 where the cumulative density functions for consumption and assets are depicted for the case where a liquidity constraint household applies the optimal consumption rule at an income process with mean income of 100 and a standard deviation of 50. It has been determined for a series with a 1000 income draws from the income process. The graphs can be used to assess the sensitivity of consumption and assets to inaccuracies in measurement. Roughly 90 percent of all consumption lies in the interval between 70 and 140. For assets holds that 90 percent of the observations lies in the interval ranging from 0 to 250, a much broader interval and hence one that is less sensitive to measurement error.

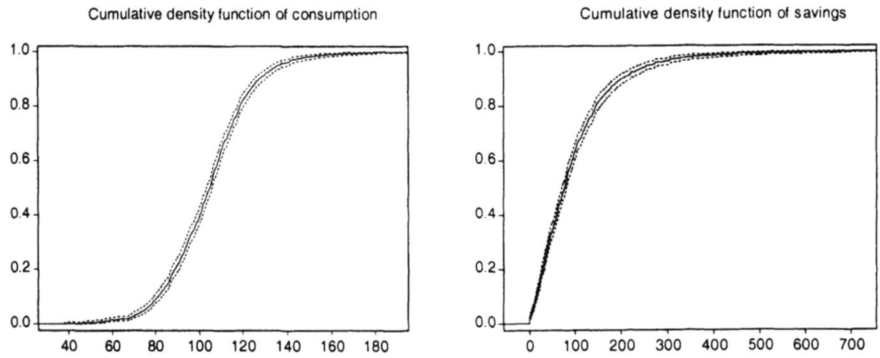


*Figure 10.7 Cumulative distribution functions of consumption and assets, of income process of 1000 draws from income with mean 100 and standard deviation 50*

Having determined the consumption rule, the extent to which households reduce the variability of consumption needs to be determined. This is illustrated in Figure 10.8 by first taking one random income draw for 100 periods for an agricultural income process with mean 100 and standard deviation 50 and by deriving the consumption that arises if the household used the consumption rule of Figure 10.6. The household is assumed to start the process without any assets, so that in the first period it is liquidity constrained and can not cushion consumption against adverse income shocks. The figure shows that the household is 'unlucky' in its first income draw so that it consumes all its income. In the next period income is much higher. This allows the household to both consume more and to build up a stock of assets. These assets turn out to be useful in the subsequent period when income drops, while consumption can be maintained. The figure shows that the household is quite successful in smoothing consumption as the latter follows a much less erratic pattern than income does.

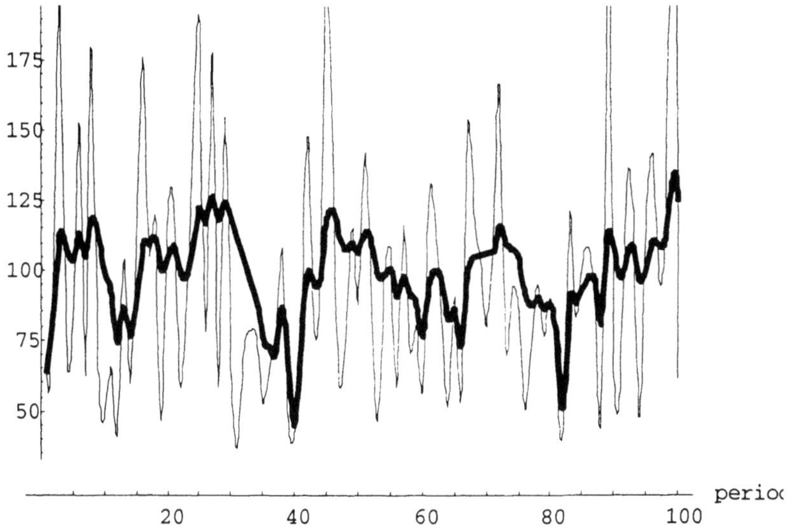


*Figure 10.8* *Stochastic income (thin line), at mean 100 and standard deviation of 50 and smoothed consumption (bold line)*

Average consumption variability can be determined with a Monte Carlo simulation, in which a 1000 times the mean and variance of income, consumption and buffer stocks are established for a sequence of 100 periods. The mean values for these variables are presented in Table 10.3.

*Table 10.3* *Mean and standard deviation for consumption, savings and income determined in Monte Carlo simulation of 1000 income draws of 100 periods*

| *Standard deviation crop income* | *50* | *80* |
|---|---|---|
| Assets in period 0 | 0 | 0 |
| Mean Consumption | 104 | 109 |
| Mean Assets | 93 | 242 |
| Mean Crop Income | 100 | 100 |
| Standard Deviation Consumption | 18 | 23 |
| Standard Deviation Assets | 70 | 166 |
| Standard Deviation Crop Income | 50 | 79 |
| Certain Value of Income (money equivalent) | 97 | 93 |
| Certain Value of Income (utility equivalent) | 91 | 83 |

The table illustrates that the average stock of asset, responds strongly to the variability of the income process but that the variability in consumption is of the same order of magnitude. In the standard deviation 80 case, average asset holdings are more than two and a half time as large than those in the relatively safe situation. In both cases average consumption is slightly above mean agricultural income, due to the fact that households earn a positive return on asset holdings. The accumulation of assets allows the household to reduce consumption variability to a large extent. Compared to a situation in which no assets are held, the coefficient of variation drops from 50 to 18 and from 80 to 23 respectively. These results, though based on stylised facts and theoretically derived, seem plausible indications of actual consumption variability. The Zimbabwe survey data for instance, show that average crop income is about Z$ 3700 per year (in 1992 prices) and that the value of livestock, the most important asset for inter-seasonal consumption smoothing stands at about Z$ 5800 or 1.6 times average income. This is of the same order of magnitude as suggested by the simulation results. The simulations also show that the variability of consumption is not sensitive to increases in income variability so that a coefficient of variation of 20 seems an acceptable point estimate of the consumption variability experienced by rural households in Zimbabwe.[13]

Doubts about the plausibility of this estimate can easily be raised. What for instance is the real value of the buffer stock if the value of livestock is negatively correlated with the outcome of the income process so that livestock is less well a buffer stock than the safe asset used in the simulation? Livestock is also kept for productive purposes suggesting that not all livestock can be considered buffer stock. And, what is the income instability of Zimbabwean farmers in general? To which extent do resettlement households represent all communal farmers in Zimbabwe. All these important qualifications need to be made. Nonetheless given that the value of livestock roughly corresponds to the predicted optimal level of buffer assets, that the variability of consumption after applying a buffer stock strategy is not very sensitive to changes in the riskiness of the income process, 20 is considered a plausible estimate of the coefficient of variation of consumption for liquidity constrained households applying a buffer stock strategy.

## 5 THE BENEFITS FROM INTRODUCING FORMAL FINANCIAL INSTITUTIONS

Despite the accumulation and depletion of buffer stocks, in the absence of formal financial institutions it is not optimal to eliminate all variation in consumption. But the consequences of the remaining variability may be grave and Figure 10.8 comprises at least 2 years which would qualify as disaster years, and a few more which are particularly bad. Households therefore are still to gain from the introduction of financial services that help them to further stabilise consumption. An ideal income insurance scheme would eliminate all consumption variability by guaranteeing the farmer his average income each year by collecting premium in years when income exceeds average and paying indemnities in all other years. But how much will farmers gain from such a service?

The price farmers are prepared to pay for full insurance can be determined by calculating the risk premium they would be prepared to pay to attain a certain income. This risk premium can be approximated by the formula:

$$\frac{\rho}{y} \approx \frac{1}{2} R \sigma^2$$

where $\rho$ indicates the risk premium, $R$ the relative rate of risk aversion (2 in our case) and $\sigma$ the coefficient of variation of consumption (Newbery and Stiglitz, 1981). Application of this formula shows that uninsured and without the use of buffer stocks, the risk premium would lie between 25 percent and 64 percent of average income. These were the figures presented in the introduction. Through self insurance farmers reduce the variability of consumption to a considerable extent but still they would be prepared to pay about 4 percent of their average income as risk premium. This is a lower bound because in the presence of formal financial institutions households are in a position to release buffer stocks and invest them more productively in other activities. The gain from this is probably limited because buffer stocks are usually kept in the form of cattle which already yield high returns (Scoones, 1992). A more substantial gain can be expected from the fact that households no longer have to opt for a combination of crops and activities to diversify their income so that they can completely specialise.

Formal financial institutions may also offer a service which is not complementary to the self insurance strategy of farmers but which substitutes it. Such would be attractive if the cost of formal insurance is lower than the opportunity costs of self insurance. The opportunity costs of self insurance are embedded in the fact that households have to accumulate assets despite the fact that the rate of time preference exceeds the rate of return on those assets. Households are nevertheless prepared to bear this opportunity cost, because of the benefit received from the reduction in consumption variability.

To determine the money metric value of the opportunity costs of accumulating assets let $y_i^{pv}$ be the certain value of income which allows the household to achieve the same discounted expected utility as the consumption stream that follows from the optimal consumption rule with buffer stocks:

[9] $$\sum_{t=0}^{N} \frac{1}{(1+\delta)^t} v(y_{i,t}^{pv}) = \sum_{t=0}^{N} \frac{1}{(1+\delta)^t} Ev(c_{i,t}) .$$

The right hand site of equation [9] is known from the Monte Carlo simulations so that $y_i^{pv}$, fixed for all t, can be determined. The certainty equivalent income values are presented in Table 10.2. They are 91 and 83 for the standard deviation 50 and 80 cases respectively implying that farmers would be prepared to forego 9 to 17 percent of their average income to obtain a stable income and to forego the necessity of accumulating buffer stocks.

## 6 DISCUSSION: CAN THE PUZZLE BE RESOLVED?

The size of the premium households (are prepared to) pay reconciles only partly the contradictory findings mentioned in the introduction to this paper. Instead of welfare costs of the order of magnitude of 25 to 64 percent of average income, income risk generates welfare costs of up to 20 percent of average income. This is a huge amount and it is doubtful whether monitoring costs and covariant risk are sufficient to explain the absence of insurance markets in rural areas. This especially because these factors need not be prohibitive. Imagine for instance an insurance scheme that only covers covariant risk. Covariant risk is most harmful to farmers, and by its very nature it can not be dealt with through income pooling. With idiosyncratic risks not included in its coverage, financial services reduce their monitoring problem substantially if only because the presence and depth of a covariant risk such as drought can easily be established. This leaves the problem of determining expected yield, its probability distribution and assessing the yield shortfall due to drought, the costs of which are substantial. But insurance even needs not to be based on actual losses. It could also pay out an amount independent of the agent's loss but dependent on the seriousness of the covariate event and the amount of premium paid. In that case, a drought of $y$ standard deviations below mean rainfall entitles the insured party to an indemnity of $x$ times the dollar premium paid.[14] Since all participants would pay the same premium and receive the same indemnity, drought insurance of this type avoids all moral hazard and adverse selection problems. Such insurance would also not have to be offered to farmers only. It might also be of interest to providers of credit who shield away from the provision of loans for consumption purposes because of the risk of default in case of the manifestation of a covariate risk. These credit providers can use the insurance to cover such risk.[15] Also the emergence of a secondary market need not be discouraged, since its existence would allow liquidity constrained households to obtain its expected value prior to the rainfall period specified in the insurance contract.

There are however, at least three potential problems with the proposed insurance (Gautam et al. 1994). First, its value as a drought coping aid depends on whether catastrophic income outcomes for most households coincide with severe droughts at nearby weather stations. Next the rainfall measurement has to be robust and foul proof. Finally, such drought insurance faces an extreme covariability problem because all the participants have to receive the same indemnity per dollar of ticket at the same time. These problems are not insurmountable however. Informal insurance mechanisms, between households or through moneylenders or traders sustain additional inter household smoothing after the formal insurer has paid out and can thus compensate potential misalloctions. Foul proof systems can be designed, leaving the covariant risk problem. The extent of this problem is a matter of scale. A covariant risk may be too large to insure for a Zimbabwean financial institution, this is unlikely from the perspective of the international financial market where reinsurance can be sought. Also the donor community might be interested since through the provision of emergency relief, it de facto already offers some kind of drought insurance (though one that leaves much to be desired[16]). Recently developed financial products such as catastrophy bonds further enhance the possibilities to cover low frequency high exposure risks through the issuance of bonds which only repay the principal in full if during the maturity of the bond the damage associated with a well defined risk remains below a certain value. This has made it possible to offer earthquake insurance up to a certain value to Tokyo and California for instance (Kuys, 1999), but could just as well be applied to drought insurance.

So, if monitoring costs and covariant risk are insufficient arguments to explain the absence of formal financial services in rural areas, why are such services not readily available? Fixed costs may provide an explanation. In the presence of sunk costs not all goods that can conceivably be produced are taken into production. Especially goods with high sunk costs will not be fabricated (Krugman, 1980; Romer, 1994). But with the introduction of new financial products, the barriers to their further development are lowered. If more experience with these products is gained and the cost of introduction subsequently lowered, it is conceivable that one day income insurance will be a feature of rain fed agriculture in Africa.

## NOTES

[1] Discussions, comments and suggestions received from Harold Alderman, Kees Burger, Chris Elbers, Jan-Willem Gunning, Peter Lanjouw, Michiel Keyzer and Ashoka Mody are highly appreciated. All errors should exclusively be attributed to the author. Much appreciation is extended to Bill Kinsey who kindly permitted the use of the data. For correspondence refer to: hhoogeveen@econ.vu.nl.

[2] In section 5 the way to derive these figures is presented.

[3] Only demand for financial services originating from a desire to smooth consumption is considered. Demand for productive investments is not under consideration.

[4] Table 10.1 already presents an ex post measure of variability namely after farmers have taken measures to reduce risk on farm by using different planting dates, diversification of plots etc.

[5] My ventures are not in one bottom trusted,
Nor to one place; nor is my whole estate
Upon the fortune of this present year;
Therefore, my merchandise makes me not sad. (Act I, Scene 1)

[6] The correlation coefficients are: maize-cotton, 0.66; maize-groundnuts,0.55; cotton-groundnuts, 0.57.

[7] An exception are the grain loans made available during the 1995 drought, but whose continuity is under threat.

[8] The following fractions have been used: cow, 0.73; heifer, 0.61; bull, 0.89; trained oxen, 1.00; young oxen, 0.57; calf, 0.30; donkey, 0.18; sheep, 0.08; goat, 0.06 and pig, 0.06.

[9] A formal derivation can be found in Obstfeld and Rogoff (1996).

[10] This is an approximation of the income distributions presented in Figure 10.5. A Shapiro-Wilks test rejects the hypothesis that income follows a lognormal distribution.

[11] To ensure that the iterative process converges, it is required that the rate of time preference exceeds the rate of return.

[12] Comparisons carried out by Duncan and Hill (1984) in the United States of employee's reports of incomes with records from their employer shows that measurement error accounts for 16 percent of he variation in reported labour income. Evidence on the size of measurement error on consumption in developing countries is not available however.

[13] To get an intuition of what this implies the series 108, 117, 83, 88, 142, 77, 126, 90, 95, 114 and 104 has a mean of 104 and a coefficient of variation of 20.

[14] This is not the first time that an insurance of the rainfall lottery type is proposed. Walker et al. (1986) for instance also do so for India. They are less optimistic about its chances of success because of the weak correlation between mean

village net household income and rainfall in the three villages they did research on. In Zimbabwe this is not the case.

[15] Precisely for this reason, the World Bank has taken the initiative for a stand by credit to be released to the government of Nicaragua in the event of a drought.

[16] Food aid provided on a massive scale during the 1992 drought and grain loans distributed in 1995 was included in the determination of total income in Section 2. Nonetheless the variation in income remained very high and widespread suffering could not be prevented.

**REFERENCES**

DEATON, A. (1989) Saving in developing countries: theory and review. In *Proceedings of the World Bank annual conference on development economics*, pp.61-108.

DERCON, S. (1992a) *Agriculture and risk*, conference paper, mimeo.

DERCON, S. (1992b) The role of assets in coping with household income fluctuations : some simulation results, mimeo.

DUNCAN, G. & D. HILL (1984) An investigation of the extent and consequences of measurement error in labor economic survey data, Survey Research Center, University of Michigan.

FAFCHAMPS, M., UDRY, C & K. CZUKAS (1998) 'Drought and saving in West-Africa: are livestock a buffer stock?' In *Journal of Development Economics*, vol 55, pp.273 – 305.

GAUTAM, M., HAZELL, P. & H. ALDERMAN (1994) *Rural demand for drought insurance*. Policy Research Working Paper 1383, The World Bank.

KINSEY, B., BURGER, K. & J.W. GUNNING (1998) 'Coping with drought in Zimbabwe: survey evidence on responses of rural households to risk'. In *World Development*, vol.1, pp.89-110.

KRUGMAN, P. (1980) 'Scale Economies, Product Differentiation, and the Pattern of Trade'. In *American Economic Review* vol. 70 no. 5, pp.950-958

KUYS, P.H.M. (1999) 'Marktconforme Verzekeringsoplossingen'. In *Economische Statistische Berichten*, vol. 4193, pp.14-18.

NEWBERY D. and J.E. STIGLITZ, Commodity Price Stabilization, Oxford University Press, 1981.

OBSFELT, M. and K. ROGOFF, Foundations of international macroeconomics, MIT Press, London, 1996.

RAVALLION, M. & S. CHAUDHURI (1997) 'Risk and insurance in village India: Comment'. In *Econometrica* vol. 65 no. 1, pp.171-184.

REARDON, T. (1997) 'Using evidence of household income diversification to inform study of the rural nonfarm labor market in Africa'. In *World Development* vol. 25, pp.735-747.

ROMER, P. (1994) 'New goods, old theory, and the welfare costs of trade restrictions'. In *Journal of Development Economics*, vol. 43, pp.5-38.

ROSENZWEIG, M. & H. BINSWANGER (1993) 'Wealth, weather risk and agricultural investments'. In *The Economic Journal*, , 1981.vol. 103, pp.56-78.

SCOONES, I. (1992) 'The Economic Value of Livestock in the Communal Areas of Southern Zimbabwe'. In *Agricultural Systems*, vol. 39, pp.339-359.

SCOONES, I. et al. (1996). *Hazards and Opportunities, Farming livelihoods in dryland Africa: Lessons from Zimbabwe*. London and New Jersey: Zed Books Ltd,.

TOWNSEND, R. (1994) 'Risk and Insurance in Village India''. In *Econometrica*, vol. 62 no. 3, pp.539-591.

WALKER, T.S., SINGH, R.P. & M. ASOKAN (1986) 'Risk Benefits, Crop Insurance, and Dryland Agriculture'. In *Economic and Political Weekly*, vol 21 nos 25 and 26 review of Agriculture, June 21-28: A81-A88.

# Part 4

## Liberalization: Institutions and Approaches

# 11 VOLUNTARY MARKETING INSTITUTIONS IN FOOD MARKETING SYSTEMS

Matthew Meulenberg

## 1 INTRODUCTION

Food marketing in Western countries has evolved from simple exchange processes at local markets into complex food marketing systems. Food marketing has also changed in other countries as a result of trends in consumer behaviour, new marketing methods, and a changing economic order.

Food marketing takes place within a specific network of institutions. In open markets the structure of this network is determined by factors related to marketing effectiveness, efficiency and equity. Occasionally enterprises have voluntarily established common marketing institutions, such as cooperatives, in order to improve marketing performance. Sometimes marketing institutions, for instance marketing boards, are imposed on a food marketing system by a government. In some countries the government enhances the establishment of common marketing institutions through subsidies or tax facilities.

This paper addresses voluntary marketing institutions established by enterprises of their own free will. We define a voluntary marketing institution (VMI) as a marketing institution which is voluntarily established by enterprises, and whose services are used by enterprises at their own discretion within the bounds of the formal relationship between enterprise and VMI. VMIs traditionally play an important role in Western food marketing systems. Examples are farmers' cooperatives, wholesaler sponsored voluntary chains and franchise organisations. The role of VMIs is changing as a result of changing markets and marketing strategies.

Marketing institutions have been studied at length in the agricultural marketing literature. Marketing textbooks (e.g. Kohls, Uhl, 1990) offer an in depth treatment of marketing boards, marketing orders/agreements and cooperatives. In particular marketing boards and cooperatives have been investigated extensively. Marketing boards have been studied regularly (e.g. Morley, 1967; Hoos (ed.), 1979; Forbes, 1982; Meulenberg, 1986; Veeman, 1997; Wallace and Schroder, 1997). There is a continuous stream of research on farmers' cooperatives (e.g. Cobia (ed.), 1989; Nilsson and Van Dijk (ed.), 1997; Van Bekkum and Van Dijk (ed.), 1997). This paper takes a more general look at these marketing institutions by focusing on VMIs as a class of marketing institutions of which marketing boards and cooperatives are important members. It is not our intention to offer an up-to-date picture of VMIs. Rather we analyse general characteristics of VMIs and present structural developments of VMIs as a result of a changing marketing environment and marketing strategies. In particular we intend to contribute to the understanding of VMIs by: (a) reviewing theories and research results from economics and marketing; (b) proposing a classification scheme of VMIs and a set of criteria to assess the viability of VMIs; (c) analysing the evolution of VMIs in western food markets. The proposed classification scheme and assessment criteria are also applicable to VMIs in non-western countries. Their relevance will increase with market liberalisation, which will enhance a further shift from public to voluntary marketing institutions.

Our paper is organised as follows. First, a concise description of VMIs in western food markets is given. Second, concepts and findings of economic theory and marketing relevant to VMIs are reviewed. A classification scheme of VMIs and a framework for assessing the viability of VMIs are proposed. Next, main trends in the Western marketing environment and in the enterprises of food marketing systems are described. The impact of these trends on VMIs in Western food markets is analysed. Some conclusions close the paper.

## 2 VOLUNTARY MARKETING INSTITUTIONS IN WESTERN MARKETS, A CONCISE DESCRIPTION[1]

VMIs are operating in many food marketing systems. They perform specific marketing functions, such as price discovery, or even perform the complete marketing operation for an agricultural/food product.

Cooperative auctions contribute to price discovery in agricultural markets. Dutch auctions (auction starts at an offer price higher than any bidder is willing to pay, which is lowered until a bidder accepts the offer) are used in markets for fresh horticultural products: flowers, fruit and vegetables and fish. English auctions (auctions start at a low offer price; bids are publicly made; the bidder who makes the highest bid receives the goods) are used among other things to sell cattle. Auctions also fulfil an important logistical function by concentrating physical supply and demand.

Product differentiation and relationship-marketing diminish opportunities for selling through auctions, since in selling through an auction no special relationship between supplier and buyer can be developed. On the other hand, better communication facilities increase the access to the auction process, and therefore the opportunity for auctioning products at an international scale.

Futures markets, sometimes established with the support of enterprises operating in the relevant agricultural sector, have been popular in the US for a long time as an institution for hedging price risks for commodities such as corn, soybeans and wheat. In Europe commodity futures exchanges are still of limited importance, but various developments are stimulating the interest in commodity futures trading:

- the trend towards big and specialised farms which operate in more open markets and receive less price support, e.g. by the CAP (Common Agricultural Policy) of the EU;
- bigger food companies, which are concerned about price volatility of their agricultural inputs.

In many European countries, such as France, Germany, Hungary, Netherlands, Spain and the United Kingdom, commodity futures markets have been started, or will start soon.

Various marketing institutions market agricultural products on behalf of participating farmers. Important institutions in this respect are agricultural cooperatives and marketing boards, see Table 11.1.

*Table 11.1 Market shares of agricultural co-operatives in the EU*[1)]

| Market shares(%) | Dairy | Fruit & vegetables | Meat | Farm inputs | Credit | Grain |
|---|---|---|---|---|---|---|
| Belgium | 50 | 70-90 | 20-30 | | | - |
| Denmark | 93 | 20-25 | 66-93 | 64-59 | | 87 |
| Germany | 55-60 | 60 | 30 | 50-60 | - | - |
| Greece | 20 | 12-51 | 5-30 | | | 49 |
| Spain | 35 | 15-40 | 20 | - | - | 20 |
| France | 49 | 35-50 | 27-88 | 50-60 | - | 75 |
| Ireland | 100 | - | 30-70 | 70 | | 69 |
| Italy | 38 | 41 | 10-15 | 15 | - | 15 |
| Luxembourg | 80 | - | 25-30 | 75-95 | - | 70 |
| Netherlands | 70-96 | 35 | 40-50 | 84 | | |
| Austria | 90 | - | 50 | - | - | 60 |
| Portugal | 83-90 | 35 | - | - | - | - |
| Finland | 94 | - | 68 | 40-60 | 34 | - |
| Sweden | 99 | 60 | 79-81 | 75 | - | 75 |
| U.K. | 98 | 35-45 | ± 20 | 20-25 | - | 20 |

[1)] *Source: Bekkum, O.F. van, and G. van Dijk (eds.), 1997, p.29.*

Agricultural cooperatives started out as organisations, which tried to improve farmers' product prices by increasing the bargaining power of farmers, or through better product quality. They are defined as user-owned and user-controlled businesses that distribute benefits on the basis of use (Barton, 1989). Three concepts distinguish cooperatives from other businesses: a) the user-owner principle: persons who own and finance the cooperative are those who use it, b) the user-control principle: control of the cooperative is by those who use the cooperative and c) the user-benefits principle: benefits of the cooperative are distributed to its users on the basis of their use. Cropp and Engelsbe (1989) indicate potential classifications of cooperatives based on functions performed, structural arrangements, organisational or financial structure.

The shift towards market orientation in food marketing requires cooperatives to adopt a customer oriented marketing policy. A cornerstone of such a policy is an agricultural supply of farmer-members which fits well with the marketing strategy of the cooperative processing/marketing company. Farmers' willingness to invest in the cooperative, in particular in its markets and R&D, is also essential for such a policy.

Adoption of modern marketing and management procedures by cooperatives influences the cooperative structure. The following organisational changes, partly due to the shift toward market orientation, can be observed:

- size of enterprise is increasing,
- quality of management is improving and the rights and responsibilities of top management are better defined,
- special financial structures are developed to generate the necessary risk-bearing capital for the cooperative enterprise,
- some cooperatives have transformed their enterprises into limited companies, whose shares, at least a majority of shares, are in the hands of the cooperative union/farmer-members.

The above developments enhance a more rational and less emotional relationship between farmers, in particular young and modern farmers, and their cooperative.

In various countries agricultural marketing boards and commodity boards contribute to the marketing of food and agricultural products of a sector. Marketing boards which are responsible for the total marketing operation of an agricultural/food product are the exception rather than the rule in Western countries. However, promotional boards which support the marketing of generic food products have been set up in many European countries. Sopexa in France and CMA in Germany are examples. Product differentiation and large size of enterprises stimulate marketing efforts for the brand of individual enterprises at the cost of marketing efforts for the generic product of a sector. On the other hand, new marketing problems emerge for some food products (e.g. a poor image of meat) which have to be tackled by marketing activities of the sector.

## 3 THEORIES AND RESEARCH RESULTS FROM ECONOMICS AND MARKETING

Voluntary marketing institutions (VMIs) perform functions and coordinate processes in food marketing systems. Their role in western food markets is changing. Theories and research results from economics and marketing are instrumental in understanding this changing role. In this section some important theories and research results are briefly reviewed from the perspective of their relevance to VMIs.

### 3.1 CONTRIBUTIONS FROM INSTITUTIONAL ECONOMICS

Economic theory defines institutions in different ways. Schotter (1994, p.5, ff.) distinguishes the following three meanings of institutions: "....conventions, sets of rules that constrain the behavior of social agents in particular situations, and an organization -usually a large, well-established organization". Coriat and Dosi give a similar classification (1998, p.6): ".. formal organizations, patterns of behaviours that are collectively shared, negative norms and constraints". VMIs are either formal, well established, organisations or sets of rules, norms or constraints.

For the analysis of VMIs it is useful to make a distinction between institutions dealing with the institutional environment (rules of the game) and institutions of governance (the play of the game), since the evolution of both types of institutions differs substantially in food marketing. The scheme of social analysis in four levels proposed by Williamson (1998, p.26) also seems helpful to understand different types of VMIs: 1) the social embeddedness level, where the norms, customs, mores, traditions, etc. are located. 2) the institutional environment..., the rules of the game within which economic activity is organised, the polity, judiciary, and bureaucracy of government 3) the institutions of governance - second order economising applies: get the governance structures (markets, hybrids, firms, bureaus) right 4) from discrete structural to marginal analysis... the level with which neo-classical economics and, more recently, agency theory has been concerned. Williamson argues that New Institutional Economics is concerned in particular with levels 2 and 3 of his scheme. The analysis of VMIs is also in particular concerned with the levels 2 and 3, of Williamson's scheme of social analysis: the institutional environment and the institutions of governance.

According to Coriat and Dosi (1998, p.6, ff.): " ....institutions not only 'parameterize' and 'constrain', but, given any one environment, also shape the 'visions of the world', the interaction networks, the behavioural patterns, and, ultimately, the very identity of the agents." They make a distinction between weak and strong institutionalism: " ...ranging from weak forms retaining a lot of the canonic microfoundations to strong forms wherein institutions have much more life of their own and also much more influence on what microentities think and do." The former type of institutionalism is characterised by: "(1) Role of institution; parameterize system variables; contain menu of strategies. (2) "Primitives" of the theory: ....rational self-seeking agents; institutions as derived entities. (3) Mechanisms of institution-formulation: Mainly intentional, "constitutional" processes, (4) Efficiency properties: Institutions perform useful coordinating and governance functions;...". Our approach to VMIs is, in the terminology of Coriat and Dosi (op cit. p.8), primarily "weak" institutionalism, since enterprises in Western food marketing systems evaluate their participation/ cooperation with a VMI increasingly on the basis of rational criteria.

Theories focusing on rules and organisational structures at macro-level may be useful in the analysis of VMIs. An example is the discussion of corporatism by Visser and Hemerijck (1997, p.65, ff.): "Corporatist governance has two analytically distinct properties: the degree of institutional integration of organized interests into the framework of public policy formation; and the degree of societal support for corporatist policies offered by organized interests." The authors classify corporatist institutional change on the basis of these two criteria. The analytical scheme for analysing corporatism by Visser and Hemerijck (op cit.) seems helpful to characterise a VMI and to derive warranted strategies.

Discussions about market- and non market regimes also offer food for thought about VMI policies. Wolf (1993, p.87) summarises market failures as externalities and public goods, increasing returns, market imperfections, distributional inequity (income and wealth). He describes non-market failures as disjunction between costs and revenues, redundant and rising costs, internalities and organisational goals, derived externalities, distributional inequity (power and privilege). The analysis of VMIs can profit from economic theories about the failures of market and non-market institutions at the national/sector level. In particular these theories may be helpful in understanding the influence of externalities of production and marketing on new opportunities for VMIs, and the influence of internalities and organisational goals, as well as distributional inequity (power and privilege) on policies and management of VMIs.

## 3.2 CONTRIBUTIONS FROM INDUSTRIAL ECONOMICS.

A centre-piece of industrial economics theory is the well known transaction cost economics theory of Williamson. This theory regards the firm as a governance structure. It is based on (e.g. Williamson, 1989, 1998) the behavioural assumptions of bounded rationality and opportunism and considers frequency, uncertainty and asset specificity to be principal dimensions of transactions. Williamson (e.g. 1989, p.146; 1998, p.37, ff.) proposes a scheme for the analysis of contracting, which starts out by determining the degree of asset specificity. In the case of asset specificity safeguards might be provided, such as penalties to prevent a breach of the contract, or the transactions might be organised under unified ownership. Information is considered a prime source of transaction costs, in particular as a result of specialisation in production and marketing (Holmstrom and Tirole, 1989, p.64).

Nooteboom (1998, p.172) argues that the transaction costs economics theory of Williamson is not complete, because it is comparative, static and not dynamic: it does not take learning into account. In order to integrate the outlooks of transaction cost economics and IMP(Industrial Marketing and Purchasing, Hakansson, 1982) Nooteboom makes a distinction between egotistic and non-egotistic sources for cooperation both at the macro- and micro level, see Figure 11.1.

| | Macro | Micro |
|---|---|---|
| Egotistic | coercion or fear of sanctions from authority (god, law) | material advantage or "interest" |
| Non-egotistic | ethics: values/norms of proper conduct | bonds of friendship, kinship or empathy |

*Figure 11.1 Sources of cooperation between companies according to Nooteboom*
*Source: Nooteboom (1998, p.174).*

He argues that while the New Institutionalist transaction cost economics focuses on the egotistic sources, Neo-Institutionalism aims at incorporating the non-egotistic sources.

In a review of transaction costs analysis in marketing research Rindfleisch and Heide (1997) conclude that firms seek to minimise transaction costs through vertical integration when faced with the need for safeguarding specific assets invested in an exchange relationship. They conclude also that the role of governance as a means of adapting to uncertain environments receives mixed support. Concepts and theories of transaction cost economics contribute to our understanding of VMIs (see for instance: Zylbersztajn, 1996). Transaction costs influence enterprises' decision about whether or not to establish a VMI. Transaction cost economics also improves our understanding of the type of relationship between enterprises and their VMI. Asset specificity and information dissemination are important in this respect. "Non egotistic" factors seem relevant to VMIs which have strong social/emotional links with participating enterprises.

Integration is associated with increasing levels of asset specificity, difficulty of performance evaluation and uncertainty (e.g. Anderson and Schmittlein, 1984; Lilien, 1979). Other reasons for vertical integration include ownership and complete control over neighbouring stages of production or distribution (e.g. Porter, 1980), the creation of barriers to entry for competitors (Bain, 1956), and the acquisition of private information (Perry, 1989, p.208).

Between vertical integration and anonymous spot market exchange there is a great variety of vertical "controls" between firms at different stages of the marketing channel. For example, vertical contractual relations, such as quantity dependent pricing, ties, royalties, requirement contracts and exclusive dealing, resale customer restraints, resale price restraints (Katz, 1989, p.656, ff.) are used to improve profits in vertical marketing relationships. Concepts and theories on integration and vertical contractual relations are relevant to enterprises' decision making about outsourcing functions to a VMI.

Models have been developed which are concerned with aspects of the relationship between a principal and his agent, e.g. models on moral hazard, incentives to agents and monitoring of agents (Holmstrom and Tirole, 1989, p.67). Models have also been proposed which search for a wage structure that stimulates agents' efforts. Free-rider problems and the division of proceeds from a joint output have been investigated. Principal-agent models, such as those concerned with moral hazard, giving incentives to agents and monitoring agents, are instrumental in analysing the relationship between VMIs and participating enterprises.

Reputation has been recognised as an important asset of enterprises, in particular when it is impossible or too expensive to sign comprehensive contracts (Holmstrom and Tirole, 1989, p.76, ff.). In that case reputation offers an implicit promise for a fair fulfilment of a contract when unexpected events occur that are not covered by contract. Since mutual trust is extremely important for the relationship between a VMI and its target enterprises, the reputation of a VMI is a valuable asset for a successful relationship with the related enterprises.

### 3.3 CONTRIBUTIONS FROM MARKETING THEORIES.

The role of marketing institutions in the channel has been a central topic of marketing theory. Meulenberg (1997) reviewed major contributions from marketing theory in this field, using the classification of marketing schools by Sheth et al. (1988).

It is not surprising that in particular the institutional school, focusing on "...the organizations that actually perform the functions required to move the goods from the producer to the consumer" (Sheth et al., 1988, p.74), contributes in particular to the theory of marketing institutions. Many scholars have analysed marketing channels and marketing institutions from the efficiency point of view (Alderson, 1954; Stigler, 1951; Bucklin, 1965, 1970). In fact, in the case of perfect competition, costs will ultimately determine channel structure and consequently the institutional set up. In other market structures, effectiveness and/or equity are also important criteria for building a structure of marketing institutions. Various studies emphasising marketing efficiency as a criterion for the channel structure have also included effectiveness as a criterion (e.g. Bucklin, 1965; Bucklin, 1966). Models from management science (see for instance Lilien et al., 1992, p.415, ff.; Stern and El Ansary, 1992, p.304, ff.) analyse channel structure by focusing on the relationship between a profit maximising manufacturer and a retailer under different assumptions about the degree of integration, the type of market structure, the profit function of the manufacturer and the profit sharing arrangement.

The organisational dynamics school of thought in marketing, which analyses marketing channels from the behavioural point of view, has also contributed to the understanding of marketing institutions. Initiated by Stern (1969), marketing channels have been analysed on the basis of power (e.g. Brown, et al., 1995), conflict and satisfaction in the relationship between channel partners. Frameworks for the analysis of marketing channels have been proposed (Stern and Reve, 1980). Marketing channels have been classified on the basis of the relationships in the channel, for example vertical marketing systems are classified in administered, contractual and corporate systems (Stern and El Ansary, 1992). Transaction cost economics has been widely used to understand marketing channels (Rindfleisch and Heide, 1997).

The coordination between marketing institutions in a marketing channel has been analysed (Celly and Frazier, 1996). Much attention is paid to franchising, for example its motivation by the "resource constraints" or by the "incentive" argument (Lafontaine and Kaufmann, 1994). Factors determining continuity in channel relationships (Anderson and Weitz, 1989), and trust in marketing channels (Geyskens, et al., 1998; Kumar, et al., 1995) have been investigated. Theories on marketing channels and marketing institutions are by definition relevant to VMIs. They offer in particular insights into the process of establishing a VMI and the relationship of VMIs with participating enterprises.

## 4 CRITERIA FOR ANALYSING VOLUNTARY MARKETING INSTITUTIONS

### 4.1 VOLUNTARY MARKETING INSTITUTIONS, A CLASSIFICATION SCHEME

VMIs take an intermediate position on the spectrum of marketing institutions:
Government-owned marketing institution --- Government controlled marketing institution --- Government-supported common marketing institution --- Voluntary marketing institution(VMI) --- Individual company. They can be classified by criteria related to organisation and policy. On the basis of the present types of VMIs we suggest the following:

*organisational criteria*

- horizontal versus vertical VMI, e.g. farmers marketing cooperative versus a wholesaler sponsored voluntary chain;
- hierarchy of VMIs, one-, two-..multilayer structure, e.g. a primary marketing cooperative set up by farmers versus a secondary national marketing cooperative established by regional cooperatives;

- purely voluntary versus hybrid (voluntary and public) VMI, e.g. a cooperative established by farmers versus a commodity board based on public law but established by majority vote of the representatives of an industry;

*policy related criteria*

- permanent versus temporary VMI, e.g. farmers' cooperative (intentionally permanent) versus a temporary marketing organisation set up by a group of enterprises to explore a new market.
- strength of the link between a VMI and its target enterprises, e.g. a relatively loose relationship between retailers and wholesalers in a wholesaler sponsored voluntary chain versus a strict relationship in a franchise organisation.
- degree of VMI-involvement in marketing a product, e.g. marketing cooperatives being in charge of the complete marketing operation of a product versus a promotional board taking responsibility for promoting the generic product only.

These classification criteria seem helpful in understanding the role of a VMI in a food marketing system. However, they are descriptive and not instrumental for assessing the viability of a VMI. In the following section we propose criteria for assessing VMIs.

## 4.2 CRITERIA FOR THE VIABILITY OF A VOLUNTARY MARKETING INSTITUTION

Enterprises can transfer their complete marketing operation to a VMI, or specific aspects of the marketing operation only, such as aspects related to product, price, information/communication, or distribution. Decision-making on that matter can use concepts and theories from economics and marketing, as reviewed in section 3. It may also profit from assessing the viability of a VMI according to the following criteria. We distinguish necessary and sufficient conditions for the viability of a VMI (Meulenberg, 1986). Necessary conditions are conditions to be fulfilled in order to make a VMI meaningful. In addition to the necessary conditions a number of sufficient conditions, not necessarily all, have to be met in order to make a VMI viable.

**Necessary conditions for the viability of a VMI.**

*Condition 1: Products/needs of enterprises served by a VMI are homogeneous.*
VMIs are viable only if they offer products and services which suit the participating enterprises, e.g. the members of a cooperative. Consequently, enterprises served by a VMI should be homogeneous with respect to the marketing object of a VMI, say the product of the enterprises or specific product attributes only, such as nutritional aspects. This condition will be met in many agricultural markets, but is increasingly a problem in food markets because of product differentiation. Nevertheless, differentiated food products may also be homogeneous for specific attributes, such as the environmental friendliness or healthiness of a product, at least from the consumers' point of view.

*Condition 2: There are market opportunities and/or threats for the "generic" product of enterprises related to a VMI.*
Without opportunities and/or threats for the generic product there is no reason for enterprises to join a VMI. In food marketing systems this condition seems almost

always to be fulfilled; markets for many generic products, such as bread, poultry meat, beef, etc., are dynamic.

*Condition 3: Marketing by a VMI is superior to marketing by individual enterprises on either efficiency, effectiveness, or equity.*
Marketing is a basic function of every enterprise. However, an enterprise may transfer the marketing function fully or partially to a VMI if this transfer increases its profits. It requires superior marketing by a VMI either on efficiency (costs), effectiveness (sales) or equity (share of profit).

These "necessary" conditions for viability of VMIs are often met in food marketing systems because of product-similarity between enterprises and the importance of small and medium sized enterprises.

**Sufficient conditions for the viability of a VMI.**

*Condition 1: Enterprises prefer to respond jointly instead of individually to market opportunities and threats.*
In a free market economy an enterprise is responsible for marketing its products and services. Whether an enterprise will outsource marketing functions to a VMI depends among other things on its size and the related economies of scale and scope in marketing. However, even when a joint marketing operation through a VMI could be more effective, an enterprise might prefer its own approach for reasons of independence and flexibility. For instance, a food company might prefer to carry an own environmentally friendly brand instead of using an industry-wide environmental label.

*Condition 2: The product marketed by a VMI is important for the income formation of the related enterprises.*
An enterprise will not be motivated to join a VMI if the product in question only contributes to a minor extent to its income.

*Condition 3: Free rider problems are negligible.*
Enterprises hesitate to join a VMI, e.g. a promotional board, if non-participants will profit from that VMI too.

## 5 ENVIRONMENTAL CHANGES RELEVANT TO VOLUNTARY MARKETING INSTITUTIONS

Developments of VMIs in Western food marketing systems, such as marketing cooperatives and marketing boards, depend on trends in the environment of western food marketing systems. We will highlight the main trends and their consequences for food marketing systems.

**Economic environment.** In spite of steadily increasing income levels in Western countries, the per capita consumption of food (in volume) remains stable. Income elasticities of the demand for various food products are small. Income elasticities of the demand for food quality and services are higher. The tendency towards a more skewed income distribution offers opportunities for market segmentation and product differentiation in many western countries.

Economic developments suggest that farmers and food enterprises should reinforce their market position by increasing added value and by focusing on specific target groups.

**Demographic environment.** The EU population is expected to increase only slightly in the period 1995 to 2000 (from 371.5 million to 375.0 million) and is even projected to decrease in the first half of the next century. On the other hand the global population is still expected to grow substantially. People in western countries are living longer. However, the food consumption pattern of old people does not differ substantially from that of the rest, except for a lower and nutrient poorer food intake, lower energy need and more need for services (Senauer, et al., 1991). The trend towards smaller families, in which both partners have a job, stimulates the demand for convenience foods and "away from home" consumption. These demographic trends also stimulate marketing strategies of increased added value (services) to the product and of focusing on specific target groups.

**Social environment.** Trends in values and lifestyles have a great impact on food marketing. From the well-known list of trends proposed by Popcorn (1992), "fantasy adventure, save our society, small indulgence and staying alive" seem particularly relevant to food consumption. New values, such as "self fulfilment ethic", "better quality of life", and "work to live" (Plummer, as quoted by Engel, et al., 1995, p.627) influence food consumption. Concern about the environment and about animal welfare affect food marketing not only through consumers' wants and needs, but also through actions of consumer groups and environmental lobbies.

Changes in lifestyle due to more women working outside the home, increasing mobility and more free time influence consumer behaviour. Trends in values and lifestyles bring product positioning, product innovation and adding more services to the product to the forefront of food marketing.

**Physical environment.** Ecological problems have become a major social issue. Many consumers are environmentally conscious, but do not behave in an environmentally friendly way. Environmentally friendly behaviour is promoted in particular by environmental lobbies and consumer groups. Ecological problems are acute in some agricultural sectors, e.g. manure problems in pig and poultry raising, use of pesticides/insecticides in horticulture, and land degradation in arable farming. The present wave of biotechnological inventions reinforces discussions about the sustainability of food systems.

Food marketing will have to pay more attention to the sustainability of the food marketing system. It will increasingly approach sustainability as an opportunity and not as a threat.

**Political environment.** Fewer trade barriers and less market intervention, combined with more involvement in environmental and health issues are trends in government policies with respect to food marketing. Member countries of GATT/WTO are committed to a reduction of market intervention, to improved market access (replacement of non-tariff barriers by "equivalent" tariffs and reduction of tariffs) and reduction of subsidised exports. The future EU membership of some middle and east European states will shift the Common Agricultural Policy (CAP) further from market intervention to income support. The CAP will pay more attention to rural and environmental policies.

Food marketing systems are becoming more market oriented and rely less on market support from government.

**Technological environment.** Technological advances in IT (information technology), computer science, biotechnology, and transportation offer new opportunities to food marketing. New information technologies improve the speed and precision of exchange processes. Computerisation of production and logistical processes reduces costs. Biotechnological inventions create opportunities for new products. However, many Western consumers are suspicious about food produced by modern biotechnological methods.

Advances in technology increase marketing efficiency and stimulate product innovation in food marketing systems. Consumer bias against new sophisticated methods of food production and marketing increases the importance of communication and company-image in food marketing.

## 6 CHANGING ENTERPRISES AND STRATEGIES IN FOOD MARKETING SYSTEMS

Changes of enterprises in food marketing systems affect the potential role of VMIs. They result from environmental trends, but also from endogenous developments in enterprises themselves, such as technological inventions and new management techniques. In spite of the variety in Western food marketing systems, the developments in retailers, food manufacturers, wholesalers and farmers, show similar trends. Major trends that are relevant to VMIs will be discussed briefly in this section.

### 6.1 CHANGING ENTERPRISES

**Food retailers.** Trends in European food retailing result from policies aimed at higher effectiveness and efficiency. The organisational structure of retail enterprises has changed in conjunction with these policies.

- Marketing effectiveness of food retailers has been increased by market oriented retail policies. Width and depth of product assortment are cornerstones of such policies. They characterise to a large extent the store type, such as hypermarket, supermarket, discount store and speciality shop. A focus on specific target markets, and a constant search for new products and services are characteristic of modern food retailing. Private labels have become an important marketing instrument of retailers. In the past a cheap alternative to national brands, they are currently positioned as products of good quality, which are substantially cheaper than national brands. Retailers also try to improve marketing effectiveness through new assortments, better services or new outlets such as food shops in petrol- and railway stations and on the Internet.
- Marketing efficiency is extremely important in a highly competitive Western food retail market. Costs of carrying a large number of items per store (Aldi has about 600 lines, Sainsbury and Tesco stores between 10,000 and 20,000 lines

(Corstjens and Corstjens, 1995, p.197)) make logistical efficiency of paramount importance. Advances in IT, such as the use of bar-codes, scanning at check-outs and electronic data interchange, contribute to efficient logistics. Contracting-out of logistical functions has decreased logistical costs too. The drive for efficiency has also stimulated a purchasing policy of obtaining price discounts. The strong bargaining power of retail chains vis-à-vis food suppliers is instrumental in this respect.

- Organisational structure of food retailing is changing. Food retailing has shifted from small independents, such as grocery stores and greengrocers, to large retail chains. Some retail chains are members of a holding enterprise. Small independents try to stay in business either as a speciality shop, as a member of a wholesaler sponsored voluntary chain, such as Spar, as a member of a retailer sponsored voluntary chain, or by participating in a franchise organisation.

**Food industry**. The European food industry has been characterised by Linda (1988) as widely diversified in the form of product extensions, geographic diversification and functional diversification. Innovation, mergers and joint ventures are means to materialise the diversification process. At present many food manufacturers try to build a competitive position in international markets on the basis of their core competencies in production and marketing.

While multinationals such as Danone, Kraft, Nestlé and Unilever have been operating on an international scale for a long time already, a great many other food enterprises, which have traditionally set up production plants and management facilities in the domestic market only, are now also internationalising their businesses. A case in point are dairy cooperatives such as MD foods in Denmark, Sodiaal in France, Campina-Melkunie in the Netherlands and Avonmore in Ireland. In spite of the concentration and internationalisation in the European food industry a great many small and medium sized food manufacturers (SME's) remain competitive by focusing on market niches, by serving regional markets or by producing under a private retailer label.

**Wholesale companies**. As a result of concentration and specialisation in food marketing systems the need for some wholesale functions, such as assembling, has diminished. Retail chains have integrated the execution of other functions, such as regrouping. For example, sorting and collection of eggs upstream in the marketing channel disappeared with the advent of specialised poultry farms. Allocation and assortment of groceries, traditional functions of wholesalers, have been taken over by distribution centres of large retail companies.

Wholesalers have responded in different ways to market opportunities and threats. Distributive wholesalers have integrated food retailing, e.g. wholesaler sponsored voluntary chains. Wholesalers in commodities, such as merchants of grain, have become global players. Other wholesalers stay in business as agents of specific food manufacturers or food chains. Specialisation in particular wholesale functions, such as Cash and Carry, is another type of response to market opportunities and threats.

Wholesalers are not only challenged by concentration and specialisation in food marketing systems. Technological innovations and new types of food outlets also create opportunities and threats.

**Farmers**. Concentration and specialisation are main trends in Western agriculture. Large specialised farms have emerged in the production of wheat, milk, pigs, broilers and horticultural products. This trend towards concentration and specialisation is still continuing. Partly in response to the predominant position of specialised commercial farms, mixed farming is experiencing a revival, particularly in the production of organic and regional foods. This development has been stimulated not only by the needs and wants of environmentally conscious consumers but also by rural planning policies.

### 6.2 STRATEGIES OF FOOD MARKETING SYSTEMS

Changes of food marketing systems and their environment go together with a shift from "selling commodities" to "marketing products". In this context food marketing systems increasingly operate on the basis of well defined strategies. Basic strategies have become:

- *Market orientation*, " ...the organization wide generation of market intelligence, pertaining to current and future customer needs, dissemination of the intelligence across departments, and organization wide responsiveness to it" (Kohli and Jaworski, 1990, p.6),
- *Market segmentation* by targeting consumers with specific needs and wants; companies often strive for market leadership in chosen target markets,
- *Product differentiation* on the basis of specific attributes of products and services,
- *Innovation* in order to serve changing consumer wants and needs and to capitalise on technological inventions,
- *Policy coordination* between decision makers of a food marketing system, chain strategies,
- *Internationalisation* of businesses in order to realise necessary economies of scale and scope in production and marketing, in particular with respect to the exploitation of core competencies.

## 7 THE IMPACT OF CHANGES IN THE STRATEGIES AND ENVIRONMENT OF FOOD MARKETING SYSTEMS ON VOLUNTARY MARKETING INSTITUTIONS

Changes in the environment, enterprises and strategies of food marketing systems have been reviewed briefly in the previous sections. Their impact on the viability of VMIs will be discussed in this section. In this context we will make use of the criteria proposed in section 4.

### 7.1 THE IMPACT OF CHANGING STRATEGIES ON VMIS.

**Market orientation**. Market orientation implies that marketing plays a central role in policies of agricultural and food companies. Therefore, VMIs responsible for marketing the products of participating enterprises should acquire a leading role in determining quality and quantity of product supply. However, the willingness of enterprises to accept a leading role of VMIs is negatively correlated with size of enterprise: bigger enterprises are more able and willing to respond in their own way to market opportunities and threats. Consequently, market orientation makes VMIs less viable vis à vis large enterprises, but reinforces their role for small enterprises, which lack the necessary marketing skills. It makes moral hazard in the relationship between a VMI and participating enterprises a more serious issue, since the reliability of enterprises' product supply to the VMI is crucial for marketing success.

**Market segmentation and product differentiation**. Marketing policies of market segmentation and product differentiation by individual enterprises increase the heterogeneity of an industry. As a result, the condition "products/needs of enterprises served by a VMI are homogeneous" will be less easily fulfilled. However, small enterprises, having limited marketing capacities, might use a VMI as a vehicle for policies of market segmentation and/or product differentiation. Producers' groups are a case in point.

**Innovation.** Many food companies have become big enough to adopt a policy of product innovation, which is some type of product modification. This development diminishes the viability of VMIs: the conditions "products/needs of enterprises served by a VMI are homogeneous" and "enterprises prefer to respond jointly instead of individually to (specific) market opportunities and threats" are not easily fulfilled.

On the other hand, enterprises become increasingly aware that basic research paves the way for a continuous stream of new products. In spite of ongoing concentration in food marketing systems, many enterprises are still too small, or even do not want to engage in basic food research. As a result, there is a common need for basic research and a willingness to consider collective action. Therefore, a VMI in charge of basic research might be viable in spite of increasing product heterogeneity due to product innovation.

**Policy coordination / chain strategies.** Serving consumers by offering specific products and services requires effective policy coordination in the food chain. In fact, chain marketing (i.e. a coordinated marketing policy of two or more subsequent enterprises in a food chain) is receiving much attention. VMIs can perform coordination tasks, but big enterprises, such as retail enterprises, often prefer to carry out such tasks themselves, also if coordination concerns one aspect of a marketing strategy, such as product quality, only. However, when many small enterprises participate in a food chain, a VMI can be instrumental in coordinating their strategies, e.g. a wholesaler sponsored voluntary chain. Under such circumstances the condition of "products/needs of enterprises served by a VMI are homogeneous" is fulfilled by the very mission of the VMI, namely the development and implementation of a uniform marketing strategy of the participating enterprises. Franchise organisations are a case in point.

**Internationalisation**. Many VMIs serve enterprises of a specific country or region. Commodity boards and farmers' cooperatives are an example. However, international enterprises with foreign subsidiaries have a broader scope and are not interested in nationally oriented VMIs. As a result the condition for a viable VMI, "products/needs of enterprises served by a VMI are homogeneous", will not be met.

A solution to this problem is internationalisation of VMIs. VMIs with an international membership are scarce at present, but will become more important, e.g. farmers' cooperatives with an international membership.

## 7.2 THE IMPACT OF ENVIRONMENTAL TRENDS ON VMIS

Trends in the environment of food marketing systems influence VMIs through their influence on food marketing strategies. From that point of view their impact has been discussed in the previous section. However, the following trends create opportunities for VMIs, irrespective of the distinguished food marketing strategies:

- increasing concern about health and safety in relation to agricultural and food products,
- increasing concern about ethical and social issues related to agriculture and food production,
- increasing concern about the sustainability of agriculture and food production.

These trends often concern the generic product. For instance: consumers are worried about the cholesterol content of eggs or about pesticide residues on fresh fruit and vegetables; society is suspicious about the ecological consequences of genetically manipulated plants and animals; many people are critical of the environmental pollution caused by some types of animal husbandry.

Food marketing systems have to respond to such trends by developing products and procedures which are socially acceptable. VMIs can be instrumental in this respect by defining criteria and controlling their implementation. Examples include VMIs which develop and implement environmental labels; VMIs in charge of decreasing the amount of package waste; or VMIs which define and control the implementation of animal welfare criteria. Such VMIs seem viable since all enterprises of an industry face the same generic problem, which can effectively be handled by a collective approach. However, instead of joining a VMI some big enterprises prefer to adopt their own individual approach to such generic problems, the reason often being to support their own enterprise and brand image.

Many VMIs that take care of marketing problems resulting from societal concern about the production and marketing of a generic product, develop and implement behavioural rules for participating enterprises. In this respect they differ from "classical" VMIs, such as farmers' cooperatives, which actively participate in the marketing process.

## 7.3 ASPECTS OF VMI'S REACTION TO CHANGING STRATEGIES AND ENVIRONMENT

VMIs, in particular big VMIs, challenged by a changing environment try to stay in business, even if their relevance for participating enterprises is decreasing. Goals and time path for transforming such VMIs have to be planned carefully, in order to avoid poor solutions, "too little, too late" or even a "mission impossible". For example, a cooperative auction cannot transform itself into a true marketing cooperative without a fundamental change in its management organisation. In fact, while big VMIs, challenged by a changing environment, appear to be well able to adopt new technologies, they often seem to have big problems in changing their management organisation and business culture.

The approach to transforming/adapting a VMI can, in analogy to the scheme of Visser and Hemerijck (1997, p.65,ff.; see section 3), be clarified by the criteria institutional importance (potential role of a VMI for the income formation of participating enterprises) and institutional capacities (capacities of a VMI in production and marketing), see Figure 11.2.

| | INSTITUTIONAL CAPACITIES<br>+ | |
|---|---|---|
| +<br>INSTITUTIONAL | INTENSIFY ACTIVITIES | IMPROVE CAPACITIES<br>or<br>QUIT |
| IMPORTANCE | ORGANISATIONAL CHANGE<br>or<br>QUIT | QUIT |

*Figure 11.2 Adaptation / transformation of a VMI to a changing environment in relation to institutional importance and institutional capacities*

Finally it is interesting to note, that VMIs in responding to fundamental environmental changes sometimes evolve towards hybrid types of VMI. Examples include a farmers' cooperative owning a majority share in a limited company which has no direct formal link with cooperative members; an industrial research institute which is doing basic research for generic products on the basis of public-private partnership.

## 8 CONCLUSIONS

The above analysis of VMIs suggests the following conclusions.

- The proposed conceptual framework is helpful for classifying VMI's, and assessing the viability of VMIs.
- Many trends in Western food marketing systems, in particular the increasing number of medium-sized and big enterprises pursuing their own marketing policies, have a negative impact on the viability of VMIs.
- VMIs still have opportunities in Western food marketing systems, by offering specific marketing services to enterprises that basically pursue their own marketing policies: a "facilitating" VMI; and by planning and implementing the

total marketing programme for the product of small enterprises: a "strategic" VMI.

- There are increasing opportunities for VMIs that address marketing problems due to the externalities of food production and marketing.
- VMIs evolve from VMIs having business relations with member enterprises only, to VMIs having business relations with both member and non-member enterprises.
- Hybrid VMIs, such as public-private partnerships or cooperatives operating through limited enterprises, are becoming more important.

**NOTE**

[1] Section 2 borrows heavily from Meulenberg and Viaene (1998)

**REFERENCES**

ALDERSON, W., 1954, Factors Governing the Development of Marketing Channels, in: Clewett, K. (ed.), 1954, *Marketing Channels for Manufactured Products*, R.D. Irwin, Homewood, Ill.

ANDERSON, E. and B. WEITZ, 1989, Determinants of Continuity in Conventional Industrial Channel Dyads, *Marketing Science*, 8, pp.310 - 323.

ANDERSON, E. and D. SCHMITTLEIN, 1984, Integration of the sales force: An empirical examination, *The Rand Journal of Economics*, 15, pp.385 - 395.

BAIN, J. 1956, *Barriers to New Competition*, Harvard University Press, Cambridge Mass.

BARTON, D., 1989, What is a cooperative? in: Cobia, D. (ed.), *Cooperatives in Agriculture*, Prentice Hall, Englewood Cliffs, N.J., pp.1 - 20.

BEKKUM, O.F. van, and G. van DIJK (eds.), 1997, *Agricultural Co-operatives in the European Union*, van Gorcum, Assen.

BROWN, J.R., JOHNSON, J.L. and H.F. KOENIG, 1995, Measuring the sources of marketing channel power: A comparison of alternative approaches, *International Journal of Research in Marketing*, 12, pp.333 - 354.

BUCKLIN, L.P., 1965, Postponement, Speculation and the Structure of Distribution Channels, *Journal of Marketing Research*, 2, pp.26 - 31.

BUCKLIN, L.P., 1966, *A Theory of Distribution Channel Structure*, University of California, Berkeley, Institute of Business and Economic Research.

BUCKLIN, L.P., 1970, The Classification of Channel Structures, in: Bucklin, L.P. (ed.), 1970, *Vertical Marketing Systems*, Scott, Foresman and Company, Glenview, Ill., pp.18 - 31.

CELLY, K.S. and G.L. FRAZIER, 1996, Outcome-based and Behavior-based Coordination Efforts in Channel Relationships, *Journal of Marketing Research*, 33, pp.200 - 210.

COBIA, D. (ed.), 1989, *Cooperatives in Agriculture*, Prentice Hall, Englewood Cliffs, N.J.

CORIAT, B. and G. DOSI, 1998, The institutional embeddedness of economic change: an appraisal of the 'evolutionary' and 'regulationist' research programmes, in: Nielsen, K. and B. Johnson (eds.), 1998, *Institutions and Economic Change*, Edward Elgar, Cheltenham, UK., pp.3 - 32.

CORSTJENS, J. and M. CORSTJENS, 1995, *Store Wars*, John Wiley and Sons, Chichester, New York.

CROPP, R. and G. ENGELSBE, 1989, Structure and scope of agricultural cooperatives, in: Cobia, D. (ed.), 1989, *Cooperatives in Agriculture*, Prentice Hall, Englewood Cliffs, N.J., pp.35 - 67.

ENGEL, J.F., BLACKWELL, R.D. and P.W. MINIARD, 1995, *Consumer Behavior*, 8th edition, the Dryden Press, Forth Worth.

FORBES, J.D., 1982, Societal Control for Producer Marketing Boards, *Journal of Macromarketing*, 2, pp.27 - 37.

GEYSKENS, I., STEENKAMP, J-B.E.M. and N. KUMAR, 1998, Generalizations about trust in marketing channel relationships using meta-analysis, *International Journal of Research in Marketing*, 15, pp.223 - 248.

HAKANSSON, H.(ed.), 1982, *International Marketing and Purchasing of Industrial Goods, an Interaction Approach*, John Wiley and Sons, Chichester, New York.

HOLMSTROM, B.R. and J. TIROLE, 1989, The Theory of the Firm, in: Schmalensee, R. and R. Willig, (ed.), 1989, *Handbook of Industrial Organization*, Volume 1, North Holland, Amsterdam-New York, pp.61 - 134.

HOOS, S. (ed.), 1979, *Agricultural Marketing Boards*: Prices, Profits and Patterns, Ballinger Publishing Company, Cambridge, Mass.

KATZ, M.L., 1989, Vertical Contractual Relations, in: Schmalensee, R. and R. Willig (eds.), 1989, *Handbook of Industrial Organization*, Volume 1, North Holland, Amsterdam-New York, pp.655 - 721.

KOHLI, A.K. and J.B. JAWORSKI, 1990, Market orientation: The construct, research propositions, and managerial implications. *Journal of Marketing*, 54, pp.1 - 20.

KOHLS, R.L. and J.N. UHL, 1990, *Marketing of Agricultural Products*, 7th edition, MacMillan Publishing Company, New York.

KUMAR, N., SCHEER, L. and J.E.B.M. STEENKAMP, 1995, The Effects of Perceived Interdependence on Dealer Attitudes, *Journal of Marketing Research*, 32, pp.348 - 356.

LAFONTAINE, F. and P.J. KAUFMANN, 1994, The Evolution of Ownership Patterns in Franchise Systems, *Journal of Retailing*, 70, pp.97 - 113.

LILIEN, G.L., 1979, Advisor 2: Modelling the market mix decision for industrial products, *Management Science*, 25, pp.191 - 204.

LILIEN, G.L., KOTLER, P. and K.S. MOORTY, 1992, *Marketing Models*, Prentice Hall International Inc., Englewood Cliffs, N.J.

LINDA, R., 1988, The food and drink industry: large firm strategies, in: de Jong, H.W.(ed.), 1988, *The Structure of European Industry*, 2nd Revised Edition, Kluwer Academic Publishers, Dordrecht, pp.127 - 156.

MEULENBERG, M.T.G. and J. VIAENE, 1998, Changing food marketing systems in western countries, in: Jongen, W.M.F. and M.T.G. Meulenberg (eds.), 1998, *Innovation of Food Production Systems*, Wageningen Pers, Wageningen, pp.5 - 36.

MEULENBERG, M.T.G., 1986, Food marketing at the sector level: the case of Dutch commodity boards, in: Kaynak, E. (ed.), 1986, *World Food Marketing Systems*, Butterworths, London, pp.209 - 221.

MEULENBERG, M.T.G., 1997, Evolution of agricultural marketing institutions, a channel approach, in: Wierenga, B., van Tilburg, A., Grunert, K., Steenkamp, J-B, E.M., and M. Wedel, 1997, *Agricultural Marketing and Consumer Behavior in a Changing World*, Kluwer Academic Publishers, Boston/London/Dordrecht, pp.95 - 108.

MORLEY, J.A.E. 1967, Marketing Boards, in: Warley, K. (ed.) 1967, *Agricultural Producers and their Markets*, Basil Blackwell, Oxford, pp.341 - 352.

NILSSON, J. and G. van DIJK, (ed.), 1997, *Strategies and Structures in the Agro-Food Industries*, van Gorcum, Assen.

NOOTEBOOM, B., 1998, Governance of transactions: a strategic process model, in: Nielsen, K. and B. Johnson (eds.), 1998, *Institutions and Economic Change*, Edward Elgar, Cheltenham, UK., pp.172 - 200.

PERRY, M.K., 1989, Vertical Integration: Determinants and Effects, in: Schmalensee, R. and R. Willig (eds.), 1989, *Handbook of Industrial Organization*, Volume 1, North Holland, Amsterdam-New York, pp.183 - 255.

POPCORN, F., 1992, *The Popcorn Report*, Harper Business, New York.

PORTER, M.E., 1980, *Competitive Strategy*, The Free Press, New York.

RINDFLEISCH, A. and J.B. HEIDE, 1997, Transaction Cost Analysis: Past, Present, and Future Applications, *Journal of Marketing*, 61, pp.30 - 54.

SCHOTTER, A., 1994, *Microeconomics*, Harper Collins.

SENAUER, B., ASP, E. and J. KINSEY, 1991, *Food Trends and the Changing Consumer*, Eagan Press, St. Paul.

SHETH, J.N., GARDNER, D.M. and D.E. GARRETT, 1988, *Marketing Theory: Evolution and Evaluation*, John Wiley and Sons, New York.

STERN, L.W. and T. REVE, 1980, Distribution Channels as Political Economies: A Framework for Comparative Analysis, *Journal of Marketing*, 44, 3, pp.52 - 64.

STERN, L.W., 1969, *Distribution Channels: Behavioral Dimensions*, Houghton Mifflin Company, Boston.

STERN, L.W. and A.I. EL ANSARY, 1992, *Marketing Channels*, 4th ed., Prentice Hall Inc., Englewood Cliffs, N.J.

STIGLER, G.J., 1951, The Division of Labor is Limited by the Extent of the Market, *Journal of Political Economy*, 54, pp.185 - 193.

VEEMAN, M.M., 1997, Marketing Boards: The Canadian Experience Revisited, *American Journal of Agricultural Economics*, 79, 5, pp.1554 - 1562.

VISSER, J.and A. HEMERIJCK, 1997, *'A Dutch Miracle', Job Growth, Welfare Reform and Corporation in the Netherlands*, Amsterdam University Press, Amsterdam.

WALLACE, L.T. and W.R. SCHRODER, 1997, *Government and the Food Industry: Economic and Political Effects of Conflict and Co-operation*, Kluwer Academic Publishers, Boston/Dordrecht/London.

WILLIAMSON, O.E., 1989, Transaction Cost Economics, in: Schmalensee, R. and R. Willig (eds.), 1989, *Handbook of Industrial Organization*, Volume 1, North Holland, Amsterdam-New York, pp.135 - 182.

WILLIAMSON, O.E., 1998, Transaction cost economics: how it works; where it is headed, *The Economist*, 146, 1, pp.23 - 58.

WOLF, C.,1993, *Markets or Governments*, 2nd edition, The MIT Press, Cambridge, Mass.

ZYLBERSZTAJN, D., 1996, Governance Structures and Agribusiness Coordination: A Transaction Cost Economics Based Approach, in Goldberg, R.A. (ed.), 1996, *Research in Domestic and International Agribusiness Management*, Volume 12, JAI Press, Inc. Greenwich, Connecticut, pp.245 - 310.

# 12 TAKING GENDER: SOCIAL INSTITUTIONS AS REGULATORS OF MARKETS

Barbara Harriss-White [1]

## 1 INTRODUCTION: MARKETS AND SOCIAL INSTITUTIONS

The "market" is conventionally seen as an atomistic realm of impersonal economic exchange of homogeneous goods carried out by means of voluntary transactions. These are mediated on an equal basis by large numbers of autonomous, fully informed entities with profit maximising behavioural motivations able to enter and leave freely. The market is thus the supreme medium for the expression of individual choice (Hodgson, 1988, p.178). Much more often than not it is assumed to be perfectly competitive. Models of other stylised market structures (monopoly, oligopoly) alter certain criteria, retain others and predict the consequences for prices and quantities.

The abstractions leave us shorn of means whereby to understand not only how supply is supplied and demand is demanded, but also the structure and behaviour of real market systems which relate demand and supply. For exactly the reason that real markets are highly diverse and complex socio-economic phenomena, real markets have proved awkward to define.

One restrictive definition of market exchange in which voluntarism, egalitarianism and informational availability are stressed has been offered by Pandya and Dholakia: "the simultaneous transaction of valued goods and services between two parties [who are] capable of accepting or rejecting the value offered ... and [who are] uncoerced ... and capable of communication and delivery" (1992, p.24). Markets are then efficient mechanisms of resource allocation resulting in an expansion of productivity.

By contrast, another definition by Fourie sees a real market as "an economically qualified, purposeful interchange of commodities on the basis of quid pro quo obligations at a mutually agreed upon exchange rate ... in a cluster of exchange and rivalry relations" (1991, p.43, p.48). Here, the social relations unique to market exchange require the combination of "horizontal" and adversarial competition between populations of buyers (and populations of sellers) on the one hand and a mass of "vertical", exclusive, mutualistic, bilateral transactions between one buyer and one seller on the other. The implications of this definition (contrary to the voluntarist definition) are that exchange rates mutually agreed on may not always be mutually beneficial, that vertical contractual arrangements may prevail over horizontal competition and that purposeful bargaining and the obligations resulting from it may rest on and reinforce a highly unequal base or fall-back position.

Without wishing to be reductionist, it would seem that market exchange will be a site of exploitation *as well as* of allocation. It has the capacity to be a site of ethnic and/or gender subordination as well as of liberation.

Much exchange which would fall within Fourie's definition of market would fall outside that of Pandya and Dholakia - repeated, relational contracting, the socially consensual determination of prices (Jorion, 1994; Guerrien, 1994) coercive interlocked contracts (Bhaduri, 1986). The extent of such non-market marketing will affect the pace at which markets react to deregulation and other policy tools of liberalisation which are normally considered to provide incentives.

Fourie's kind of market is conventionally distinguished from other types of economic activity, for example that within firms. Although some theorists have depicted firms as clusters of individual market-like contractual relationships, firms can also be seen as "a command economy in microcosm" (Folbre, 1994, p.45). Their internal structure of authority is understood by some primarily to minimise transactions costs and by others primarily as a coercive mechanism. Upon empirical scrutiny, it is apparent that firms cannot be reduced to bundles of micro-markets. Firms are a distinct type of economic institution. A conventional characterisation contrasts institutions/firms with markets. If that contrast is accepted, then markets would have to be understood as "*not-institutions*" (see Folbre, 1994, p.24 for one example).

But markets cannot exist in a deinstitutionalised form: no economic phenomena do. There are logical and historical aspects to this contention. First, it is only possible to construct supply and demand schedules on the assumption that buyers and sellers react as though any price could be the equilibrium price. Prices are thus formed in logical time as if expectations had vanished and memories were eliminated but as if complete information existed about other prices at the moment of price formation. This is a necessary condition for perfect competition. But perfect competition not only does not exist, it would not be viable for long if it did exist because, entry, exit, investment and disinvestment depend in the actual world upon the belief or the fact that information regarding opportunities is restricted. Two central tenets of the neoclassical project: the "methodological individualism" involving voluntarist, individualist subjective preference and instrumental rationality on the one hand and the market (as an actually existing bundle of "legal, customary, political and other social arrangements" (Hodgson, 1988, p.174) on the other are famously incompatible. Markets are institutions.

In his treatise on 'The Great Transformation', Karl Polanyi sought to explain the origin and purpose of such arrangements. He argued that throughout most of time, space and society, material goods have been valued in so far as they serve social standing, claims and assets (1957, p.48). By this he meant that an economic system is usually run for non-economic motives. The market, where it exists at all, has been a mere appendage to other systems of exchange. Polanyi simplified these latter to consist of 'redistribution' and 'reciprocity', the power relations of, and rules for which are regulated by social institutions or arrangements. When, as in our contemporary era, the economy is dominated by the market 'instead of the economy being embedded in social relations, social relations are embedded in the economic system.... Market economy can only flourish in market society' (op.cit., p.57). Market society, however, is something of a contradiction in terms, for the 'self regulating market... could not exist for any length of time without annihilating the human and natural substance of society' (op.cit., p.3; see also Altvater, 1993 for a recent restatement and development with respect to the environment). The measures through which a market is construed are understood by Polanyi to be the way in which societies protect themselves from the destructive force of markets. Insodoing they 'disorganise' the pure and abstract market in two ways: first by structuring and contesting the division of returns, particularly distributive shares; second, by resisting commercialisation *per se*. They thereby give all markets their continually evolving economic, social and political characters.

Following Polanyi, markets are thus not *only* controlled and regulated by 'themselves', through formal (associational) and informal (normative, contractual) institutions of firms, organisations and capital. They are not only regulated by the laws and institutions of states. They are also regulated and controlled by social institutions.

"Institution" is a notion used in at least three rather loose ways. Economists distance their use of the concept from that of sociologists by carving out two distinct territories for 'institution'. The first is micro-economic. Institutions are understood in "special case" terms. North (1990) would have us recognise organisations as distinct from (social normative) institutions and examine the tensions between them. Organisations are groups of individuals bound by some common purpose to achieve objectives (op. cit., p.5). The organisations or micro-economic institutions of interest to economists are those concerned with production: firms and contracts. The second sense of institution is macro-economic and legal, encompassing the definition of rights, the scope of economic behaviour, the mechanisms to protect exchange, penalise miscreants and through taxation ensure state legitimacy (Giddens, 1992; Shaffer, 1979).

The third usage of 'institution' is sociological. Any behavioural regularity or patterning is the manifestation of an institution and may have implications for markets (Fourie, 1991, p.52). Thus a conference is an institution but so also is the way in which biological sex becomes social gender or norms of justice, values of money or goods and other aspects of ideology and social rules are developed and reproduce. Exchange processes are constituted by, and constitute in turn, a wide set of social institutions: state, locality, class, ethnic group, caste, religion, kin, age and gender. These institutions can be ascribed (e.g. caste), acquired (e.g. 'Lionesses'[2], alumni ), multiple, not co-ordinated (patriarchal behaviour, for instance, may be in direct conflict with egalitarian developmental beliefs; families are sites of both cooperation and conflict) and not monitored (Folbre, 1994). Contrary to the new home economics, returns to participation in such social institutions are not consciously calculated, nor is it likely that they are calculable at all. By means of such institutions the terms and conditions of contractual participation and resistance are defined, the structure of wants is constituted, value and motive are confirmed, status operates and worth, identity and honour are expressed according to unique and specific configurations. Markets thereby get social meaning.

It is one thing to state this but quite another to examine these complex relationships empirically. Yet, since all social institutions work in non-universalistic ways, specific to time and place, since they inter-relate and change over time and place and since the social configurations of markets vary in significant ways, there can be no escape from empirical enquiry.

My project (still methodologically in its infancy) is to explore the evolution and significance of institutions of social regulation and resistance as an economy passess from 'being embedded' to 'embedding' and possibly to 'engulfing'. Even if institutions were strictly speaking 'constraints' to the operation of freely engulfing self-regulated markets, they are falsely regarded as constraints on the working of real historical markets. If 'market society' would self-destruct there can be no sustainable economy run by absolute, pure, commercial logic. Research from anthropology and psychology reveals two important attributes that could not have been known by Polanyi. The first is that even in advanced quasi-market societies, the vast majority of economic transactions are personalised and socialised (Emler, 1996). The second is the pervasiveness of reciprocity and redistribution and many other forms of exchange even in the most advanced OECD countries (Davis, 1992).

Social institutions can be of two types, first, social organisations or groups (e.g. family, caste and ethnic group), second, social norms, ideologies and conventions. The transition from the market's being embedded in social relations to social relations' being embedded in the market will not necessarily result in change to groups and norms in consistent or predictable directions. Older forms of family may dissolve under the impact of commercialisation but patriarchal ideology may be reinforced for the while. Nor is there a simple reading-off of increased material advantage or social functioning or capability from such a transition. Just as they may be destructive of non-market values and ideologies, markets *per se* contain no inherent safeguards of life or livelihood.

In order to elaborate these ideas, the rest of the discussion will be devoted to the intersection of commodity markets and the social institution of gender (and to a lesser extent, class[3]). Almost everywhere grain forms the basis of social reproduction and is therefore a useful analytical 'prism'. Here, we try to build an analysis of the regulatory role of a) gender on grain markets and of b) grain markets on gender in the context of liberalisation using case material from Guinea in West Africa and Tamil Nadu in South India.

## 2 HOW GENDER REGULATES MARKETS

Marketplaces and staple food markets are highly gendered complexes of social institutions. But (although there are no religious textual proscriptions of women from trading) gender works in these markets in ways differing radically from society to society. The central questions concern institutional autonomy: in what ways do the social institutions of production, household reproduction and the reproduction of gender ideologies shape female participation in such markets.

For the highly original economist Esther Boserup, gender divisions of labour in production were related straightforwardly to those in markets (1970, pp.87-92). The gendering of property ownership or usufructuary rights over land was reflected in those over products. Female farming systems had female traders and marketplaces. Male farming systems, male traders and marketplaces. However, this is not, nor was it at the time, an accurate generalisation.

In West Africa, for instance, while agricultural production may be controlled by women as well as by men, local trade in basic consumption goods has been widely observed as a public domain for women[4] in which not only can competition be fierce but also accumulation can happen. Even so, women are portrayed as tending to occupy particular niches in female grain marketing systems [5] more often than not defined by:

1. commodity (staple food, cooked food, beer) because access to, or production of, these goods can be carried out domestically; these goods may also be a female gendered subset where commodities are gendered according to notions of rank, status and purity (where female goods are inferior);
2. points in the market system (small scale processing and retailing) because of highly gendered concepts of the 'strenuous work' required by other marketing activities or because of the prior requirements of the reproductive burden;
3. by organisational forms (individuated) and forms of reproduction of the firm (simple/oriented to subsistence) - the result of gendered inheritance customs and access to capital and of gendered constraints on spatial mobility;
4. by motivation (household subsistence/target incomes) because of the gendered division of intrahousehold expenditure responsibilities and gender specific norms of ostentation in private consumption, savings and investment;
5. by territoriality and spatial mobility (local, non mobile) because of male gendered transport ownership and gendered notions of propriety in the public territories of marketplaces;
6. by season (most active in, if not fully confined to, the post harvest months with a rapid turnover time and lower capital requirements).

The irresistible implication is that other forms of local trade will tend not to be controlled by women, even in 'female' marketing systems.

Pujo has thoughtfully reviewed speculation about the female gendering of West African trade (1997, pp.41-44). All the explanations put forward are problematical. One materialist hypothesis - of 'residualisation' stresses the role of land scarcity under conditions of rapid commercialisation which results in production's becoming increasingly dominated by men. Women then turn to the control of commercialisation and the sphere of distribution, giving them a domain independent of men[6]. But for this there is little empirical support. In particular, the division by gender of male production and female distribution is not rigid in any society and trade is far from a residual activity, being capable of generating comparatively great wealth. A second hypothesis - that of 'compatibility' - rehearses the idea that trade is easy to combine with household activities (which are understood as prior claims on female labour). Where social reproduction requires market exchange as well as labour within the household, then women are found in markets. Again this is toppled by empirical evidence. A third 'cultural' argument sees female identity as involving both fertility and independence. Thus trading and the separate budget associated with it are just as much required of women as a large number of children. But not all women trade and marketplaces and processes are not so simply arranged[7]. A general explanation has not proved possible. As with so many social phenomena[8] the explanations for patterns of gendering will be both specific and complex. To reveal how social institutions and economic behaviour interact it is useful to be able to examine gender ideologies and then to look at the contrariness of their outcomes in the 'actual' observable gendering of tasks, control over technologies and capital in markets, and in turn at their impact on structural aspects of markets: entry, concentration, competition and information and thus on performance. To my knowledge, Pujo's work in the forest interior of Guinea in 1992 is the only study that has succeeded systematically to look at the rice market anywhere in this way. Performance has even been specified in gendered industrial organisation terms[9]: the gendered distribution of returns to trade, employment and comparative productive efficiency (Pujo, op.cit., pp.239-74).

While the labour process in rice production in Guinee Forestiere is mostly gender-sequential, that of marketing is gender-segregated. "The idea that processing transforms 'male' paddy into 'female' rice is used by many peasant women to obtain access to the output" (op.cit., p.182) . But from then onwards the rice market is segmented by gender, operational scale and technology. The vast bulk of trading firms are individuated. Eighty per cent of firms in Guinee Forestiere were run by women and all of them were individuated. The classification of activity in the marketing system therefore corresponds with types of firm (which is not the case in India). In practice it coincides closely with the ideologies of appropriate activity. Marketplace trade is indeed female. These women traders are relatively younger than male traders (average age 37 versus 50) and their firms are smaller (average 32 tonnes versus 134). Female traders tend to operate through personalised 'network' trading contracts and are price takers. By contrast shop-based trade is controlled by men. Men control all derived markets (e.g. the rental markets for transport and storage) and all materially productive activity (mechanical milling). They are price makers and can and do practise collusive price formation. Quite exclusive proscriptions prevent women from being mill mechanics. Trading capital and skills result from gender-specific social relations. While young men acquire capital and skill by labouring (seasonally) in the firms of relatives and/or by commercial loans, young women (when released from reproductive chores) watch female relatives in the marketplace but their access to trading capital depends entirely on loans from parents, siblings or husbands. These loans are more restricted in size than those to men. Further they carry with them the obligation to share the returns from trade - on average half of their earnings.

Thus the gender construction of trade limits the access of women to capital and causes an unequal distribution of costs, returns and obligations. Gender intersects with other social institutions:

1. class (found to be particularly awkward. Though the commoditisation of labour is at an incipient stage, males dominate capital accumulation; the household is a site of exploitation through exchange; dependent women traders can also be exploited by means of usurious interest on credit from male wholesalers. Indeed in these circumstances gender has been argued to be *the* class division (Pujo, op.cit., 291-4));

2. locality (information and access to capital is heavily urban biased);
3. ethnic group ( the historically trading *ethnies* put up barriers to entry);
4. religion (allegiance to which consolidates network transactions and may also lead to forms of contract which mask interest payments);
5. family (marriage patterns affect entry to marketing systems);
6. age (women gain independence with age and can accumulate at a faster pace; however the great majority of women stop trading altogether when their dependents leave the households) and
7. the state (whose political oppression provoked the emigration of (male) traders and encouraged accumulation through smuggling and where current new entrants to the large scale wholesale trade in rice are male officials who have often derived their starting capital from bribery) (Pujo, op.cit., pp.291-301).

So not only can certain men be far poorer than certain women (in the Kissi region, male peasants are universally poorer than female bulkers of rice) but also female traders are themselves far from being a homogeneous category. In Guinea, the returns to female traders are more highly unequal than those to men (differing by a factor of four between female retailers and wholesalers in Gueckedou (ibid. p.253)).[10]

In South Asia, by contrast, local staple food markets would appear to be much more highly commercialised and economically differentiated. Depending on location, private wholesaling firms are up to 100 times larger in terms of gross output than their counterparts in Guinea. There is a significant wage labour force in the marketing systems (working for the owners of firms and in gangs in marketplaces) which is absent in the accounts of Guinee Forestiere. The classification of activity in markets does not correspond with types of firm. Firms can be very much bigger and more internally differentiated with tendencies towards uniqueness in the complex combinations of activities that each performs (Harriss-White, 1996). The patriarchal labour process in grain marketing is complex in ways different from that in West Africa. And female *traders* in India are much more homogeneous than their counterparts in West Africa.

Our field evidence from Tamil Nadu in south east India (1980, 1993-95)[11] shows that in a typical firm owned by the males of a joint family and practising a diversity of trading and transformation activities, there are multiple relations of male labour. Male family labour (up to 13 members have been encountered) whose work is loosely specified, may work part-time or seasonally or as part of a multiple occupational profile. Permanent workers (averaging 3 but up to 7, whose task specification may be quite refined but whose terms and conditions are vague) may work at the simultaneous performance of more than one activity within the firm or on own account. Rates of pay are unsystematised, and accentuated by both patronage and debt bondage. Male casual labour (averaging 9 but up to 40) may be permanently attached to a trading firm but employed on a daily, weekly, seasonal, group contract or piece rate basis for manual work. While tasks are highly specified, contracts, terms and conditions, and rates of pay vary greatly. Lastly, male child labour is underestimated[12], used at key points in the grain marketing system (messages, carrying food and drink for negotiations, cleaning), and may be paid secure though very low wages. For some children, such work is an apprenticeship (though there is no reason why such apprenticeships should replace formal school because the children of rich traders participate in both activities).

Labour "markets" in grain marketing are heavily structured by the assets distribution and by networks of gender, caste and locality. Markets are segmented spatially and socially. Large differences in contractual forms, in modes and levels of payment can obtain over short distances such that the very notion of a market in the restricted definition of Pandya and Dholakia is challenged.

Women participate in this grain trade in four ways according to their caste and class position.

1. Directly (as in Guinea, but from a more restricted and poorer population segment), women from pauperised, female-headed and/or low caste households are confined to petty and often seasonal operation, to subsistence orientation and simple reproduction, to particular positions and activities within the system (especially processing and retailing), local territorial linkages, weekly marketplace sites and unlicensed and/or illegal transactions in cash. Their participation is conditioned by the life cycle and when children are no longer dependent, these women tend to stop trading.
2. Female casual wage workers from the assetless class form the large substratum of labour in grain milling and premilling processing. Marketing systems rest on the backs of these women. The average mill employs 15, but up to 70 have been encountered[13]. Outcaste women are allowed to turn paddy on the large drying yards because the kernel is still protected from ritual pollution by its husk. As in Guinea, women are debarred from being mill mechanics and it is unusual to find them handling heavy consignments of scalding paddy during the parboiling process. Female "coolie"[14] is prevalently but incorrectly regarded by mill owners as a household supplement for their employees. Wage differentials of two thirds to a half that of male wages in rice mills in no way reflect productivity (which in any case would be tantamount to impossible accurately to measure since the division of tasks in milling is sex sequential). Female mill work is deliberately casualised and I have never encountered a unionised female labour force. The sexual exploitation of the mill work force by management is not unknown.
3. In the recent past in smaller family firms, unwaged female family members have provided that part of the wage to labour in trading firms which takes the form of prepared food (though with the commercialisation of labour, this practice of payment in tea and meals is dying, or itself being commercialised). Female labour will then subsidise the firm. It has to be added that a large rice milling and trading firm will almost certainly "subsidise" some of the costs of reproduction of their male labour force such that the social reproduction of male labour is not entirely born by female labour within working class households. Accidents on site are usually compensated and medical expenses often paid at times of sickness. Both male and female labour receive at least one month's extra pay at a major festival. Often this is given in kind as cloth. It can also be argued however that such payments both retard the formation of labour markets (because of their informational opacity) and reflect "backward" relationships of patronage rather than market exchange (by dint of their discretionary element).

4. Women in accumulating firm-families are used for the caste-based reproduction and expansion of firms first by means of their dowries on marriage and second through the (rare) practice of fictitious ("benami") registration of a trading company in a woman's name generally for the purposes of tax avoidance. In the first case, the higher education of such women is a good example of the economic inefficiency of gender institutions. For such women, education is a status good and leads neither to economic participation nor control over assets or over major economic decisions [15]. Well educated mothers are thought better to educate their own children than less educated women; but the structure of ownership of large mercantile companies, framed by the preemption of tax laws, frequently requires strong male control of young male adults and discourages migration for advanced education, so there exist constraints to education other than gender and other than its costs.

So the South Asian grains trade is also seen to be highly gendered. In turn gender intersects with caste and class position. The lack of ownership or control over property, or any means of circulation (which determine creditworthiness) makes the economic role of women belonging to the accumulating oligopolies somewhat indirect, in contrast to the role of poor women in petty trade and the casual labour force.

Thus male marketing systems are dynamised by female wage labour and female marketing systems have male traders at their power points even if they also have a profusion of apparently independent female traders. The classification of marketing systems as male or female might be appealing but it is inaccurate as a generalisation and it obscures increasingly common characteristics of rice markets in these two regions. Both types of marketing system provide more livelihoods to women than they do to men, but women's livelihoods are on the whole relatively humble. We now turn to the thorny question whether even these humble livelihoods can be emancipating.

## 3 MARKETS AND AMBIVALENCE IN GENDER RELATIONS

We have seen how marketplaces in certain parts of the world have been identified as female domains, as sites for the exercise of female autonomy. But equally, we have seen that market exchange is both complex and inherently riddled with ambiguity. It does not follow that processes of commodity marketing either ensure female autonomy and emancipation (within either the marketing system or the domestic unit) or remove the shackles from female property ownership or accumulation. Just as the household is a site for relationships which subordinate women so also are the 'markets' for wage labour and for grain.

In West African *rural* markets, despite the fame of the "market queens", "mama benz" or "market mammies" and despite the undoubted greater freedom to trade and autonomous control over the returns to trade than in India, it seems the upward economic mobility of women is rarely achieved[16]. In Guinea, the access of women to market places and to commerce is restricted more often than not by the decisions of men about market supplies, which are frequently generated out of production processes centring upon the household. Women in marketplace trade may thus be prevented from buying and selling and constrained to selling. While market information is segmented by gender (and women are socialised and 'apprenticed' from childhood to appreciate better than men the subtle conventions surrounding volumetric measurements[17]), the economic resources necessary for the spatial and social mobility to amass wholesale consignments, command transport and/or own processing facilities are in the hands of men. Inheritance down the female line is not possible, so the intergenerational transfer of commercial resources from mother to daughter has to be made early in the latter's maturity. Only petty quantities can be transferred in this way. The average capital of male traders is 27 times that of female traders. In Pujo's study only one woman trader gained returns greater than any of the male traders.

Once entry into trade is achieved, then gender-stratified access to technology and money means that processing costs for women are double those for men. Women's wholesale trading costs are 12 per cent higher than those of men; their transport costs are 55 per cent higher per unit distance; their storage costs are greater by a factor of three. Even the way trade is taxed disadvantages women in Guinee, the mechanism being gender specific because of the gender-geography of trade. On men tax is levied at a fixed rate per shop. On average it amounts to about 1 per cent of income. Women, by contrast, pay a tax every day they occupy a site in a marketplace - amounting to about 5 per cent of earnings. As a result male traders in Gueckedou make $3,500 per annum to a weighted average of $820 for women (Pujo, 1997, p.239-274).

Men exert other kinds of indirect control over female trading. In the forest interior of Guinea, rotating credit associations among women traders facilitate accumulation and yet women traders' investments in cattle, houses and jewellery, and their savings, are still frequently controlled by men. The extent of economic and functional differentiation means that it is not to be assumed that women traders have any particular collective interest on account of their gender.

In South India, the same phenomena are writ large and much more intensely. Roles within market systems are structured by non-market criteria. Control over capital is exclusively male while 55 per cent of the casual livelihoods comprising the physical work of marketing and post harvest transformation are done by women. Casual female labour gets take-home pay based on piece rates which is on the average about two thirds that of casual male labour. The subordination of women is extremely crude depending on the type and class of commercial firm. Petty trade admits low caste and poor women. Their 'independent' returns are rarely much more than can be got by wage labour in commercial firms and are on the poverty line: 2 per cent of the estimated average returns of a rice wholesale firm. But gender constrains entry into the permanent wage labour force where take-home wages including perqs in kind average 20 to 300 per cent higher than earnings in the casual female labour segments (Harriss-White, 1996).

Gender also constrains accumulation which is overwhelmingly by men. In elite firms, the role of women is very rarely productive or managerial. Nor are women family members deployed in the firm. Instead women are constructed socially as vehicles for the transfer of commercial resources between kinship groups. The exploitation of low-paid casual female labour by this commercial class not only fuels capitalist profits but also in turn reproduces the educated, somewhat secluded, leisured state of women in the households of the 'rurban' elite. Indian grain markets reflect in vivid ways the gender subordination characteristic of Indian society as a whole. Resistance to these arrangements is rare and meets intense male hostility.

Not only is gender far from being a homogeneous category, but the extent to which gender divisions and ideologies structure and constrain individual or collective agency varies. While it looks as though what Folbre terms the 'structures of constraint' are less fortified in rural and 'rurban' West Africa, they are formidable in India. There, businesswomen and female entrepreneurs can be found. But a) not in rural or even 'rurban' areas and b) not in the grain trade.

## 4 THE CONSEQUENCE OF GENDER EMBEDDEDNESS FOR MARKET EFFICIENCY

The commonest question asked by economists of markets concerns their efficiency. By virtue of there being no de-institutionalised rice marketing systems anywhere, efficiency is extremely difficult to evaluate. For the purposes of this discussion we can distinguish productive efficiency in which outputs from given inputs is maximised, from adaptive efficiency (North 1990, 80-2) which involves judgements about the flexibility of the norms and institutions shaping the way an economy develops.The point to be made here is that the economic consequences of the gendering of rice marketing system reduces both sorts of efficiency in the examples we have used.

In Guinea, the gender division of labour has been calculated to cause losses in productive efficiency. Women face reduced incentives in production because of male control of the product of their labour, in processing because of the gendered stratification of technology and in trading because of the higher costs which were discussed earlier, which reduces their returns by 20 per cent[18]. The inefficient impact of these gendered relations is to raise consumer prices, leading to malnutrition, to raise the competitive advantage of imported over local rice (and imported rice is completely controlled by men) and to lead to massive female underemployment in the scarce season, which does not bode well for the nutritional status of their dependents (Pujo, op.cit., pp.239-274).

In liberalising India, while gender roles, ideologies and the division of tasks in metropolitan conurbations may show considerable new elasticity (particularly among the 'middle classes') are viscous outside such social territories, and are 'adaptively inefficient'. In a South Indian market town of some 100,000 people whose economy has been studied every decade from 1973, the adaptive inefficiency of the gendering of markets works differently from the West African case. Considerable strides have been made in female education and there is a small core of women doctors and lawyers. But women cannot own or manage commercial firms. The reasons given by the few women who want to enter trade are instructive. The state, in the guise of tax officials, block licencing on suspicion of bad faith and of intention to avoid tax by the family concerned. The state, in the guise of banks, also refuses to regard women as eligible for credit on the grounds that their registered collateral is 'benami' (bogus). With respect to the acquisition of skills and contacts, the process of learning has to be confined to the domestic arena (in which a special room can be a private haven for male business activity - passively observed by women) and such women have faced male sanction on travel for purposes of commerce (Basile and Harriss-White, 1999).

While there are no gender barriers to entry into high skill wage work such as computer programming in metropolitan cities and while the next stage of technical change in agro-commerce - automation - urgently requires computerisation, women in the marketing town who have somehow obtained relevant qualifications from the district capital's educational institutions are still debarred (on declared grounds of gender) from computer operation on commercial premises in the market town. The agro-commercial elite reject the female gendering of high-tech work, located socially (and physically) above their male labour force. The diffusion of computerisation is currently hindered and confined to the kind of simple accounting operations, which can be learned by trial and error by under-educated male family labour.

The commercial elite boasts many examples of what Sunder Rajan (1993) has ironically termed the 'new Indian woman': much more highly educated than their agro-commercial magnate husbands, these women have been imported for purposes of marriage from the metropolis. Their rare and always frustrated wish to enter commerce is seen as a natural outcome of a merchant capitalism construed everywhere as benevolent to them and massively reinforced in the popular English language media. Such 'modern women', confined to teaching the children of their elite peers and to philanthropic social work, with their roots in a carefully sanitised version of 'female tradition' accentuate the gap between their own apparent modernity and the alienated mass of women socially imprisoned by gender ideologies and by gendered material poverty.

Meanwhile in this male marketing system the petty retail trade in staples which 20 years ago was an arena for women in the precincts of the Municipal Market has been taken over by men. Remaining women retailers are the most economically disadvantaged - widows and abandoned women with children. The older generation of women traders ceased this activity once their children were no longer economically dependent upon them. While a few elite women see large scale commerce principally in terms of its profitability and contemplate entry, poor women experience marketplaces as public, physically dirty and ritually polluting places and contemplate exit. For this latter class the non-participation of women is what they understand as development and progress. For the rich and poor aspirant to commerce alike, information and credit are **sine qua non** to enable contingent acts of individual agency to consolidate themselves into a successful patterned challenge to gender roles.

## 5 THE IMPACT OF MARKETS ON GENDER: TECHNICAL CHANGE AND LIVELIHOODS

Staple food marketing systems provide huge numbers of livelihoods. Eighty percent of firms in Guinee Forestiere were run by women. One hundred and forty nine agrocommercial firms studied in South India generated jobs for 5,500 adults, of whom 72 per cent (or 3,960) were for women (Harriss-White, 1996). That proportion would be higher if the invisible and unvalorised role of women from the families of elite firms were included.

Technical change in foodgrains (rice and oil) processing has involved i) foreign imports of technology developed for the radically different ' factor endowments and ratios' of the U.S. and Japan, ii) foreign technical assistance, the quality of whose scientific advice left a great deal to be desired and iii) protective domestic legislation outlawing the 'indigenous' technologies (Harriss and Kelly, 1982). Adoption of such imported packages was pressed on state institutions (co-operatives and parastatals) which were isolated from the other state and private institutions of finance, pricing, procurement and logistics upon which the package depended. Low capacity utilisation and higher per unit marketing and processing costs have resulted in chronic state-subsidisation. 'The market' has resisted technical change in package form (a rational response) but has over the years adopted techniques in stages and by component. This has resulted in economies of scale and accentuated the structural concentration of assets in marketing systems. These assets are male-controlled.

Each wave of incremental technical change has led to the net displacement of labour, but has been biased hugely against casual female labour. Whether or not this displacement destroys livelihoods or 'merely' reduces drudgery does not depend only on the degree of commercialisation of the product, it also depends on the extent to which the labour relations of post-harvest processing have become commercialised. The first mechanised technology (the 'Lewis Grant huller, adapted in the early 20th century from a coffee grinder and now very widespread, cost effective but nevertheless made illegal in India) is the most female-labour-displacing of all because it replaces hand- or foot- operated machines worked by women [19]. The latest components (husk-fired driers) substituting for the public good of sunshine pose very serious threats to large quantities of female livelihoods in marketing systems. While Douglass North has hypothesised that technology is generally adopted so as to maximise the use of lesser skilled workers who do not have the bargaining power to disrupt production (1990, p 65), here, technology is adopted which does away with precisely those people least capable of bargaining or of witholding labour - the cheapest factor of marketing - female labour. The newest technical components are only cost-effective at market prices under conditions of high capacity utilisation which are difficult to achieve. Yet, inexorably, they are being adopted in the very reverse of a green-revolutionary development[20]. This is hard to explain. Perhaps the employment of armies of low caste people who happen to be women is more of a status-reducing expression of the contaminating caste relations of a merchant than it is an status enhancing expression of patriarchy. Technical change can not only concentrate the economic asset structures of the marketing system but also reinforce their gender power relations.

## 6 LIBERALISATION AND MASCULINISATION

If liberalisation is successful, it results in the development of more integrated national markets, which in turn entail not just the commercialisation of production and the scaling-up of post harvest technology but also an increase in specialisation and interregional wholesale trade, not the least in staples. The development of long-distance interregional trading in female marketing systems has been observed to be dominated by men. In parts of West Africa, masculinisation may substitute for female trading (e.g. northern Nigeria (Robson, forthcoming)) while elsewhere they may be complements (e.g. Guinee Forestiere). Female marketing systems may be articulated with national territorial markets without a reduction in female livelihoods. Indeed they may very well expand. The point is that the integrated system as a whole is dominated, economically if not in numbers, by men. This domination places a new constraint upon female commercial accumulation. It is not necessary that a change in the gender division of work occur. It is enough that there is expansion in precisely that type of work historically dominated by, if not entirely restricted to, men . It is also the case that forces other than liberalisation and the formation of national markets may masculinise trade: the expansion of Islamic seclusion ideology and practice is one, the gender ideologies of the colonial and post colonial state are another.

The working women forming the foundation of increasingly 'male' marketing systems are caught in a pincer. On the one side technical change, on the other a reinvigorated patriarchal ideology coupled with the crudest of commercial oppression threaten their livelihoods and their capacity collectively to improve their working conditions. Both in 'female' and 'male' marketing systems and for a wide variety of different reasons, it is no exaggeration to conclude that millions of female livelihoods worldwide are in jeopardy from the reworking of the social institution of gender during the transition from embedded to embedding grain markets. This is the corollary of liberalisation.

## NOTES

[1] This is a reworked version of 'Female and Male Grain Marketing Systems: analytical and policy issues for West Africa and India' pp 171-189 in (eds) C. Jackson and R. Pearson: 'Feminist Visions of Development: Gender, Analysis and Policy', London, Routledge (1998). These ideas have been developed in dialogues springing from a comparison of my work in South Asia with that of doctoral students Laurence Pujo, working in Guinea and Elsbeth Robson, working in Nigerian Hausaland, to both of whom I am very grateful. The response of QEH's Centre for Cross Cultural Research on Women's seminar, where an earlier version was presented, of Cecile Jackson at UEA and the suggestions of the editors have also been very helpful. The remaining weaknesses are entirely my own.

[2] 'Lionesses' comprise the women's wing of the global network of Lions' Clubs which are social and philanthropic organisations of local notables.

[3] Elaborated in Harriss-White, 1998.

[4] Bohannan and Dalton, 1962; Lawson, 1971; Hill, 1977; Christiansen, 1993.

[5] Koopmans, 1990; Bauer, 1992

[6] In societies where divorce is common,such independence is required of women.

[7] The first view has been advanced and discussed by Cordonnier,1987, the second and third by Hopkins, 1973.

[8] C.f. Bain, 1959

[9] For example, the regional patterns of the sex ratio spring to mind

[10] At the other extremity in Hausa Nigeria, though the grain trade is dominated by men and women are generally secluded, there is very considerable economic differentiation among female traders (Robson, forthcoming).

[11] Harriss-White, 1995, 1996 (see also Harriss, 1993 for West Bengal in the north west of India)

[12] Male and female child labour is crucial for the perpetuation of secluded women's trade in Nigerian Hausaland. Their education may be denied them for this reason, or they may trade with other pupils at school (Robson, forthcoming).

[13] The female casual labour force can number up to 700 in cotton ginning and wholesaling firms (see Harriss-White, 1996, ch7).

[14] 'Coolie' in Tamil means wages for casual labour.

[15] Female education leads to the lowering of birth rates ( though not to reduced gender discrimination in regions of South Asia where this is practised (Das Gupta,1987; Jeffery and Jeffery, 1998)). But it is primary rather than tertiary education which achieves this result.

[16] Robson's work in Nigerian Hausaland rams home the points that, there, men do not share their control of agricultural production. Men have over the recent past come to engulf and to dominate periodic marketplace trade and even to share the village market with secluded women traders. The latter, for reasons of inheritance custom, access to capital and periodic taboos on trading based on life cycle events, are unable to accumulate (Robson, forthcoming)

[17] See Christiansen, 1993, for a pioneering study of the volumetric conventions in price formation by women in marketplaces in Benin.

[18] Not more because male traders have a wage labour cost component which female traders do not have (Pujo, 1997, p263)

[19] The process of displacement is most backward (and therefore acute) in the north east of the subcontinent, see Greeley, 1987.

[20] A revolutionary component should increase output per unit of input and thus lower total costs per unit of output.

**REFERENCES**

ALTVATER E. (1993) *The Future of the Market*, London, Verso,

BAIN J.S. (1959) *Industrial Organisation* New York, John Wiley

BASILE E. AND B. HARRISS-WHITE (1999) Corporative Capitalism: Civil Society and the Politics of Accumulation in Small Town India

Paper for the Gordon White Memorial Conference, Institute of Development Studies, Sussex University, Brighton

BAUER S. (1992) Action to Assist Rural Women: Marketing Approach for Income Generating Activities in Four African Countries in L. Cammann (ed) *Traditional Marketing Systems* Feldafing, German Foundation for International Development pp.10-20

BHADURI A. (1986) Forced Commerce and Agrarian Growth *World Development* 14,2,267-72

BOHANNAN P. AND G. DALTON (1962) *Markets in Africa* Illinois, Northwestern University,

BOSERUP E. (1970) *Women's Role in Economic Development* London, Allen and Unwin

CHRISTIANSEN B. (1993) *Effizienz des Vermarktungssystems von Mais, dargestellt am Beispiel der Provinz Atlantique, Benin* Kiel, Wissenschaftsverlag

CORDONNIER R. (1987) *Femmes Africaines et Commerce: Les Revendeuses de Tissu de la Ville de Lome* Paris, L'Harmattan, Serie Villes et Entreprises

DAS GUPTA M. (1987) Selective Discrimination against Children in Punjab *Population and Development Review* 13

DAVIS, J. (1992) *Exchange* Buckingham, Open University Press

FOLBRE N. (1994) *Who pays for the kids? Gender and the Structures of Constraint* London, Routledge

FOURIE, F.C. VON N. (1991) A Structural Analysis of Markets in G. Hodgson and E. Screpanti (eds)

GIDDENS A. (1992) *Sociology: A Brief but Critical Introduction* London, Macmillan

GREELEY M. (1987) *Post Harvest Losses, Technology and Employment: the Case of Rice in Bangladesh* Boulder, Westview Press

GRIFFON M (ed) (1990) *Economie des Filieres en Regions Chaudes: Formation des Prix et Echanges Agricoles* Monpellier, CIRAD

GUERRIEN B. (1994) L'Introuvable Theorie du Marche in Caille et al (eds)

HARRISS B. (1993) *Markets, Society and the State: Problems of Marketing under Conditions of Small Holder Agriculture in West Bengal* Development Policy and Practice Monograph no 1, Milton Keynes, Open University

HARRISS B AND C. KELLY (1982) Food Processing; Policy for Rice and Oil technology *Bulletin Institute of Development Studies* 13,3, 32-44

HARRISS J., J. HUNTER AND C. LEWIS (eds) (1995) *The New Institutional Economics and Third World Development* London, Routledge

HARRISS-WHITE B. (1995) Maps and Landscapes of grain Markets in South Asia in (eds) J. Harriss et al pp 87-108

HARRISS-WHITE B. (1996) *A political Economy of Agricultural Markets in South India: Masters of the Countryside* Sage, New Delhi

HARRISS-WHITE, B. (1998) Female and Male Grain Marketing Systems: analytical and policy issues for West Africa and India, pp 171-189 in (eds) C. Jackson and R Pearson

HILL P. (1977) *Population, Prosperity and Poverty: Rural Kano 1900 and 1970* Cambridge, Cambridge University Press

HODGSON, G. (1988) *Economics and Institutions* London, Polity

HODGSON G. AND E. SCREPANTI (eds) (1991) *Rethinking Economics* London, Elgar

HOPKINS A. (1973) *An Economic History of West Africa* London, Longmans

JACKSON C. AND R. PEARSON (eds) (1998) *Feminist Visions of Development: Gender, Analysis and Policy*. London, Routledge

JEFFERY P. AND R. JEFFERY (1998) Silver Bullet or Passing Fancy: Girls Schooling and Population Policy' pp 239-258 in (eds) C. Jackson and R Pearson

JORION P. (1994) L'Economie comme Science de l'Interaction Humaine vue sous l'angle des Prix: vers une Physique Sociale in Callie et al (eds)

KOOPMANS J. (1990) Gender Issues in Food Production and Marketing: Core Issues in Africa's Food Crisis in (ed) M. Griffon pp 519-531

LAWSON R. (1971) The Supply Response of Retail trading Services to Urban Population Growth in Ghana in (ed) C. Meillassoux pp 377-399

NORTH, D. (1990) *Institutions, Institutional Change and Economic Performance* Cambridge, Cambridge University Press

MEILLASSOUX C. (ed) (1971) *L'Evolution du Commerce en Afrique de l'Ouest* Oxford, Oxford University Press

MOSER C.O.N. (1993) *Gender, Planning and Development: Theory, Practice and Training* London, Routledge

PANDYA A. AND N. DHOLAKIA (1992) An Institutional Theory of exchange in Marketing *European Journal of Marketing* 26,12,19-41

PENNY, D. (1985) *Starvation: A Political Economy* Canberra, Australian National University

POLANYI, K. (1957/1985) *The Great Transformation: The Political and Economic Origins of our Time* Boston, Beacon Press

PUJO L. (1997) Towards a Methodology for the Analysis of the Embeddedness of Markets in Social Institutions: Application to Gender and the Market for local Rice in Eastern Guinea, Oxford University, D. Phil Thesis

ROBSON E. (forthcoming) Gender, Households and Markets: Social Reproduction in a Hausa Village, Northern Nigeria, Oxford University, D. Phil Thesis

RODGERS G. (ed) (1989) *Urban Poverty and the Labour Market: Access to Jobs and Incomes in Asian and Latin American Cities* Geneva, International Labour Office

SHAFFER J.D. (1979) Observations on the Political Economics of Regulation *American Journal of Agricultural economics* 11, 2, 766-74

SUNDER RAJAN R. (1993) *Real and Imagined Women: Gender, Culture and Postcolonialism* London, Routledge

# 13 RICE AND CORN, AND THE ASEAN FREE TRADE AREA AGREEMENT

Lilibeth A. Acosta[1]

## 1 INTRODUCTION

The Association of Southeast Asian Nations (ASEAN), founded in 1967, is probably one of the oldest regional blocs among developing countries which continuously move ahead in strengthening the economic ties that bind the member countries. Although the perspective of attaining closer economic cooperation was very auspicious, the way towards this end was not always smooth. Regional economic cooperation formally started only in 1976, almost a decade after the creation of ASEAN. In the area of trade and industrial cooperation, the ASEAN member countries implemented various programs which included ASEAN Preferential Trading Arrangement (PTA), ASEAN Industrial Joint Venture (AIJV) Scheme, the ASEAN Industrial Complementation (AICs), etc.

PTA represented the first major commitment on the part of all the ASEAN countries to make joint efforts to liberalize intra-regional trade, mainly through reduction of tariff barriers. ASEAN economic cooperation moved at a lackluster pace and several problems came on the way that left the economic cooperation with little accomplishments. The ASEAN members thus saw the need for stricter guidelines of trade liberalization in lieu of the loose implementation of PTA. The strategy that superseded PTA, the ASEAN Free Trade Area (AFTA) Agreement, was announced in 1992. The mechanism for tariff reduction, which is necessary for the formation of a free trade area, is the so-called Common Effective Preferential Tariff (CEPT) Scheme. The failures experienced in PTA served as important guiding rules for the ASEAN member states in formulating AFTA. Yet, the framework for the implementation of AFTA received criticisms from many experts; the ASEAN members have thus continuously elaborated and further developed the principles guiding the AFTA Agreement, specifically the CEPT Scheme, since its inception in 1992.

The success and viability of an economic integration in the form of a free trade area in the ASEAN region have been doubted if not denied by several authors in the past on the basis of several premises. Among which are the following: (a) the diversity of resources, histories, culture and legal traditions; (b) the differences in the level and pace of economic and industrial development; and (c) the lack of an appropriate institutional framework. Partly due to this pessimism, the ASEAN member states opted to exclude primary agricultural products in the CEPT product coverage in 1992. Teh (1992) mentioned important reasons for excluding agricultural products in the 1992 CEPT agreement. Firstly, primary agricultural products are politically sensitive for most ASEAN member states for national security reasons. Secondly,

ASEAN agriculture is competitive rather than complementary due to similarity in natural endowments and climatic conditions. This is believed to be an impediment to greater specialization and trade in agriculture. Lastly, there is a strong presumption that scale effects are of little importance to agriculture. Notwithstanding the negative views raised regarding the inclusion of primary agricultural products in AFTA, the ASEAN members decided to include these products in the CEPT Scheme in 1995.

It remains a fact that the bulk of the ASEAN population is still engaged in agriculture and agriculture-related employment, despite a quite impressive development in the industry sector in the ASEAN region. Moreover, many primary agricultural commodities continue to hold a substantial share in ASEAN trade. Among the primary agricultural commodities, rice and corn continue to play important roles in the ASEAN agricultural sector as major staple food and feed commodities. Recognizing this fact, an important question is whether the creation of a free trade area in the ASEAN region will bring about the desired effect, namely to bring about growth in these commodity sectors. This paper thus concentrates on the analysis of the likely impact of AFTA on rice and corn industries in ASEAN countries through application of a Spatial Price Equilibrium (SPE) model. This model determines the changes in demand and supply quantities of these two primary agricultural products, as well as the changes in prices. The spatial characteristic of the model allows not only the quantities of imports and exports to be endogenously derived, but also their direction. One can thus draw from the results of the SPE model the changes in intra-ASEAN and extra-ASEAN trade as a result of the implementation of AFTA.

## 2 THE ASEAN FREE TRADE AREA AND THE STRATEGY OF IMPLEMENTATION

### 2.1 THE COMMON EFFECTIVE PREFERENTIAL TARIFF SCHEME

The CEPT Scheme provides the procedure for trade liberalization in the ASEAN region. The 1992 CEPT framework envisaged zero to five percent tariff rates within 15 years, or in year 2008. The 1992 CEPT Agreement, however, indicates a very general procedure for implementing AFTA and lacks details as to the specific product coverage. Despite the scantiness of information contained in the 1992 CEPT Agreement, it served as a working guideline for the succeeding negotiations on trade liberalization and the turning point for the ASEAN member countries to open their economies to a broader spectrum of economic integration.

According to the 1992 CEPT Scheme, the tariff reduction will apply to all manufactured products, including capital goods and processed agricultural products.[2] Primary agricultural products will be excluded from the CEPT Scheme. This scheme also categorized the products into two groups, namely products falling under the "Fast Track" program and those under the "Normal Track" program. There were 16 products selected for the "Fast Track" program (ASEAN Secretariat, 1996). The rest of the products in the 1992 CEPT thus fall under the "Normal Track" program. Figure 13.1 shows the difference in the tariff reduction schedule of the two product groups (hint: refer to the 1992 old schedule). Products falling under the "Normal Track" program are expected to reach 0-5 percent tariff levels by the year 2008, while those under the "Fast Track" program are expected to reach this in

2003, hence 5 years earlier than the former program. The schedule of tariff reductions depends not only on the type of reduction program but also on the level of tariff during the initial implementation of the CEPT Scheme. Both the reduction programs (Fast Track and Normal Track) and the stages of reduction (i.e. initial tariff levels) are not binding in the sense that the member countries are allowed to accelerate the reduction of their tariff rates faster than the 1992 CEPT Agreement.[3] Moreover, the member countries can opt to include the products initially under the exclusion lists in the CEPT Scheme.

The period from 1992-1994 saw further developments in regional and world trade, i.e. the plan of creating free trade area in Asia Pacific region and the conclusion of the Uruguay Round. In view of these changes, a Protocol to amend the "Framework Agreement on Enhancing ASEAN Economic Cooperation" was signed during the Fifth Summit Meeting in Bangkok in 1995, which aims to accelerate the implementation of AFTA to ten years. This means that AFTA tariff level of 0-5 percent is to be achieved in 2003 instead of 2008. The new tariff reduction schedule based on the Fifth ASEAN Summit is likewise shown in Figure 13.1 (hint: refer to the 1995 new schedule). Based on this figure, tariff reduction schedules can be summarized as follows:

Fast Track program

- products with tariff rates above 20 percent must be reduced to 20 percent by 1998 and reach 0-5 percent tariff target by 2000; and
- products with tariff rates of 20 percent and below must go down to 0-5 percent by 1998.

Normal Track program

- products with tariff rates above 20 percent must reach 20 percent by 2000; and
- products with tariff rates of 20 percent or below must go down to 0-5 percent by 2003.

The above tariff reduction schedule was used as the basis of the simulation scenarios in this study as it reflects the most recent developments in the AFTA Agreement.

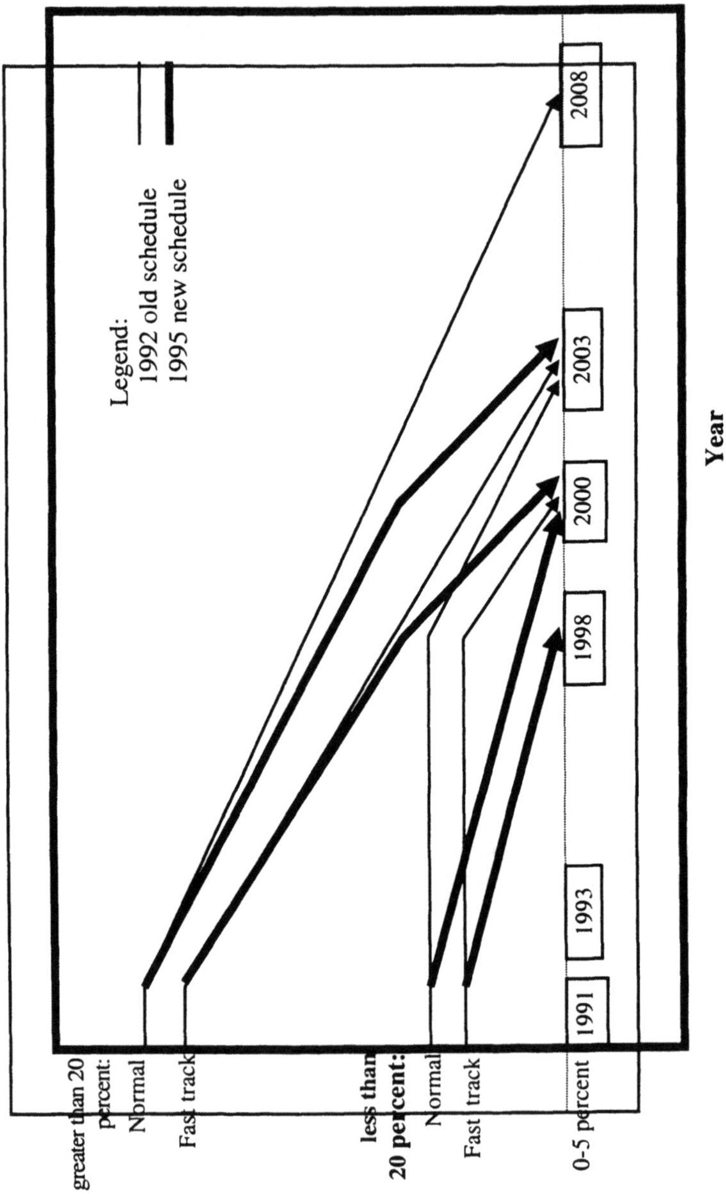


*Figure 13.1 Comparison of Schedule of Tariff Reductions between the Original (1992) and Modified (1995) CEPT Scheme.*
*Source: Author's own presentation.*

## 2.2 THE INCLUSION OF AGRICULTURAL COMMODITIES IN THE CEPT SCHEME

In addition to the acceleration of the tariff reduction schedule, the product coverage of CEPT was modified. The 1995 CEPT Scheme covers not only the manufactured and processed agricultural products, but also the primary agricultural products or unprocessed agricultural products (UAPs). This modification in the CEPT product coverage has proved the volition of the ASEAN member countries to finally open their markets to regional competition.

The 1995 CEPT Scheme differentiates the UAPs into three categories, i.e. Immediate Inclusion List, Temporary Exclusion List and Sensitive List (ASEAN Secretariat, 1996). The UAPs in the first category, Immediate Inclusion List, were included in the CEPT Scheme on 1 January 1996. These products have already begun the process of tariff reduction, so that by 2003, the tariffs on these products will be within the 0-5 % range. The UAPs in this category thus follow the 1995 new schedule of tariff reduction (Figure 13.1). Member countries are also required to eliminate quantitative restrictions and other non-tariff barriers on these products. The UAPs in the second category, Temporary Exclusion List, can be kept out of the CEPT Scheme only for a limited time. Each member country was required to phase in, at equal installments, those unprocessed agricultural products in their Temporary Exclusion List in January 1997. All the products in the Temporary Exclusion List should be in the CEPT Scheme by 1 January 2003. This means that these products should have tariffs at 0-5 percent in 2003, and the quantitative restrictions and other non-tariff barriers should have been removed.

The third category of the UAPs is the Sensitive List and the products included in this List are treated differently. A Special Arrangement with features that need not be the same as the CEPT will be created for these products. But even for the sensitive UAPs, the ASEAN member countries have agreed that there is a deadline by which these products have to be phased into the CEPT Scheme. It was agreed upon that the deadline for the inclusion of the sensitive UAPs in the CEPT Scheme is on 1 January 2010. Hence, the special arrangement for the products in the Sensitive List represents a transitional and not a permanent special treatment of UAPs.

There are around 2,025 tariff lines of UAPs in the CEPT Scheme as of 1994 (ASEAN Secretariat, 1996). Bulk of the UAPs in the 1994 CEPT Scheme falls under the Immediate Inclusion List, 68 percent. Around 19 percent of the total tariff lines of the UAPs are in the Temporary Exclusion List. Since the UAPs in this second category will be phased into the CEPT Scheme in 1997, approximately 87 percent of the total UAPs will be having tariffs at 0-5 percent in 2003. The remaining 13 percent of the UAPs are in the Sensitive List.

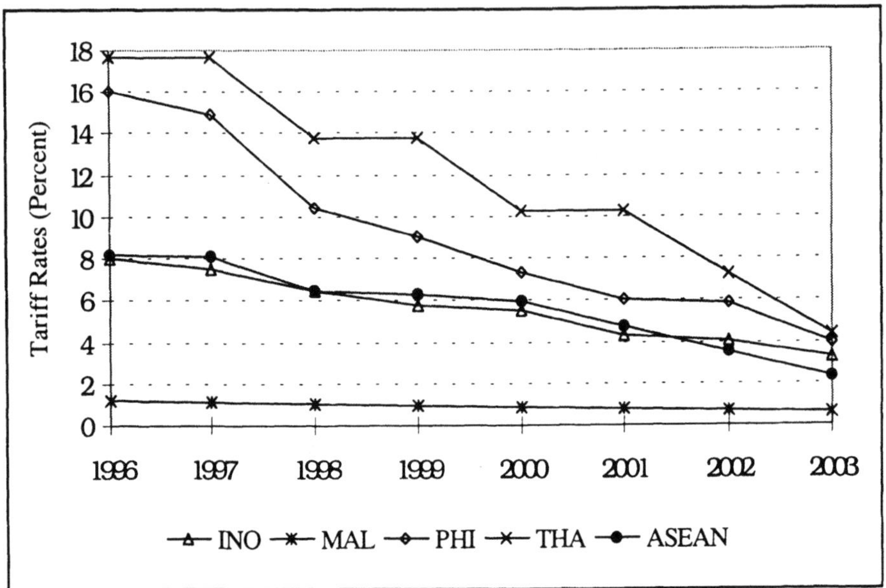


*Figure 13.2 Average Tariff Rates of Unprocessed Agricultural Products in the Immediate Inclusion List of ASEAN Countries, 1996-2003. For codes see Table 13.1.*
*Source: ASEAN Secretariat, 1996.*

The average tariff rate of the UAPs in the Immediate Inclusion List is given in Figure 13.2. Thailand and the Philippines have obviously the highest initial level of tariffs for the UAPs, approximately 18 and 16 percent, respectively. These countries are expected to reduce the tariff levels at a higher rate per year as compared to ASEAN countries with lower level of tariffs in 1996 (e.g. Malaysia). The average tariff level of the UAPs in Indonesia approaches the average tariff level in the ASEAN region. For the whole region, the average tariff rate for UAPs was about 8 percent in 1996 and is expected to be down to 2.1 percent in 2003.

## 3 DESCRIPTION OF THE MODEL CONCEPT AND METHODOLOGY

### 3.1 THE SPATIAL PRICE EQUILIBRIUM (SPE) MODEL[4]

The economic analysis that was employed to meet the objectives of the study is a Spatial Price Equilibrium (SPE) model. The theory underlying this model was pioneered by Enke (1951) and its mathematical solution was first suggested by Samuelson (1952). Takayama and Judge (1971) extended Samuelson's "net social payoff" (NSP) to include transportation activities. They used an empirically oriented "quasi-welfare" quadratic function to generate a spatial structure of prices, production, consumption and allocation for all commodities. The attributes of the SPE model applied in this study are partial equilibrium and comparative-static.

To understand the theoretical concepts of the spatial price equilibrium model, consider a case of two-region and one-commodity trade as graphically depicted in Figure 13.3. The basic assumptions for a competitive market in micro-economic principles and for a free trade in international trade theory remain valid in the subsequent analysis, e.g. homogeneity of commodities across regions, rational production and consumption decisions, etc. In addition, it is assumed that each region is a single and distinct market, separated by unit transportation cost. Supposed $\mathbf{D}_1^{t_0}$, $\mathbf{S}_1^{t_0}$ and $\mathbf{D}_2^{t_0}$, $\mathbf{S}_2^{t_0}$ are the regional demand and supply functions of regions 1 and 2, respectively, in a particular time period $\mathbf{t_0}$. Without commodity exchange, the equilibrium condition in the two regions is where the supply and demand functions intersect, namely $\mathbf{E}_1^{t_0}$ and $\mathbf{E}_2^{t_0}$. At these initial equilibrium conditions, the price $\mathbf{P_1}$ of the commodity in region 1 is much lower than the price $\mathbf{P_2}$ in region 2.

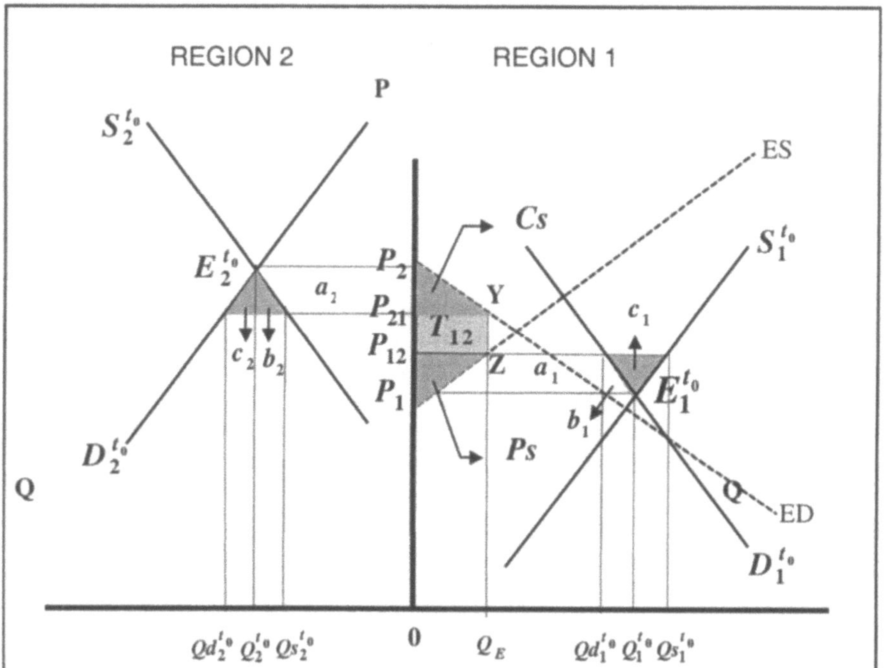


*Figure 13.3 Spatial Price Equilibrium Model for Two-Regions and One-Commodity. Source: Author's own presentation based on literature on international trade.*

Having assumed the absence of trade barriers, the price difference between these two regions gives a signal for a commodity exchange. The movement of commodity from one region to another entails some costs, generally referred to as transfer costs ($\mathbf{T_{12}}$). An important prerequisite for a commodity exchange to occur is that the difference between the trade prices $\mathbf{P_{21}}$ and $\mathbf{P_{12}}$ equates to the transfer costs $\mathbf{T_{12}}$. If the margin between the trade prices does not match the costs to be incurred in moving the commodity, then there exists no commodity exchange despite the price difference between the two regions. Assuming the existence of an appropriate price margin, trade is likely to result to new equilibrium conditions in both regions, to wit:

Region 1:

- increase in quantity supply from $Q_1^{t_0}$ to $Qs_1^{t_0}$ due to increase demand for export at higher price;
- decrease in quantity demand from $Q_1^{t_0}$ to $Qd_1^{t_0}$ since the exportation of the product decreases the domestic supply and, hence, increases price;
- a surplus or excess supply of the amount $Qs_1^{t_0}$-$Qd_1^{t_0}$ represents the quantity exported at price $P_{12}$.

Region 2:

- increase in quantity demand from $Q_2^{t_0}$ to $Qd_2^{t_0}$ due to availability of the imported commodity at a lower price;
- decrease in quantity supply from $Q_2^{t_0}$ to $Qs_2^{t_0}$ since inefficient producers cannot compete with the low-price imported commodity;
- a deficit supply or excess demand of the amount $Qd_2^{t_0}$ -$Qs_2^{t_0}$ represents the quantity imported at price $P_{21}$.

The gains from trade resulting from the shifts in the initial equilibrium conditions can be depicted from the geometric model in Figure 13.3. A drop in price from $P_2$ to $P_{21}$ enables the consumers in region 2 to gain the areas $a_2 + b_2 + c_2$. The net gain, however, is equivalent only to the areas $b_2 + c_2$ since $a_2$ represents producers′ loss from the reduction in price and quantity supply. On the contrary, producers in region 1 gain the areas $a_1 + b_1 + c_1$ due to price increase from $P_1$ to $P_{12}$. Note, however, that the consumers lose the areas $a_1 + b_1$ yielding only a net gain of $c_1$ for the region. The areas $a_1$ and $b_1$ represent the loss due to increase in price and decrease in demand. The sum of the areas $c_2 + b_2$ (consumer surplus) in region 2 and area $c_1$ (producer surplus) in region 1 are defined as the net social payoff or, equivalently, the net welfare. The sum of these "surpluses" is the object to be maximized in the SPE model.

## 3.2 PROGRAM FLOW OF THE SPE MODEL

To provide a general overview on the steps that were adopted in running the SPE model, the program flow of the model is presented in Figure 13.4. It can be seen from the figure that the core of the SPE model is a net quasi-welfare function. The program flow shows that four groups of exogenous variables are needed as inputs to the SPE model, namely demand and supply functions, transfer costs and tariff rates. The nature of the problem that has been dealt with in this study is a comparative-static.[5] It deals with comparison of different equilibrium states brought about by the changes in the parameters and exogenous variables of the model. The initial equilibrium state is compared with the final equilibrium. The different states of equilibrium considered in this study are the pre-AFTA (initial equilibrium state) and post-AFTA periods (the final equilibrium state), the former refers to the base year period (1992) and the latter refers to the implementation of AFTA in year 2002. The post-AFTA periods include both the reference run and the AFTA scenarios.

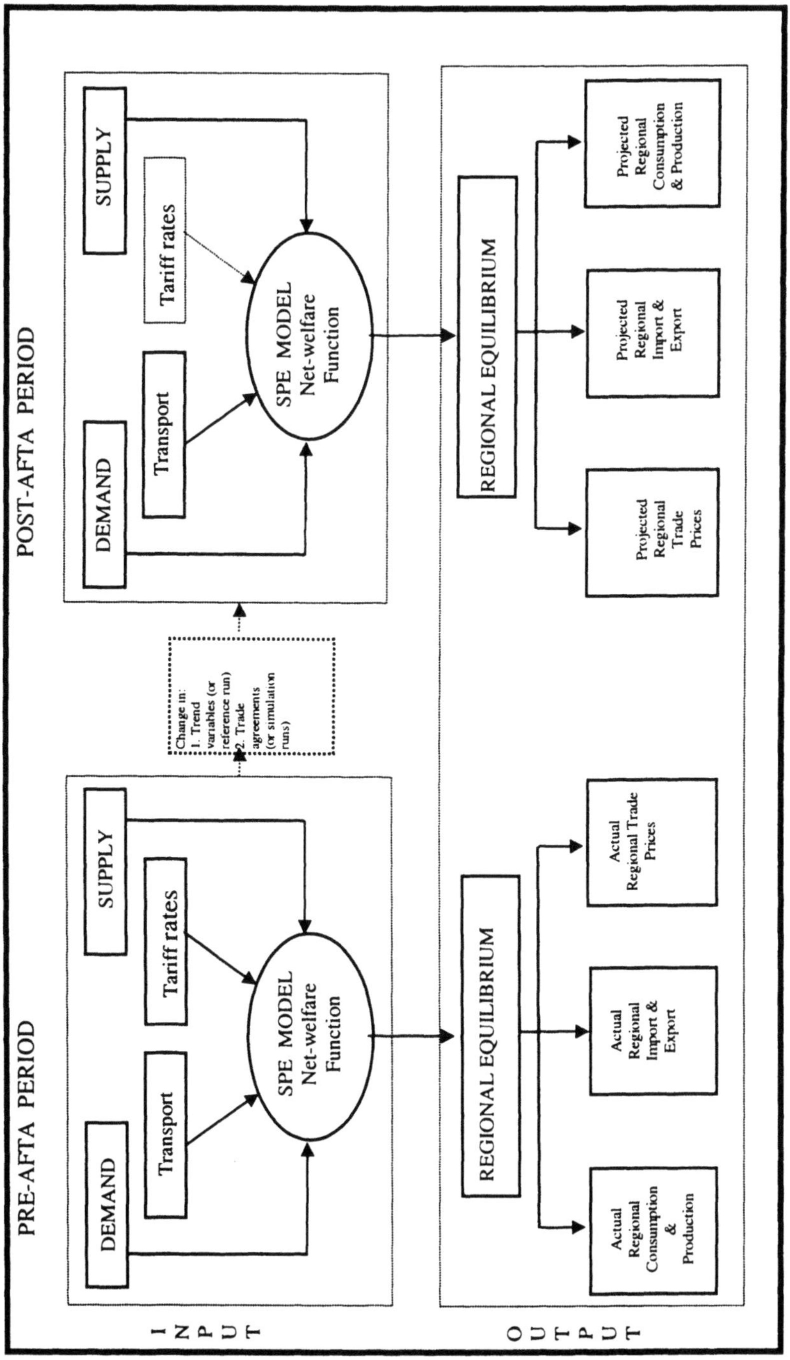


*Figure 13.4 Program Flow of the Spatial Price Equilibrium (SPE) Model*
*Source: Author's own presentation.*

In between the two boxes corresponding to the two scenario cases is a box in broken lines. It contains the list of factors that will cause changes in the group of variables that enter the SPE model in the post-AFTA period. The first aspect to be considered for the post-AFTA period is the effect of the changes in trend variables (e.g., increase in population and income in case of the demand functions; technology advancement and hectarage expansion in case of supply function, etc.) up to the year 2002 and the corresponding SPE model will be referred to as "reference run".

The second aspect is the incorporation of the free trade agreements among ASEAN countries in the SPE model, which will be referred to as "simulation runs". In these simulation scenarios, the tariffs will thus be reduced or phased out within the ASEAN region but not against the non-ASEAN countries. The results of the simulation runs are compared to the results of the reference run in order to quantify the effect of the free trade agreements.

The SPE model analysis will give emphasis on the five major agricultural countries in the ASEAN region (Table 13.1)[6]. Since Japan, European Union and the United States are ASEAN's major trading partners, these countries were also included in the model analysis. The rest of the countries were aggregated as one country and will be referred to as ROW. While non-ASEAN countries were included in running the SPE model to derive a world trade equilibrium, only the results for ASEAN countries are discussed in details in this paper.

*Table 13.1. Countries and the Corresponding Codes used in the SPE Model.*

| ASEAN Countries | Codes | Non-ASEAN Countries | Codes |
|---|---|---|---|
| Indonesia | INO | Japan | JAP |
| Malaysia | MAL | European Union | EEC |
| Philippines | PHI | United States | USA |
| Thailand | THA | Rest of the World | ROW |
| Vietnam | VIE | | |

# 4 APPLICATION OF THE SPE MODEL

## 4.1 DATA USED FOR THE EMPIRICAL ANALYSIS

The Institute of Agricultural Policy at the University of Bonn, Germany has developed a comprehensive data bank system which provided many of the data necessary for building the model for the ASEAN region. Among others, the data bank includes economic variables such as production, consumption, export, import, yield coefficients, etc., covering the period 1970-1992. These data are available for a large number of countries and agricultural products. The data were taken from FAO's Supply and Utilization Accounts (SUA's) and Trade Statistics. For the 1992 base year run, variables like the quantity of supply and demand and the trade prices were obtained from the data bank. Another important input to the SPE model are the slope coefficients. The slope coefficients were derived from the price elasticities of demand and supply. There is ample literature that provides elasticity data for the ten agricultural products considered in this study. Mueller (1995) collected and calibrated elasticities of selected agricultural products for the ASEAN countries. These elasticities were used in the derivation of the slope coefficients for Indonesia, Malaysia, the Philippines and Thailand. For the remaining countries like Vietnam, EU, Japan and the United States, the elasticities were obtained from the data bank of the Institute of Agricultural Policy.

The trend factors were also derived from the data bank of the Institute in Bonn. The Institute utilized time-series regression analysis to compute these trend factors. If the trend factors for particular variables were not available in the data bank, additional time-series analyses were undertaken in this study to fill the data gap. Some of the data relating to domestic policies (i.e. PSEs and CSEs) in ASEAN were lifted from Mueller (1995), while those for Japan, EU and the US were taken from the Institute's data bank. Data relating to trade policies of the ASEAN countries were obtained from various issues of WTO/GATT publications. Very crucial parameters in the computation of a spatial model are the transportation costs data. However, there was no comprehensive secondary data available for this particular variable. Thus efforts were made to collect transportation costs from primary sources (e.g. shipping industries, export-import companies). Due to scantiness of the transportation costs gathered from the shipping companies, a regression analysis was undertaken to fill in the missing data.

## 4.2 THE BASE YEAR AND REFERENCE RUN

The base year is the initial state where the ASEAN countries trade with each other and with their major trading partners in the presence of trade restrictions. In short, prior to the free trade agreement among ASEAN countries or pre-AFTA. The purpose of the "base year run" is to generate trade prices, quantities of supply and demand, and volume of trade flows that approximate the actual data in 1992. The net quasi-welfare (NW) function for the base year run was computed from the following equation [1]:

$$\text{Max. } NW^{sim}_{i,c}(P^{end}_{i,c}, P^{end}_{j,c}) = \sum_{i,j=1}^{n} \sum_{c=1}^{m} \begin{bmatrix} \left(\alpha^{sim}_{i,c} P^{end}_{i,c} - (\tfrac{1}{2} P^{end}_{i,c} \beta^{sim}_{i,c} P^{end}_{i,c} + \beta^{sim}_{i,c} CP^{sim}_{i,c} P^{end}_{i,c})\right) \\ - \left(\theta^{sim}_{r,c} P^{end}_{j,c} + (\tfrac{1}{2} P^{end}_{j,c} \gamma^{sim}_{j,c} P^{end}_{j,c} + \gamma^{sim}_{j,c} PP^{sim}_{j,c} P^{end}_{j,c}\right) \\ - \left(Q^{sim}_{q,c} P^{end}_{q,c}\right) \end{bmatrix}$$

$$\text{s.t. } G^{sim}_{i,c} P^{end}_{i,c} - G^{sim}_{j,c} P^{end}_{j,c} - G^{sim}_{q,c} P^{end}_{q,c} \geq t^{sim}_{ij,c} + ST^{sim}_{ij,c} + AT^{sim}_{ij,c}$$

$$P^{end}_{i,c}, P^{end}_{j,c}, P^{end}_{q,c} \geq 0$$

where:

$\mathbf{P^{end}_{i,c}}$ and $\mathbf{P^{end}_{j,c}}$ are the endogenous trade prices, respectively, such that $P_{i,c} = P_{j,c}$ if $i = j$

$\alpha^{sim}_{i,c}$ and $\theta^{sim}_{j,c}$ are intercepts of the demand and supply functions, respectively, such that $\alpha_i > 0$, $\theta_j \geq 0$ or $\leq 0$

$\beta_{i,c}^{sim}$ and $\gamma_{j,c}^{sim}$ are slopes of coefficients of the demand and supply functions, respectively, such that $\beta_i, \gamma_j > 0$.

$\mathbf{Q}_{q,c}^{sim}$ and $\mathbf{P}_{q,c}^{end}$ refer to the quantity of import quota and the endogenous price associated to this quota, respectively, in region $q$ (where $q \in i,j$)

$\mathbf{ST}_{ij,c}^{sim}$ and $\mathbf{AT}_{ij,c}^{sim}$ are the specific and ad valorem tariffs, respectively[7]

$\mathbf{PP}_{j,c}^{sim}$ and $\mathbf{CP}_{i,c}^{sim}$ are the per unit producer and consumer incentive prices, respectively $i$ and $j$ correspond to the importing and exporting countries, respectively $c$ refers to the agricultural commodities

There are several factors that can affect the developments of production (i.e. prices of the product and related products, prices of inputs, technological advancement, etc.) and consumption (i.e. prices of the product and the related products, increase in income and population, etc.) over time. These factors, except for prices of the products and related products, were assumed to be constant for the base year estimation. Over the long run, however, these factors will change and affect the initial equilibrium conditions of supply and demand. The development trends in area and yield will shift the level of production, while trends in population and income will shift the level of consumption. The reference run is the equilibrium state where the effects of trend variables are taken into consideration. Since only the effects of the trend variables are taken into account in the reference run, the values of the exogenous variables like quota ($\mathbf{Q}_{q,c}^{sim}$), transportation costs ($\mathbf{t}_{ij,c}^{sim}$), and tariffs ($\mathbf{ST}_{ij,c}^{sim}$ and $\mathbf{AT}_{ij,c}^{sim}$) are similar to the base year run. The difference between the base year and the reference run is reflected in the intercepts of the quantity of demand $\alpha_{i,c}^{sim}$ and the quantity of supply $\theta_{j,c}^{sim}$. The intercept parameters were derived from the following equation for the reference run:

[2]
$$\alpha_{i,c}^{sim} = y_{i,c}^{ref} + \beta_{i,c}^{ref}\left(\mathbf{P}_{i,c}^{ref} + \mathbf{CP}_{i,c}^{ref}\right)$$
$$\theta_{j,c}^{simf} = x_{j,c}^{ref} - \gamma_{j,c}^{ref}\left(\mathbf{P}_{j,c}^{ref} + \mathbf{PP}_{j,c}^{ref}\right)$$

It is assumed that the domestic policy supports remain unchanged during the reference run. The market prices are also assumed to be constant. The terms $y_{i,c}^{ref}$ and $x_{j,c}^{ref}$ reflect the increase in the quantities of demand and supply due to shifting of the demand and supply curves as a result of the changes in trend variables.[8]

The results of the reference run show that Vietnam and Indonesia will experience the largest expansion in supply of rice, approximately 25 and 30 percent respectively, in year 2002 (Figure 13.5). The Philippines is also expected to show significant increase in rice production at more then 20 percent. The projected increase in rice production in Thailand is 10 percent lower than in Vietnam, which implies that the latter country will continue to overtake the former in rice production in the next ten years. Although Vietnam will register the highest growth of rice production, Indonesia will remain the largest producer of rice in the region. As for rice demand, the largest increase will be seen in Vietnam (28 percent), followed by the

Philippines and Thailand. The rates of increase in rice demand in these three ASEAN countries will tend to match the rate of increase in rice supply. Meanwhile, the percentage increase in rice demand in Malaysia will be much higher than the percentage increase in rice supply. Indonesia is the only ASEAN country whose percentage increase in rice demand is projected to significantly lag behind the percentage increase in rice supply. Nonetheless, having the largest population in ASEAN, Indonesia will continue to dominate the region in terms of volume of rice consumption. With respect to the projected changes in price of rice, the Philippines will show the highest percentage increase at more than 10 percent. Vietnam will experience the lowest increase in price of rice, which will contribute to the very high percentage increase in rice demand in this country.

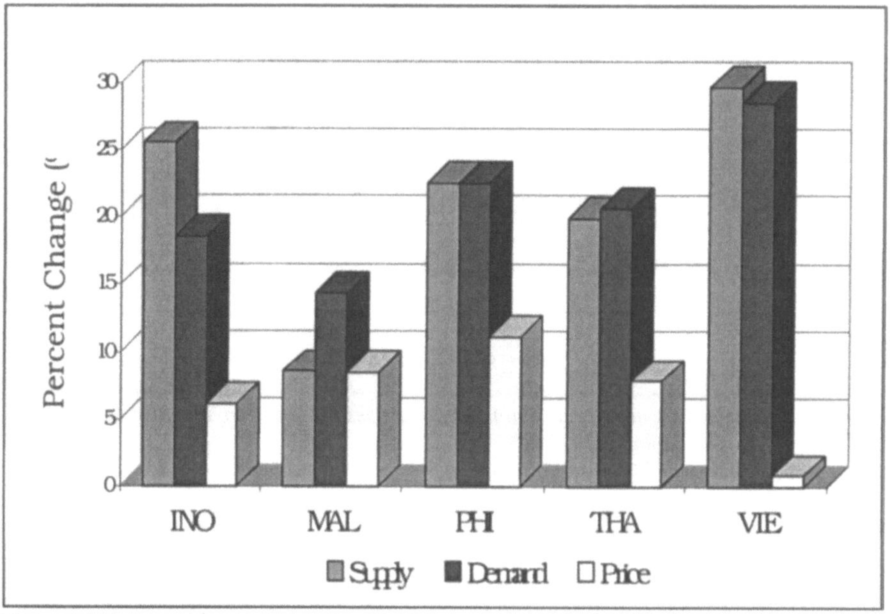


*Figure 13.5 Projected Changes in Quantities and Prices of Rice from the Base Year Run, by ASEAN Country, 1992-2002*
*Source: Results computed from the SPE model.*

As for corn, Malaysia and the Philippines will have the largest percentage increase in supply at more than 30 percent from 1992 to 2002 (Figure 13.6). The percentage increase in the demand for corn in the Philippines will tend to catch up with the percentage increase in supply. On the contrary, the percentage increase in the demand for corn in Malaysia will tend to lag behind the percentage increase in supply. Despite more than 35 percent increase in corn supply, however, Malaysia will continue to produce the smallest volume of corn in the region. Like in Malaysia, the percentage increase in demand for corn in Indonesia and Thailand will also be lower than the percentage increase in supply. But unlike in Malaysia, the volume of corn production in Indonesia and Thailand will be much higher than consumption, making these countries significant exporters of corn in year 2002. Indonesia, Malaysia, and Thailand are all projected to experience a fall in the price of corn. Vietnam will tend to show the largest expansion in the demand for corn in year

2002, about 35 percent. Although the percentage increase in demand for corn in Vietnam will be larger than in the Philippines, the percentage increase in the price of corn will be relatively higher in the latter than in the former country. This is because the absolute volume of corn supply in Vietnam will remain larger than corn demand.

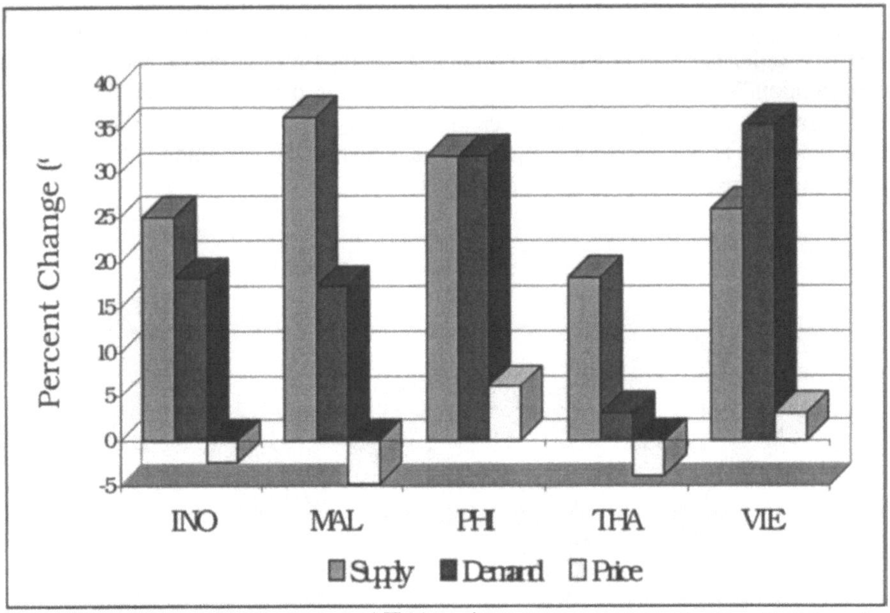


*Figure 13.6 Projected Changes in Quantities and Prices of Corn from the Base Year Run, by ASEAN Country, 1992-2002*
*Source: Results computed from the SPE model.*

**The AFTA Scenarios**

The simulation runs for the AFTA scenarios attempt to measure the changes in the quantities, prices and trade flows if the trade barriers within the ASEAN are reduced or abolished through the AFTA Agreement. The exogenous variables in the simulation runs are similar to the reference run with the exception of the trade-related policies like import quota and import tariffs. There are two simulation runs undertaken in this study; the first simulation reduced the tariffs $\mathbf{ST_{ij,c}^{sim}}$ and $\mathbf{AT_{ij,c}^{sim}}$ within the ASEAN region based on the 1995 modified CEPT schedule and the second simulation assumed that the tariffs $\mathbf{ST_{ij,c}^{sim}}$ and $\mathbf{AT_{ij,c}^{sim}}$ are totally abolished in year 2002. The first simulation (the partial abolition of tariffs) will be referred to as "*AFTA run1*" and the second simulation (the total abolition of tariffs) as "*AFTA run2*".

For AFTA run1, the tariffs applied for the base and reference runs were reduced according to the computed percentage reduction in the average tariff level of UAPs (see Figure 13.2). Alternatively, the reduction formula of CEPT were applied if the tariff levels remained too high. This second reduction procedure is based on the formula: $\frac{(\mathbf{X}-20)}{5}$ per annual reduction where $\mathbf{X}$ is the initial tariff level in 1992.

This formula was used if the initial tariff level of the agricultural product is higher than 20 percent in 1992. As a late comer to the AFTA Agreement, Vietnam was given special treatment in terms of tariff reduction. The tariff rates of Vietnam for the agricultural products in this scenario were thus assumed to be similar to the reference run. For the rest of the ASEAN countries, rice and corn are assumed to fall under the Immediate Inclusion List or Temporary Exclusion List, unless submitted for inclusion in the Sensitive List. The average tariff levels of rice and corn in the ASEAN region will be around 10 and 5 percent, respectively, in year 2002 (Acosta, 1998). The average tariff level for rice remained high in the ASEAN region because the Philippines included this product in the Sensitive List. The Philippines is also claiming for the inclusion of corn in the Sensitive List, hence a relatively high tariff rate for corn (12.71 percent) was assumed for this country.

In the second scenario, AFTA run2, it was assumed that all the ASEAN countries have agreed to bring down the tariff levels of the agricultural products in the region to zero percent in year 2002, while maintaining the existing tariff levels against non-ASEAN countries. The aim of this scenario is to analyze the probable effects of introducing a complete liberalization of agricultural trade in 10 years period. This assumption of zero tariff level of the agricultural products in year 2002 may not be very unrealistic considering that the ASEAN countries are continuously improving and accelerating the implementation of AFTA. It is assumed in AFTA run2 that Vietnam has agreed to be integrated in the CEPT Scheme, devoid of any special treatment. It was also assumed in this scenario that the Philippines agreed to exclude rice and corn from the Sensitive List.

**Changes in Quantities and Prices:**

The results of the AFTA scenarios for the supply and demand of rice are shown in Figure 13.7. The Philippines is expected to experience the largest percentage decrease in the supply of rice both in AFTA run1 (1.3 percent) and AFTA run2 (2.5 percent). Recall that the Philippines included rice in the Sensitive List, which appears to be justified by the fact that the inclusion of this product in the CEPT in year 2002 will have a larger negative effect on the level of domestic production. If rice however is temporary excluded from the AFTA Agreement, the contraction in rice supply in the Philippines will be minimized since a relatively higher price will be maintained. Meanwhile, the demand for rice in the same country will drop if AFTA is partially implemented, but will increase significantly by 3.5 percent if AFTA is fully implemented. Thus, while total liberalization of rice will hurt the producers (i.e. small farmers) in the Philippines, the rice consumers will benefit from AFTA as shown by the high level of consumption. The significant expansion in the level of rice demand in the country will be encouraged by the substantial fall in the level of price of rice by about 5 percent, the largest percentage decrease in price of rice in the region (Figure 13.8).

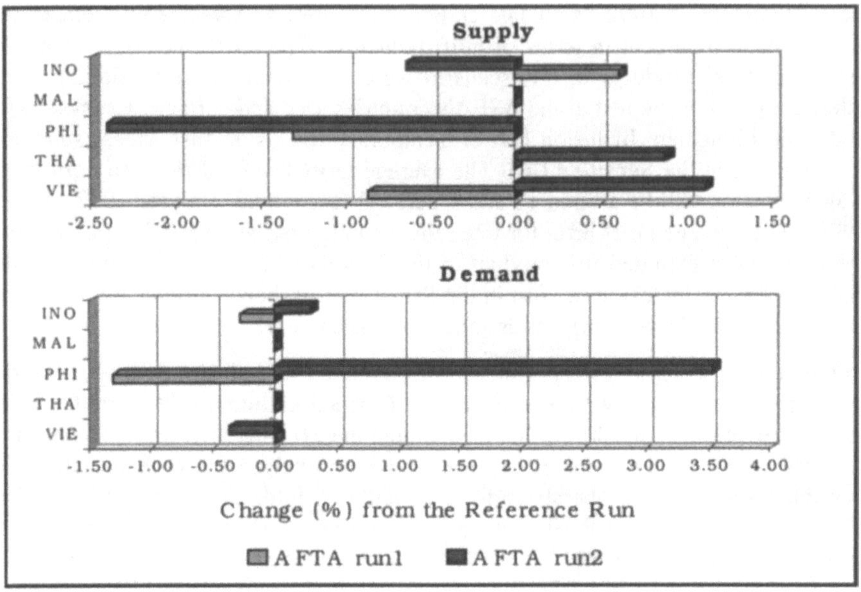


*Figure 13.7 Relative Deviation of Rice Supply and Demand in AFTA Scenarios, by ASEAN Country*
*Source: Results computed from the SPE model.*

Vietnam is expected to experience the second largest percentage changes in quantity of rice supply due to AFTA implementation (Figure 13.7). The two AFTA scenarios will have, however, contradicting effects on rice supply in Vietnam: a partial implementation of AFTA is expected to decrease rice supply, while full implementation will increase rice supply. Note that Vietnam was assumed to retain its current tariff levels for the agricultural products in AFTA run1 scenario as a late comer to AFTA. The exclusion of Vietnam from the AFTA Agreement will thus have a negative impact on the level of rice production. Figure 13.8 shows that the price of rice in Vietnam in AFTA run1 scenario will slightly fall, which will tend to discourage domestic production of rice. The consumers will not be, however, very responsive to the drop in price of rice as shown by the very tiny increase in demand for the product. Meanwhile, the full integration of Vietnam in AFTA already in year 2002 will have positive impact on the level of rice production but a negative impact on the level of rice consumption, which will be attributed to the increase in price of rice by almost 1.5 percent in AFTA run2.

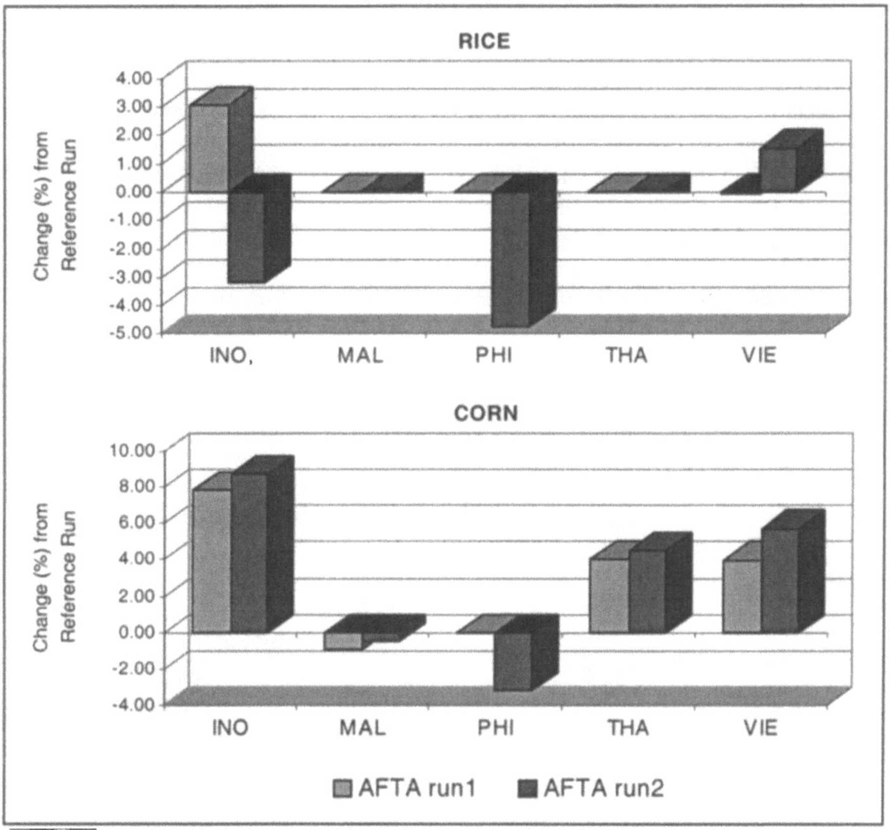


*Figure 13.8 Changes in Price from the Reference Run, by ASEAN Country*
*Source: Results computed from the SPE model.*

The partial implementation of AFTA in Indonesia will tend to increase the level of rice supply by about 0.60 percent (Figure 13.7). The model results show that Indonesia is the only ASEAN country that is expected to experience an increase in rice production in AFTA run1. This can be explained by the dramatic increase in price of rice in Indonesia in this scenario, while the prices in the rest of the ASEAN countries will remain stagnant or slightly decrease (Figure 13.8). In AFTA run2 rice supply in Indonesia is expected to contract by 0.70 percent, and rice demand to increase by 0.30 percent. The rice consumers will thus generally benefit from the full implementation of AFTA in the form of increased consumption, which will be stimulated by the drastic drop in price of rice. Finally, only full implementation of AFTA will bring about changes in the quantity of rice supply in Thailand, increasing by approximately 0.80 percent.

As compared to rice, AFTA implementation will have greater influence on the supply of and demand for corn in the ASEAN region (Figure 13.9). For instance, Indonesia is expected to experience about 3.5 percent and almost 4.0 percent increase in corn supply in AFTA run1 and AFTA run2, respectively. These figures are so far the highest percentage changes computed for the corn supply in the region. This impressive development in corn supply in Indonesia may be explained by the upward trend in price of corn. The price of corn will go up by at least 8 percent in both AFTA scenarios (Figure 13.8). Thailand and Vietnam are also expected to experience increase in corn supply in AFTA run1, i.e. between 2.0 and 2.5 percent. The percentage change in corn supply in Vietnam will increase by another 0.8 percent in AFTA run2. Indonesia, Vietnam and Thailand will exhibit declining trend in corn demand for both scenarios. Although the percentage increase in price of corn will be highest in Indonesia, the contraction in the demand for corn in this country will not be largest. Thailand will show the largest fall in corn demand, more than 3.5 percent in AFTA run1 and about 4.0 percent in AFTA run2 scenario.

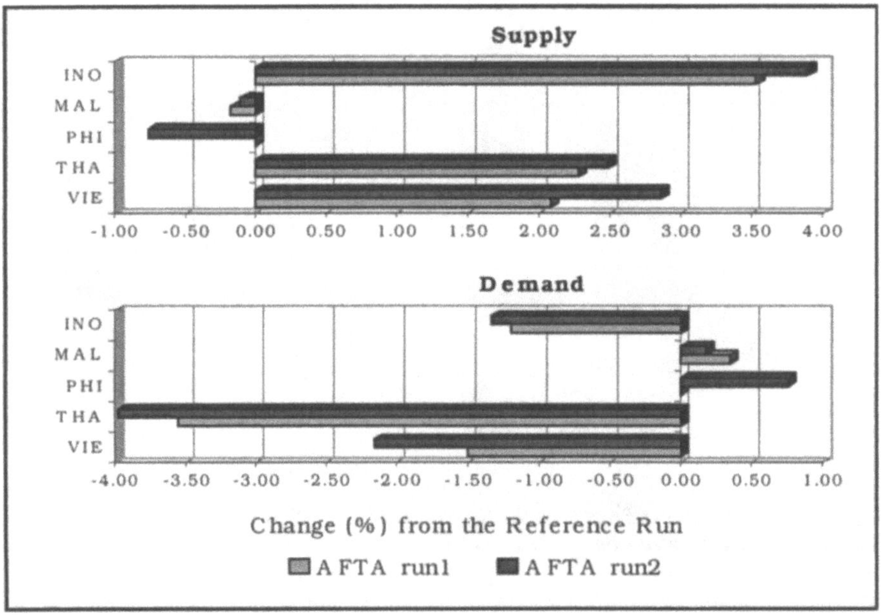


*Figure 13.9 Relative Deviation of Corn Supply and Demand in AFTA Scenarios, by ASEAN Country*
*Source: Results computed from the SPE model.*

The rest of the ASEAN countries, on the other hand, will face a downward trend in corn supply, although not very large at less than one percent (Figure 13.9). Worth mentioning, however, is that while partial implementation of AFTA will have no effect on the quantity of corn supply in the Philippines due to the relatively high tariff level of corn for this scenario, the introduction of zero tariff (AFTA run2 scenario) for this product will cause a 0.75 percent contraction in corn supply. The quantity of demand for corn in the Philippines in AFTA run2 will accordingly expand at almost the same rate. This development in supply of and demand for corn in the Philippines will be accompanied with a decline in domestic price of around 3 percent (Figure 13.8). The total liberalization of trade in corn will thus be to the disadvantage of the corn producers, but to the benefit of the corn consumers --- mainly livestock producers. The corn demand in Malaysia will also increase in

AFTA run2, but the percentage increase is not as large as in the Philippines. The rate of increase in demand for corn in Malaysia in AFTA run2 will be even smaller than that computed in AFTA run1 scenario.

**Direction of Trade Flows:**

Figure 13.10 shows the direction of trade flows of rice in the ASEAN region and the major trading partners in year 2002. Indonesia will remain the largest producer of rice in the region, and yet will not be a significant exporter due to large domestic demand for this product. Thailand and Vietnam will be the two major rice exporters in the region. In AFTA run1, Thailand will direct most of its rice exports to the rest of the world, 83 percent, and the remaining 17 percent to Malaysia. Meanwhile, Vietnam will export 94 percent of its surplus domestic production to Indonesia, and the remaining 6 percent to ROW. The volume of trade between the ASEAN countries will increase in AFTA run2. Indonesia will experience an expansion in the volume of imports due to the increase in the level of rice demand in this country, not to mention the drop in price of rice in ASEAN rice-exporting countries. Thailand will export rice to Indonesia in AFTA run2, amounting to 2 percent of total export volume. The volume of trade flows from Thailand to the Philippines will be larger at 8 percent. Nevertheless, Malaysia and ROW will remain the main destination of rice exports from Thailand.

The results of the SPE model show that some ASEAN countries like Indonesia, Thailand and Vietnam will improve the level of corn production in year 2002. Based on AFTA run1 scenario, all these three ASEAN countries will become exporters of corn in the region, specifically to Malaysia (Figure 13.11). The volume of corn exports from Indonesia to Malaysia will be the largest at 1175 thousand metric ton. The second AFTA scenario slightly altered the direction of corn trade flows in the ASEAN region. Vietnam will begin to export corn to the Philippines and will stop exporting to Malaysia. However, Indonesia and Thailand will continue to export to Malaysia. The corn exports from Thailand to Malaysia will increase by 26 thousand metric tons, while that from Indonesia will expand by 49 thousand metric tons. The increase in the volume of corn exports from Thailand and Indonesia will compensate for the cut in corn exports from Vietnam, hence the total corn imports of Malaysia will not significantly fall in AFTA run2 vis-à-vis AFTA run1.

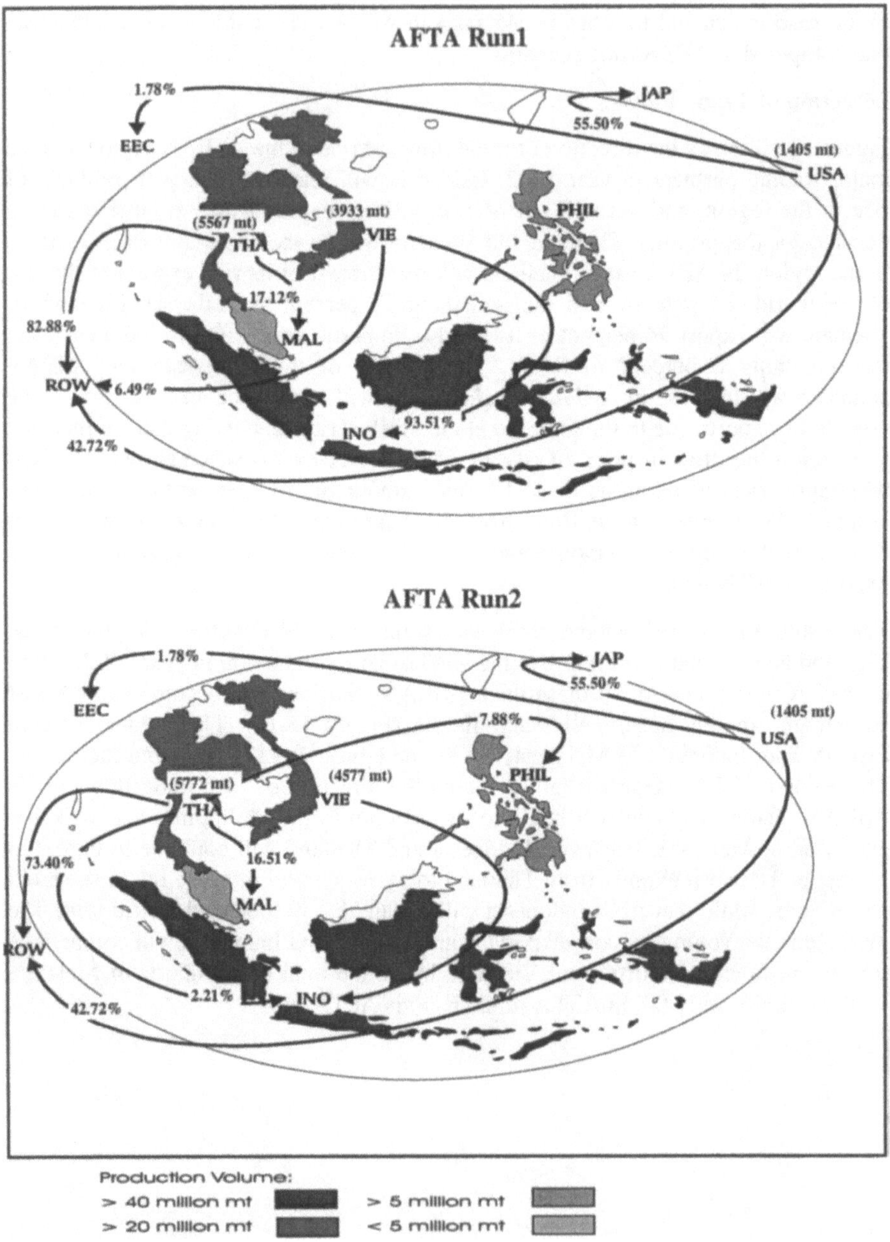


*Figure 13.10 Trade Flows of Rice in the ASEAN Countries and Major Trading Partners, by AFTA Scenarios.*

*Note: Trade quantities in parentheses are in thousand metric tons.*

*Source: Results computed from the SPE model.*

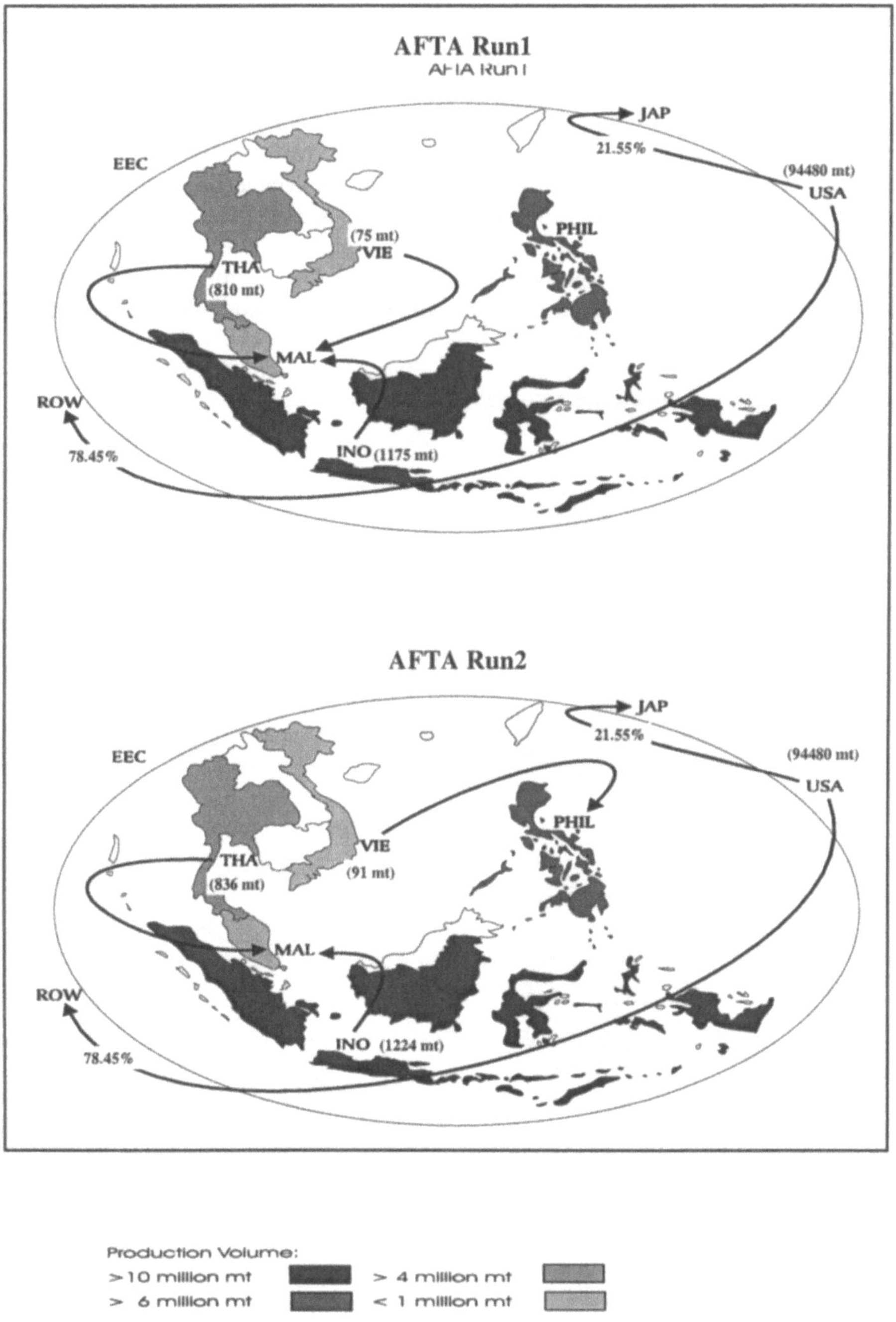


*Figure 13.11 Trade Flows of Corn in the ASEAN Countries and Major Trading Partners, by AFTA Scenarios*

*Note: Trade quantities in parentheses are in thousand metric tons*

*Source: Results computed from the SPE model*

**Change in Intra- and Extra-ASEAN Trade:**

Figure 13.12 reveals an important aspect of ASEAN trade which is not very obvious from Figures 13.10 and 13.11. According to the former figure, the partial implementation of AFTA will cause around 10 percent contraction in intra-ASEAN trade of rice. Rice exports to non-ASEAN countries (extra-ASEAN exports) will increase however by around 5 percent. The full implementation of AFTA will significantly boost the volume of intra-ASEAN trade in rice. Rice exports within the ASEAN region will increase by almost 20 percent, but will be accompanied with an 8 percent fall in extra-ASEAN exports. This implies that the total liberalization of rice trade in the ASEAN region will have some trade diversion effects. In terms of extra-ASEAN imports, both the partial and total implementation of AFTA will not affect the quantity of rice to be imported by the ASEAN countries from the major trading partners and the rest of the world.

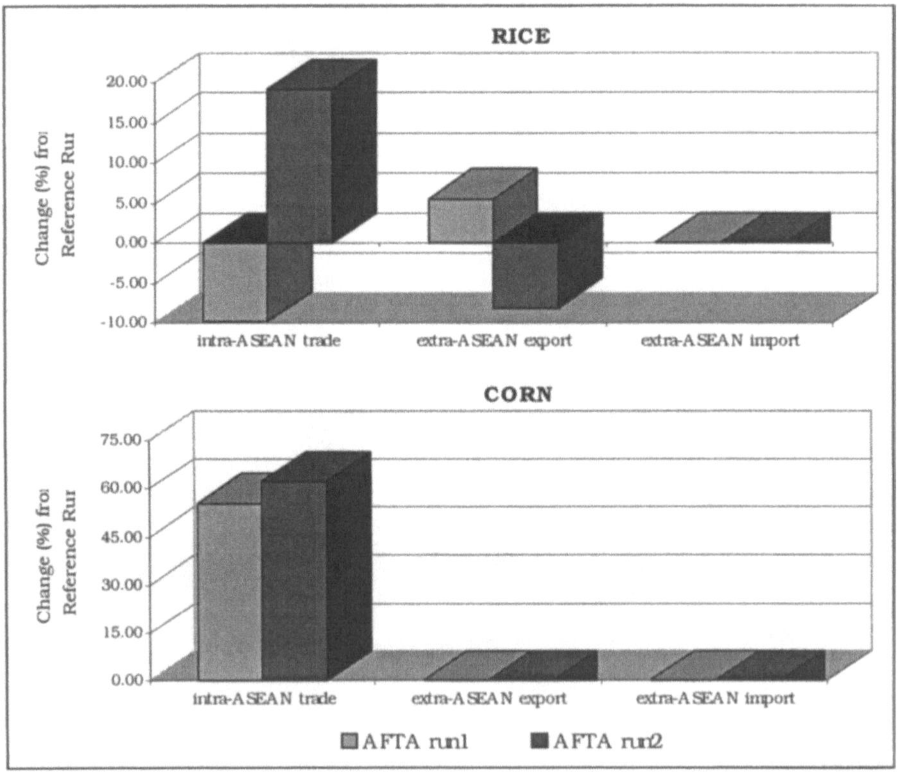


*Figure 13.12 Changes in Intra-ASEAN and Extra-ASEAN Trade of Rice and Corn, by AFTA Scenarios*
*Source: Results computed from the SPE model.*

As for the trade in corn, both partial and total implementation of AFTA will have a positive effect on intra-ASEAN trade (Figure 13.12). The results of the AFTA run1 scenario showed that there will be more than 50 percent increase in intra-ASEAN trade, while those of the AFTA run2 scenario indicated a growth of more than 60 percent. The impact of AFTA implementation on trade of corn within the region is thus more promising than that of rice. The volume of corn exports as well as corn imports of the ASEAN from the non-ASEAN countries will remain unaltered notwithstanding the reduction or abolition of corn tariffs within the region. As the volume of corn production within the region is expected to increase, the available corn supply within the region in year 2002 even in the absence of the AFTA agreement (i.e. reference run), both AFTA scenarios will not be expected to cause any diversion in corn trade. There will be bright prospects for trade creation from implementing AFTA in the ASEAN region.

## 5 CONCLUDING REMARKS

The positive impact of the AFTA agreement on the volume of production in the ASEAN region will generally be more promising for corn than for rice. Almost all the ASEAN countries will be expected to experience an increase in the level of corn production if the tariffs for this product are partially or totally eliminated within the region. Indonesia will reap the largest benefit from AFTA in terms of growth in corn production. Corn producers in the Philippines will stand to lose the most from full implementation of AFTA. The negative effects of ASEAN trade liberalization on the Philippine rice sector will be even greater, causing significant decline in the level of rice production. The results of the model thus suggest that the Philippines was justified in including both rice and corn in the Sensitive List of the CEPT Scheme. In contrast, Vietnam will be better off if rice is not excluded from the AFTA agreement. Vietnam's comparative advantage in production should thus be taken into account in the inclusion of rice in the CEPT Scheme, although this country was a late comer to ASEAN.

The demand for corn will be more responsive to AFTA than the demand for rice. The implementation of AFTA will cause a drop in demand for corn in Indonesia, Thailand and Vietnam. All these ASEAN countries are expected to experience increases in corn prices if AFTA is partially or fully implemented. Malaysia, the major corn importing country in the ASEAN region, and the Philippines will show an increase in demand for corn. These two countries will contribute to the expansion in intra-ASEAN trade of corn in the region. While there will be bright prospects for trade creation from liberalization of corn trade in ASEAN, liberalization of rice trade will be expected to have some trade diversion effects.

The SPE model is a good tool for simulating regional demand and supply quantities. Specifically, the computed values from the base year run have small relative deviation from the actual values. A peculiar pattern of world trade is that individual countries may import and export a product in the same time period. The SPE model used in this study cannot capture, however, this complexity in world trading system. The acceptability of the results thus considered the net trade position of a country in the world market. There is a dearth of parameters that can represent some of the policies affecting trade flows (e.g. prohibitive import licensing and registration requirements for rice, seasonal import-inducing policy for corn to support livestock industry). Increasing the transparency of trade policies in many ASEAN countries would help improve the results of trade models, particularly in terms of better approximation of trade volumes.

## NOTES

[1] This paper is part of the author's doctoral research project and she would like to acknowledge the kind support of her supervisor, Prof. Dr. Wilhelm Henrichsmeyer, at the Institute of Agricultural Policy, University of Bonn in Germany.

[2] In practice, the products included in PTA but not yet in CEPT Scheme will continue to enjoy the tariff preferences offered under PTA. Hence, AFTA is complementary rather than a complete replacement to PTA.

[3] However, only those who adopt the accelerated schedule shall be entitled to the concession.

[4] For detailed theoretical framework refer to Takayama and Judge (1971), Martin (1981) and Acosta (1998).

[5] This should be differentiated from dynamics analysis which considers the time paths through which the variables attain different equilibrium states.

[6] Brunei and Singapore are also members of ASEAN but were excluded in the model analysis due to lack of agriculture-related data. Laos and Myanmar joined ASEAN only in 1997, hence these countries were also omitted in the model.

[7] The $AT^{sim}_{ij,c}$ was computed as follows: $AT^{sim}_{ij,c} = \left( P^{end}_{j,c} * \frac{AVTR^{sim}_{ij,c}}{100} \right)$ where $AVTR^{sim}_{ij,c}$ is the ad valorem tariff rate imposed by the importing country $i$ against the exporting country $j$ on the its export product $c$.

[8] These quantities are defined as follows for the reference run:

$$y^{ref}_{i,c} = y^{bas}_{i,c}\left(1 + y^{trend}_{i,c}\right)$$

$$x^{ref}_{j,c} = x^{bas}_{j,c}\left(1 + x^{trend}_{j,c}\right)$$

The terms $y^{bas}_{i,c}$ and $x^{bas}_{j,c}$ refer to the quantities of demand and supply for the base year, while $y^{trend}_{i,c}$ and $x^{trend}_{j,c}$ represent the extent of increase (or decrease) in demand and supply quantities from the base year period (1992) to the reference period (2002). The trend factors for the latter terms were derived from time-series regression analysis.

**REFERENCES**

ACOSTA, L. (1998) *The Impact of ASEAN Free Trade Area (AFTA) on Selected Agricultural Products in ASEAN Countries.* Frankfurt: Peter Lang GmbH.

ASEAN secretariat (1996) AFTA Reader: *The Fifth ASEAN Summit.* Jakarta, Indonesia.

ENKE, S. (1951) 'Equilibrium Among Spatially Separated Markets: Solution By Electric Analog.' In *Econometrica*, no. 19, pp.40-47.

GENERAL AGREEMENT ON TRADE AND TARIFFS (1991a) *Indonesia: Trade Policy Review.* Geneva: GATT Publication Services.

GENERAL AGREEMENT ON TRADE AND TARIFFS (1991b). *Thailand: Trade Policy Review.* Geneva: GATT Publication Services.

GENERAL AGREEMENT ON TRADE AND TARIFFS (1993a). *Malaysia: Trade Policy Review.* Geneva: GATT Publication Services.

GENERAL AGREEMENT ON TRADE AND TARIFFS (1993b). *Philippines: Trade Policy Review.* Geneva: GATT Publication Services.

GENERAL AGREEMENT ON TRADE AND TARIFFS (1995a). *Indonesia: Trade Policy Review.* Geneva: GATT Publication Services.

GENERAL AGREEMENT ON TRADE AND TARIFFS (1995b). *Thailand: Trade Policy Review.* Geneva: GATT Publication Services.

HENRICHSMEYER, W., W. BRITZ & M. VON LAMPE (1995) *SPEL-Trade Final report.* Bonn: Institute of the Agricultural Policy.

MARTIN L. J. (1981) 'Quadratic Single and Multi-Commodity Models of Spatial Equilibrium: A Simplified Exposition.' In *Canadian Journal Of Agricultural Economics*, no. 29, pp.21-48.

MUELLER, K. (1995) *The Political Economy of Agricultural Trade of the ASEAN Countries.* Bonn: Verlag M. Wehle.

SAMUELSON, P. A. (1952) 'Spatial Price Equilibrium and Linear Programming.' In *American Economic Review*, no. 42, pp.283-303.

TAKAYAMA, T. & G. G. JUDGE (1971) *Spatial and Temporal Price and Allocation Models.* Amsterdam: North-Holland Publishing Company.

TEH, R. (1992) *Implementing the Common Effective Preferential Tariff Mechanism and Philippine Trade and Resource Flows.* Laguna: Center for Policy and Development Studies.

# 14 MODELING PUBLIC GOODS PROVISION IN AGRICULTURE

Ernst-August Nuppenau and Ousmane Badiane

## 1 INTRODUCTION

### 1.1 PROBLEM STATEMENT

As a market oriented approach, economic reforms under structural adjustment programs were intended to speed-up structural changes in the agricultural sector together with a stimulation of growth, in general, and an urgently needed intensification of agricultural production, in particular. Essentially, since the agricultural sector produces tradable goods, it was hoped that trade policy reforms should have a major impact on the intensification and modernization of an export oriented agriculture. Improvements in the export oriented sector in agriculture, in particular, should have a trickle-down-effect on food production and food security. External and internal trade liberalization and, in detail, the abolition of any government interference in prices, are supposed to be a prerequisite for the encouragement of agricultural exports and, simultaneously, modernization of agriculture. Stimulated by the tradable good sectors, agricultural intensification should appear and a technical upgrading of farms was expected.

Moreover, it was hypothesized that market orientation will not only lead to modernization in high potential areas, but also lead to an urgently need sustainable intensification in fragile areas (Scherr et al., 1996). It was argued that limited market access, inter alia, has contributed to a poor performance of agriculture and hampered intensification, especially, in areas of high population growth (Barbier, 1996). However, progress of market development has been slow in many countries (Hartmann, 1992), especially, in African (Jaeger, 1992) and new paradigms of agricultural development beyond passive liberalization policy are of big interest (Lall, 1996).

Three major shortcomings have been detected. First, changes in macroeconomic policies and liberalization policies do often not translate into significant and rapid changes in agricultural incentives on the farm level due to underdeveloped local markets. (Platteau, 1996; Badiane, 1992). A slow response of private traders in undertaking the necessary investments in marketing augments this finding (Badiane & Shively, 1996). Second, public spending on extension in agriculture has been cut and the progress of technological change has been declined considerably (Savadogo et al., 1994; Reardon et al., 1996). Third, aggravation in soil quality problems and alarming rates of soil degradation have been reported in the second phase of market liberalization. Since fertilizer prices and prices for improved seeds have risen more than proportionally, intolerable low application rates of fertilizer and improved seeds are becoming more and more a serious problem in many reforming countries, especially in Sub-Saharan Africa (Reardon, T., et al., 1997). Investments in soil quality are increasingly hampered by ill-functioning input markets due to reduced public spending in fertilizer and seed development programs and, again, the lack of marketing infrastructure. From the point of view of positive external effects of improved input provision at affordable prices, a new situation of nearly inaccessibility to modern inputs for small-scale farming has risen.

Despite the undoubted success of structural adjustment policies (necessary condition), additional public sector policies (sufficient condition) are unavoidable to stimulate intensification. By modeling the consequences of public good provision, the effects of missing public goods can be evaluated (Putterman, 1996). The complementary of price liberalization and public good provision has been stressed more recently even by protagonists of liberalization (Schiff & Montenegro, 1997) and is now seen as pivotal to the success of liberalization programs. But, the urgent question is how much and how can the failures of the past been avoided in a policy of minimal intervention. Though it seems to be accepted now that no single policy can solve complementary problems in policy, policy interaction has not been described sufficiently.

This paper provides a starting point in describing elements for policy design that shall solve the previously mentioned conflict between structural adjustment (liberal, getting prices right) policy and structural modernization (interventionist, government led innovation) policy, in a sense, that the paper offers theoretical and conceptual thoughts for active policy interaction. Though, the pretension of such a project apparently is high, there seems to be, in our opinion, scope for stepwise approaches. To solve first hand problems, we start from a theoretical and modeling perspective and give suggestions how to continue with research. It is our hope that we can come up with a framework for modeling of policy interaction. That framework shall enable future empirical work to built upon testable theories. We are fully aware, that we can not solve any operative problem emerging while policies are carried out. But, we are admittedly ambitious in the sense that we want to make a contribution to solving some fundamental conceptual problems.

## 1.2 SCOPE OF THE PAPER

This paper suggests a modeling approach for the design of public good provision policy in agriculture. That policy shall be complementary to structural adjustment policies and built upon liberalized market structures. In particular, the intended policy will have a focus on public goods provision in infrastructure for marketing, knowledge dissemination for improved agricultural practice and, additionally, a link to policies that foster investments in soil quality. The paper explicitly outlines how one can link public goods provision and price liberalization policies in an agricultural sector. Apparently, it is assumed that the sector is constrained by a lack of infrastructure. If this holds, we demonstrate how additional production effects can be created by higher farm gate price due to improved infrastructure and how improved soil fertility can be created by investments in farmers' knowledge on sustainable practice. In order to investigate synergetic effects from policy interaction, the model explicitly considers the strains of improved market access and soil fertility simultaneously. However, since any public goods provision strategy requires active policy devises, such policies can not avoid to come up with prescriptions on public investment paths. But, to make it a complementary strategy to liberalization, these paths should also include phasing out public investments when they have reached a sufficient level and private activities have been built up around public infrastructure.

In the paper we use control theory to show how one can analytically achieve optimal policies of structural change under given information. We start with the outline of such concept and present the underlying methodology. Most importantly, in an explicit chapter on model building, it will be shown, how one can derive an objective function for public investment, presuming that the intensifying sub-sector is constrained by market imperfections on the land market. Furthermore, the marketing sector is constrained in infrastructure, and we will show, how one can express mathematically intertemporal constraints, that policy makers most likely are facing in cases of lack of infrastructure and knowledge deficits. This chapter is supplemented by a chapter on the intertemporal optimization using control theory. However, the major focus is on the design of an agricultural sector model that satisfies the need for an appropriate treatment of theoretical and practical aspects, subject to deficits in public goods provision in agriculture. That model will help to improve policy design in conjunction with structural adjustments and gear agricultural policies towards intensification and modernization.

## 2 CONCEPT AND METHODOLOGY

The process of structural transformation and desired intensification in agriculture, as perceived in this paper, is reflected in a gradual move from a "traditional", resource extraction oriented and subsistence based form of agriculture, called stagnant agriculture, to a "modern", science-based and market oriented form of agriculture, called intensifying agriculture. This process is driven by production and marketing decisions of farm households and traders that provide marketing services to agriculture. Policy changes and complementary public investments undertaken in the context of market reform and liberalization programs affect farmers' output sales and input purchase decisions as well as decisions by local traders to engage in arbitrage activities on output markets and provision of inputs. Moreover, decisions on technical innovations by farmers are subject to extension work and price incentives, most importantly specified as farm gate prices, not as market price, i.e. transaction costs are explicitly considered.

The reform-induced changes in farmer incentives and in the geographic coverage of marketing activities influence the cost of transacting in local markets and the propensity of farmers to enter the exchange system. To capture the effect of price reforms and investments on the process of transformation, the model therefore defines two sectors, one subsistence-oriented and one modernizing sector, and uses the amount of land transferred from the former to the latter as an indicator of the speed of transformation. Furthermore, it uses the concept of central and local markets in analogy to the literature on market integration in order to model the impact on changes in arbitrage activities and costs on local markets.

As has been said, soil fertility problems have become a major issue in current structural adjustment programs. Capturing this issue, the intensifying sector is characterized by responses to soil quality preservation needs (fertilizer), improved technology (seeds), extension and knowledge dissemination (soil quality preserving practices), and improved marketing (cash crops). In contrast, the stagnant sector is characterized by the vicious cycles of a continuation of current practices of wealth extraction from soils due to backward technology, low investment in maintaining soil quality due to low income, and primarily food crop orientation due to subsistence needs.

## 3 MODELING THE OBJECTIVE FUNCTION

Research on behavioral aspects of structural changes in agriculture requires to start with an appropriate formulation of the objective functions of participating units. The specification and modeling of the government objective function draws on the foundation of micro based objective functions and their sectoral aggregation. From a methodological point of view (methodological individualism), this type of modeling refers to behavioral distinctions in separate sectoral units of agriculture. Farmers in the two sectors, identified as stagnant and intensifying farmers, are assumed to maximize profit functions such as the ones given in equation [1] and [2.1] and [2.2].

Similar approaches of modeling structural change, for example, have worked with a distinction of agriculture and forestry (Ehui & Hertel, 1989). Consequently, objective functions in distinct sectors have to be specified in order to capture intensification as a process. Then, the societal objective function has to be derived from summarizing sector objective functions. Moreover, adding government costs to individual sector costs and benefits will provide the final objective function (i.e. no weighting for sub-sectors except equal weighting). Sector objective functions will be discussed first, and the aggregate function will be provided subsequently.

### 3.1 SECTOR SPECIFIC OBJECTIVE FUNCTIONS

#### *3.1.1 DUAL FARM SECTOR*

Both objective functions of the farm sub-sectors (i.e. profit functions for stagnant and intensifying), are constrained profit functions, and they exhibit standard properties of profit functions (Chambers, 1988, Varian, 1992; later-on quadratic functional forms will be applied). The profit function of the stagnant farm sector is a dual function $\Pi^t_t(.)$ of farm prices for cash crops $P^{c,l}_t$ and food crops $P^{f,l}_t$, a constraint in land allocation $L^*\text{-}L_t$ , soil degradation $Z^t_t$ , and other variables $v^t_t$. The last variable (vector) summarizes exogenous variables such as output and input prices.

$$[1] \qquad \Pi^t_t = \Pi^t_t(\underset{(+)}{P^{c,l}_t}, \underset{(+)}{P^{f,l}_t}, \underset{(+)}{L^*\text{-}L_t}, \underset{(-)}{Z^t_t}, v^t_t)$$

In practice the land constraint may be not too much binding at the first hand, since most of the land is initially in the stagnant sector. However, the identification of land constraints in empirical work is crucial for the discussion on intensification, as will be shown soon, since it appears again in the profit function of the intensifying sector. Applications of constrained profit functions and their estimation have been developed, for example, by Jayne (et al, 1994). For further explanation: Total used land L* is fixed, and the stagnant farm sector is restricted to the difference between total land L* and land in the intensifying sector $L_t$, discussed next. Initially the constraining effect "$L^*-L_t$" will be with total land L*, i.e. $L_t$ is minimal, because, presumably, most land will be in the stagnant sector. In the process of intensification certain amounts of land $l_t$ will be exchanged (land market) and the low level of land constraint, initially with the stagnant sector, will be reallocated to the intensifying sector, making the stagnant sector disappearing.

Moreover, almost equally important for the discussion of soil quality problems, the equation [1] contains, as a new element in applied research on production economics, a constraint on soil degradation $Z^t_t$. The introduction of a soil degradation measure $Z^t_t$ into profit functions has been most recently suggested by Gretton & Salma (1997). As has been demonstrated by these authors, applying Australian data from a degrading region, profits are highly effected by degradation of agricultural productivity, and soil degradation can be a good index for the measurement of negative effects that have been accumulated in farm communities after quite a time of soil mining. As a negative proxy for soil quality, soil degradation can serve as a reasonable good measurable constraint, that has been developed for depicting decline in soil quality (Gretton & Salma make reference of the actual construction of an overall index using 10 indicator variables such as irrigation, salinity, induced soil acidity, soil texture etc.). Suggesting a similar approach in this paper, soil degradation should have a considerable negative impact on profitability in the stagnant farming sector, even as negative external effects. The decline of the performance, stagnation or degrading in the part of the stagnant agricultural sector will be measured and directly integrated into the approach. This perception is precisely in line with another strain of thinking that has been advanced by Coxhead (1995), who perceives soil quality as an augmenting factor for land. He suggests a land constraint that is multiplied by a productivity factor $(L^*-L_t)\cdot Z^t_t$ and further explains factors that contribute to the productivity index (see §4.5).

In contrast, the profit "$\Pi^i_t$" in the intensifying and modernizing sector (equation [2]) shall be flexibly corresponding to factors that foster intensification. Initially this sector is strongly constrained by land $L_t$ (low level). The situation improves only by private land transfers from the stagnant sector to this sector. (Land transfers can be practically thought of as either handing over of land between farmers, a land market, or farmers changes position from stagnant to intensifying). Land exchange is a secondary effect. Primarily, the design of intensification as a process and its recognition in the profit function determines higher productivity level "$h^i(.)$" and higher productivity increases competitiveness on land market. The initial limitation of productivity is a constraint in the profit function, but, its level can be increased by extension and inputs. Prices $P^{c,l}_t$ and $P^{f,l}_t$, exogenous to profits, but endogenous to the system (see §4.4); play an important incentive role.

$$[2.1] \qquad \Pi^i_t = \Pi^i_t(\underset{(+)}{P^{c,l}_t}, \underset{(+)}{P^{f,l}_t}, \underset{(+)}{L_t}, \underset{(+)}{h^i}, v^i_t)$$

The wish for higher productivity $h^i(.)$ in equation [2.1] is deliberately only modeled as sectoral problem of modernizing agriculture. It is a function of the inter-annual state of knowledge $N_t$, state of infrastructure provision $I_t$, the annual fertilizer $f_t$ and seed proliferation $s_t$, and most importantly, soil degradation. In the beginning, the productivity constraint to the intensification in agriculture should be high, but depending on adoption of technology (a dynamic process) it will relax over time, dependent on complementary government investment and private farm adjustments. Evidently, the status *of* and changes *in* soil degradation $Z^i_t$, play a major role in productivity changes, if synergetic effects between the factors, that determine productivity, are reasonably recognized. A functional relationship [2.2] that is accessible to empirical tests specifies the case:

$$[2.2] \qquad h^i = h^i(\underset{(+)}{N_t}, \underset{(+)}{I_t}, \underset{(-)}{Z^i_t}, \underset{(+)}{f_t}, \underset{(+)}{s_t})$$

Moreover, the dynamics of soil degradation will influence equation [2.2] (The dynamics itself will be discussed in the next chapter). For the moment let us assume we can measure only an overall index for soil degradation $Z_t$, and that index does not distinguish between the effects of soil degradation as effecting the stagnant $Z^t_t$ and the intensifying $Z^i_t$ sector. By that constellation we end up in a dilemma of either measuring one uniform soil degradation index or satisfying the need for two separate variables, though we should treat soil degradation individually for the stagnant and the intensifying sector. In order to solve that problem we could think of an independent decline of soil quality in the stagnant sector. This specification is reasonable, since it can be imagined that the stagnant sector is soil mining whatever the incentives are. In other words, regardless what the production is about, applying almost no fertilizer is by definition the very characteristic of this sector. It means that the sector will continue to practice soil mining, at least, with population growth (Barbier, 1996). Hence $Z^t_t$ changes autonomously: $Z^t_t = z_o \exp\{z_1 t\}$. Exploring further the distribution of land as a weigh for average $Z = L_t/L^* Z^t_t + [L^*-L_t]/L^* Z^i_t$, we receive after Taylor approximation $Z = g_0 L_t/L^* + g_1 Z^t_t + g_1 Z^i_t$; solving for the autonomous development of $Z^t_t$:

$$Z^i_t = g_0/g_1/L^* L_t + 1/g_1 Z_t + g_1/g_1 z_o \exp\{z_1 t\}$$

Hence, the individual soil degradation, as relevant in the stagnant and intensifying agricultural sectors, can be translated into the overall soil degradation including the reallocation of land and the development of "natural" soil degradation in the stagnant sector. For final notice in this chapter, the two sectors compete on the "land market". Increases in the share of intensifying agriculture in land $L_t/(L^*-L_t)$ can be indirectly geared by making this sector more competitive. Technically speaking, changing (improving) the factor demand function for land derived from the profit function, increases the competitiveness of those that are willing to modernize, and those that respond to knowledge dissemination and soil quality improvements. Finally, for empirical evaluation of coefficients, explicit quadratic or corresponding profit functions are used (Appendix).

### *3.1.2 MARKETING AND CONSUMER SECTOR*

Price incentives are transmitted by traders. Traders carry out their marketing activities to maximize profits according to equation [3]. The profit function of the marketing sector "$\Pi^m$" is based on central market prices $P^{c,c}_t$, $P^{f,c}_t$ and local market prices $P^{f,l}_t$, $P^{c,l}_t$. Presumably, marketing is constrained in infrastructure $I_t$, and servicing area $L_t$ (Diewert, 1986). Moreover, price differences between the central and the local market determine profits as well as factors $z^m_t$.

$$[3] \qquad \Pi^m_t = \Pi^m_t(P^{c,c}_t - P^{c,l}_t,\ P^{f,c}_t - P^{f,l}_t,\ I_t,\ L_t,\ f_t,\ s_t,\ v^m_t)$$

Finally the demand side has to be analyzed. In order to simplify, representative consumers "c" maximizes their welfare function (consumer surplus) subject to the central market price for food crops, primarily a staple food crop, $P^{f,c}_t$ and other exogenous factors $v^c_t$. Additionally, population growth enters with an exponential function $c_o \exp\{c_1 t\}$

$$[4.1] \qquad \Omega^c_t = \Omega^c_t(P^{f,c}_t,\ v^c_t)\ c_o \exp\{c_1 t\} \qquad \text{(consumers)}$$

Second, the cash crop shall be primarily marketed overseas as export $x^c$, resulting in the earning of foreign currency at the central market

$$[4.2] \qquad \Omega^x_t = P^{c,c}_t\ x^c \qquad \text{(exports)}$$

However, in case of more complex formulations, necessary with different marketing channels, a third level of export and import parity prices could be easily introduced.

Inasmuch as the specifics of the formulation of a sector objective function covering all participants are concerned, it has to be mentioned that the model [1] to [4] can be easily extended to world trade and multi-product specification. If food imports and cash crop exports become directly involved in the sector approach, a reformulation on the base of world market prices either requires an inclusion of export earnings and import expenditures in a separate sub-sector unit, or, alternatively, a balance of currency flows of exports and imports has to be established. (In general, the aspect of trade surplus, i.e. cash crop export earnings minus import expenditures for food crops, may substitute equation [4]). Since exports "x", for example, earn foreign currency, and potentially enable food imports "m", export earnings and import expenditures, both, at world market prices for cash crop exports "$p^{w,c,c}$" and food imports "$p^{w,f,c}$", become part of the objective analysis.

Hence, concerning coverage of costs and benefits, the approach caters for standard arguments in modernization of agriculture, i.e. the food-cash crops debate! Most easily, this can be done on the base of components that distinguish import and export parity prices and model export marketing as a fifth component in the sector coverage of individual sector objective functions. Prices in a vector presentation $P^{w,c}_t = [p^{w,c,c}, p^{w,f,c}]$ are listed according to the trade situation. Hence, the objective function of the foreign trade sector $\Pi^x_t$ can be modeled by surpluses or margins, if no additional costs occur in shipping beside fixed export margins $c^x$ and import margins $c^m$.

$$[5] \qquad \Pi^{x/m}_t = [p^{w,c,c} - p^{c,c} - c^x]\ x^c + [p^{w,c,f} - p^{c,f} - c^m] m^c$$

Apparently, a fully expression of this last part of the sector objective function requires an internal solution of "$x^c$" and "$m^c$". Such solution of "$x^c$" and "$m^c$", most easily can be approached if linear supply and demand functions are used. Analytically, no problem exists if farm objective functions are specified as quadratic surplus functions. Then inserting of "$x^{c}$" and "$m^{c}$" in market equilibria provides the sufficient type of trade constraints in the model.

## 3.2 SOCIETAL AND INTERTEMPORAL OBJECTIVE FUNCTION

For a full coverage of intertemporal costs and benefits the next step is to integrate the sector objective functions (previous Section) into the societal objective function $\Pi^s(.)$, and provide the base for the following intertemporal decision making on sector level. In the given context, that presumes equal weighs to producers, consumers and traders, the societal welfare summarizes to:

$$[6] \qquad \Pi^s(.)=\int_0^T \exp\{\rho t\}[\Pi^t_t(.)+\Pi^i_t(.)+\Pi^m_t(.)+\Omega^c_t(.)+\Pi^x_t(.)-w_1 i_t-w_2 n_t-w_3 f_t-w_4 s_t+c_o\exp\{c_1 t\}]dt$$

In equation [6] the first part contains surplus functions of the traditional sector $\Pi^t_t(.)$, the intensifying sector $\Pi^i_t(.)$, the marketing sector $\Pi^m_t(.)$, the consumer $\Omega^c_t(.)$, and the trade sector $\Pi^x_t(.)$ and the second part caters for government costs of new infrastructure $i_t$, extension service expenditure $n_t$, fertilizer $f_t$ and seed subsidy $i_t$. Beside given elements of the previous specification, the full coverage of intertemporal costs and benefits as a sector specification finally uses a discount rate $\rho$, and it considers additional temporal government costs as per unit costs of investment in infrastructure"$w_1$", per unit costs of investment in knowledge (extension) "$w_2$", and the subsidy costs per unit of provision of fertilizer "$w_3$" and seeds "$w_4$". The complete objective function, made treatable for intertemporal optimization, shall explicitly exhibit a quadratic form (Appendix):

$$[7] \qquad \Pi^s(.)=\int_0^T \exp\{\rho t\}\{0.5y(t)\Pi_1 y(t)+y(t)\Pi_2 u(t)+u(t)\Pi_3 u(t)+z(t)\Pi_4 y(t)+v(t)\Pi_5 u(t)+\pi_0' y(t)$$
$$+p^W{}_0'\Pi_6 y(t)-w'u(t)\}dt$$

Whereby vectors are read as follows: $y'(t)$ is the vector $[\mathbf{L}_t(t),\mathbf{N}_t(t),\mathbf{I}_t(t),\mathbf{Z}_t(t),\mathbf{P}^{c,l}(t),\mathbf{P}^{f,l}(t)]$ of the state variables; $u'(t)$ is the vector $[\mathbf{l}(t),\mathbf{n}(t),\mathbf{i}(t),\mathbf{f}(t),\mathbf{s}(t)]$ of the control variables applied to one particular period and $v'(t)$ is a vector $[v^{t'}(t),v^{f'}(t),v^{m'}(t),v^{c'}(t)]$ of exogenous variables as associated with individual characteristics the four participating sub-sectors. Moreover, $p^W{}_0'$ is the world market price vector, initially constant but suitable for flexibility; and $w'$ is a vector of government costs$[w_1,w_2,w_3,w_4]$.The individual coefficients $\Pi_i$ in the objective function of [7] are an aggregation of the coefficients of the separate elements of the objective function (Appendix):

$$[8.a] \qquad \Pi 1 = \pi^{t,1}{}_{i,j} + \pi^{i,1}{}_{i,j}, + \pi^{m,1}{}_{i,j}+ \pi^{c,1}{}_{i,j} + \pi^{x/m,1}{}_{i,j}$$

$$[8.b] \qquad \Pi 2 = \pi^{t,2}{}_{i,j} + \pi^{i,2}{}_{i,j}, + \pi^{m,2}{}_{i,j}+ \pi^{c,2}{}_{i,j} + \pi^{x/m,2}{}_{i,j}$$

$$[8.c] \qquad \Pi 3 = \pi^{t,3}{}_{i,j} + \pi^{i,3}{}_{i,j}, + \pi^{m,3}{}_{i,j} + \pi^{c,3}{}_{i,j} + \pi^{x/m,2}{}_{i,j}$$

[8.d] $\Pi 6 = \pi^{x/m,6}{}_{i,j}$

The particular coefficients $\pi^{i,j}{}_{i,j}$ and their aggregation in equation [8] are directly linked to supply and demand conditions of the sub-sectors in agriculture on local level, the respective marketing activities between local and central level, and the demand conditions on the central market level. Note further that the coefficients can be either calculated on the base of econometric estimations of constrained profit functions (Jayne et al. 1994) or on the base of generalized sector models that reflect shadow price determination and constraints (Howitt, 1995) using, for example, most recently developed maximum entropy techniques (Golan, 1996). Hence, an estimation of coefficients or parameters, necessary to evaluate the model, can draw on well established methods.

## 4 INTERTEMPORAL CONSTRAINTS

By determining the objective function as dependent on stage variables "y(t)", changeable only in the long run and being structural, and control variables "u(t)", being changeable in the short run, the objective function for public optimization has been established. However, before we will proceed in describing the kind of optimization technique to be applied in this case, the objective function has to be supplemented by the intertemporal constraints the government faces. To establish a reasonable dynamic problem, in our opinion, six constraints have to be recognized and formulation to be discussed. Any of the six constraints has basically one dynamic element.

### 4.1 THE LAND ALLOCATION CONSTRAINT

The land allocation constraint sums up the reallocation of farm land from the stagnant farm sector to the intensifying farm sector. By definition "L(t) dot" (i.e. the change in L) is equal to the annual redistribution of land l(t). The amount of land redistributed can be summed up over various times to the already achieved reallocation (Ehui and Hertel, 1989)

[9] $$\dot{\mathbf{L}}(t) = -\,\mathbf{l}(t) \quad \text{since, } \mathbf{L}(t) = \int_0^t -\,\mathbf{l}(\tau)\, d\tau$$

The value of L(t) shows the volume of land remaining in stagnant agriculture. It will most likely temporally decline. Assuming a given land availability L*, the difference to total land (L*-L(t)) will most likely increase. This difference displays the land transferred into commercial agriculture. Land reallocation serves as the prime indicator for intensification.

### 4.2 GOVERNMENT INFRASTRUCTURE PROVISION:

The provision of infrastructure I(t) by governments includes a depreciation and a new investment component. It developments as follows (Arrow and Kurz, 1970):

$$[10] \quad \dot{\mathbf{I}}(t) = \underset{\substack{(-)\\ \text{depreciation}}}{\chi_{2,2}}\, \mathbf{I}(t) + \underset{\substack{(=1)\\ \text{physical identity}}}{\chi_{2,7}}\, \mathbf{i}(t)$$

This is a technical equation generally applied in investment theory. It assures that the status of infrastructure can be maintained. The coefficient $\chi_{2,2}$ measures the wearing out of infrastructure, an important aspect of investment problems of infrastructure, especially, in Africa.

## 4.3 KNOWLEDGE FORMATION AND GOVERNMENT INVESTMENT IN EXTENSION

As can be deducted from knowledge acquisition theory, a self-learning component and a knowledge teaching component by external sources may best establish a knowledge improving function (Arrow, 1992). In it simplest form intertemporal knowledge formation or acquisition may exhibit a first order differential equation such as:

$$[11] \quad \dot{\mathbf{N}}(t) = \underset{(-)}{\chi_{3,4}}\, \mathbf{N}(t) + \underset{(+)}{\chi_{3,8}}\, \mathbf{n}(t) + \underset{(+)}{\chi_{3,2}}\, \mathbf{I}(t) + \underset{(+)}{\chi_{3,9}}\, \mathbf{f}(t) + \underset{(+)}{\chi_{3,10}}\, \mathbf{s}(t)$$

| | | | | |
|---|---|---|---|---|
| Depreciation of knowledge (if set at –1 no formation) | Impact of knowledge in extension expenditure | Knowledge and infrastructure link | Knowledge and fertilizer provision link | Knowledge and seed provision link |

This equation shows how the intertemporal building up of agricultural expertise and experience N(t) is depending on extension n(t), fertilizer f(t) and seed proliferation s(t), and infrastructure I(t). The behavioral equation can be empirically determined as described in Nuppenau (1995)

## 4.4 PRICE FORMATION AND MARKETING DEPENDENCE ON INFRASTRUCTURE

Finally, the price transmission must be explicitly discussed in order to establish the link between infrastructure on the one side and price margin behavior in marketing on the other side. In other words, a theoretically sound or empirically accessible building bloc on the degree of price transmission must be incorporated. By doing this, increases in price incentives due to policy reform in developing countries can be linked to intensification. In particular, it was expected that due to market liberalization price transmission is immediate and producers could directly benefit from reform. Unfortunately, farm gate prices on the local level, retail prices on the central level, and consumer purchase prices are not totally linked to each other, allowing for margin increases and low price transmission. The following approach integrates current literature on measurement of imperfect price transmission with structural research on marketing margin investigations.

First, we can assume a "linear" spread between consumer prices $P^c(t)$ and producer $P^l(t)$. (See equation [12.1] referred to as margin, if $B^m{}_l = 1$. For each element, that implies a pure additive margin; Badiane & Shively, 1996. Or assume $B^{m}_{1} \neq 1$. That implies proportional pricing.) Moreover, this type of margin can be easily proved to be the outcome of a profit maximizing production and marketing sectors assuming a quadratic objective function.

$$[12.1] \quad \mathbf{P^c}(t) = \beta^{m}_{0} + B^{m}_{1}\,\mathbf{P^l}(t) + \Gamma^{m}\,\mathbf{R}^{m}$$

The variable $\mathbf{R}^{m}$, i.e. actually a vector, $\mathbf{R^m}=[\mathbf{I}(t), \mathbf{L}(t), v^m(t)]'$, contains variables that shift the marketing cost. These variables are most importantly infrastructure $\mathbf{I}(t)$, land area served $\mathbf{L}(t)$, and other exogenous variables, while $\beta^{m}_{0}$ caters for the constant element of marginal costs due to an estimation function that determines marketing costs.

Moreover, written as a first order differential equation, we can use equation [12.1] for the purpose of amalgamating price margin behavior and price transmission.

$$[12.2] \quad \dot{\mathbf{P}}^{\mathbf{c}}(t) = B^{m}_{1}\,\dot{\mathbf{P}}^{\mathbf{l}}(t) + \Gamma^{m}\,\dot{\mathbf{R}}^{m}(t)$$

This means combined with the condition for price transmission (long run multiplier notation, derived from the fundamentals of the price transmission analysis (Badiane and Shively, 1996).

$$[12.3] \quad \dot{\mathbf{P}}^{\mathbf{l}}(t) = M_1\,\dot{\mathbf{P}}^{\mathbf{c}} + M_2\,\mathbf{P^c}$$

a first order differential equation, that describes price movements in the model on the base of producer prices, can be found:

$$[12.4] \quad \dot{\mathbf{P}}^{\mathbf{l}}(t) = M_1[B^{m}_{1}[\dot{\mathbf{P}}^{\mathbf{l}}(t)] + M_2[\beta^{m}_{0} + B^{m}_{1}\,\mathbf{P^l}(t)] + M_2\Gamma^{m}\,\mathbf{R^m}(t) + M_1\,\Gamma^{m}\dot{\mathbf{R}}^{\mathbf{m}}(t)$$

Solving for $\mathbf{p^l}(t)$ yields:

$$[12.5] \quad \dot{P}^{l}(t) = [1 - M_1 B^{m}_{1}]^{-1}[M_2[B^{m}_{1}\,\mathbf{P^l}(t) + \beta^{m}_{0} + M_2\Gamma^{m}\mathbf{R^m}(t) + M_1\Gamma^{m}\dot{\mathbf{R}}^{\mathbf{m}}(t)]$$

Then, if coefficients are sorted, merged and it is recognized that the marketing cost Rm(t) are de-pendent on infrastructure I and land L, differential equations [12] and [13] can be obtained.

$$[12] \quad \dot{\mathbf{p}}^{\mathbf{c,l}}(t) = \chi_{4,4}\,\mathbf{p^{c,l}}(t) + \chi_{4,5}\,\mathbf{p^{f,l}}(t) + \chi_{4,1}\,\mathbf{L}(t) + \chi_{4,2}\,\mathbf{I}(t) + \chi_{4,6}\,\mathbf{l}(t) + \chi_{4,7}\,\mathbf{i}(t) + \chi_{4,0}$$

$$[13] \quad \dot{\mathbf{p}}^{\mathbf{f,l}}(t) = \chi_{5,4}\,\mathbf{p^{c,l}}(t) + \chi_{5,5}\,\mathbf{p^{f,l}}(t) + \chi_{5,1}\,\mathbf{L}(t) + \chi_{5,2}\,\mathbf{I}(t) + \chi_{5,6}\,\mathbf{l}(t) + \chi_{5,7}\,\mathbf{i}(t) + \chi_{5,0}$$

Apparently, the coefficients of equation [12 and 13], $\chi_{i,j}$, are a combination of the underlying elements mostly important and recognizable in incentive proliferation:

Margin decrease (equation [12.2]) and price transmission equation [12.3]. Equation [12] and [13] contain the hypothesis that improved price incentives (higher prices), benefiting farmers in the short run and society in the long run, can only be achieved by the supplementary provision of infrastructure by governments.

## 4.5 SOIL DEGRADATION

By its very nature, soil degradation should be treated as a dynamic phenomena. Four strains of thinking in the current literature on soil degradation and decline of quality of agricultural land can be translated into a formal representation of soil degradation. From this four strains three strains are directly relevant for the endogenous modeling of long run soil degradation. They are: organic matter build up, agro-ecological suitability of crop composition and appropriate management techniques of natural resources. The fourth argument is a short run argument looking at direct improvement on soil quality from increased fertilizer and seed application.

First, soil scientists have frequently argued that organic matter plays a crucial role in sustaining or even improving fertility of soil. Tracing the argument further back, enriching soils with organic matter is primarily an aspect of green manure that is obtained by increased current productivity (Burt, 1981). Increased current productivity or the provision of surpluses of green manure, however, can only be achieved by reduced extraction of agricultural produces (i.e. planting green manure crops instead of food and cash crops) or by higher provision of external fertilizer (Barbier, 1990). Hence, even integrated soil management projects, very often rely heavily on increased artificial fertilizer application (Breman, 1997).

Second, the composition of agricultural crops seem to be a determinant for soil degradation (Goetz, 1997). As Coxhead (1995) has argued, mostly cash crops are more environmentally suitable in degradation prone areas than food crops. Presumably, cash crops are produced where comparative advantages for cash crops and long term environmental digestibility coincide. Moreover, cash recovery allows reinvestments in soils if farmers are liquidity constrained on rural credit market. For example, with regard to perennial crops, these crops are mostly more suitable to guarantee sustainability, including preservation of soil fertility than food crops, especially under various tropical agro-climatical conditions with high rainfall and temperature. With regard to a comparison between annual cash crops, the facts are not that clear, but, for example, cotton can be very suitable for dry areas. Hence, the share of a cash crop plays a crucial role in soil degradation containment (Coxhead, 1995). In regress, the share of cash crops in production is determined by the prices of crops. Hence, prices enter the degradation function.

Third, improvements in the managerial practices are vital for decreases in soil degradation in many underdeveloped rural economies. Coxhead (1995, p.15) directly links management (knowledge of integrated soil fertility management) aspects to an index of "land fertility". Unfortunately, he does not further explore the importance of and need for *increased* managerial capacities in soil conserving techniques, practical skills in private maintenance of on farm investments in soil conserving measures (such as terracing, ditch irrigation facilities etc.), and improved handling capacities of modern technology (zero tillage, drip irrigation, etc.). Rather, he focuses on pricing aspects only, and negative effects from protection of food crops. However, in this paper both aspects, the pricing and the complementary impact of managerial improvements are of concern and the aspects should be treated simultaneously.

Finally and fourth, external improvements in technology, primarily embodied in seeds are suggested by Coxhead and Jayasuriya (1995, p.15). As a major factor influencing soil degradation in the short run improved seeds are important. Unfortunately, Coxhead and Jayasuriya do not discuss aspects of a divergence between private and social returns on input use and spread of fertilizer and seeds as a prerequisite for long run technology improvements. Since they argue in a steady state general equilibrium framework, they miss the crucial point of temporally big difference between potentials for technology improvements and obstacles to technology improvements, though they are everywhere prevalent in developing low income countries.

Summarizing the given four aspects into a representation of state and control variables, such as used in the previous approach of this chapter, knowledge N(t), prices of cash $P^{c,l}(t)$and food crops $P^{f,l}(t)$, fertilizer f(t) and seed s(t), a reduced form of the deliberations can be written as the sixth intertemporal constraint. Moreover, in this presentation (equation [14]), soil degradation under agricultural production is regarded as a "natural" process that can be halted only by the positive activities listed above. This presumes that the crowding out of pre-existing eco- systems has already been taken place, which is not an unrealistic assumption with regard to the situation of many small-holder agricultural sectors in developing countries.

$$[14] \quad \dot{\mathbf{Z}}(t)=\chi_{6,6}\,\mathbf{Z}(t)+\chi_{6,3}\,N(t)+\chi_{6,4}\,\mathbf{P}^{c,l}(t)+\chi_{6,5}\,\mathbf{P}^{f,l}(t)+\chi_{6,7}\,\mathbf{f}(t)+\chi_{6,8}\,s(t)+\chi_{6,0}$$

| (+) | (-) | (-) | (+) | (-) | (-) |
|---|---|---|---|---|---|
| Appreciation of soil degradation (if set=1 no degredation increase if>1) | Impact of knowledge in extension service that is less degrading | Price of cash crops stimulates production that is less degrading | Price of local food production encourages degradation | Fertilizer application improves organic matter | Modern seeds improves soil quality by lower extraction |

For further notice, the equation could become even more realistic and flexible, but at the same time more complicated and less treatable from a mathematical point of view, if non-linearity becomes involved. For example, it could be modeled that increases in fertilizer use, that compensate for soil degradation, have a positive impact unless a certain threshold is reached. Above that threshold, as observed already in more economically advanced regions of some developing countries, a negative impact of increased fertilizer on soil quality in the long run becomes more pronounced than a positive impact. The current presentation of the interactions between soil degradation in equation [14] is deliberately limited to the case of "under-use" of fertilizer as suggested, for instance, in many African countries in the Sahel (Reardon, et al., 1997). The dynamic equations [9] to [14] can be expressed in a differential equation system:

[15] $\dot{y}(t) = \Psi\, y(t) + N\, u(t) + \Gamma\, e(t)$

where **y**(t) is again a vector of the state variables **y**(t)=[**L**(t),**I**(t),**N**(t),$\mathbf{p}^{c,l}$(t),$\mathbf{p}^{f,l}$(t),**Z**(t)] and **u**(t) is a vector of control variables **u**(t)= [**l**(t), **i**(t), **n**(t), **f**(t), **s**(t)]. The system of dynamic constraints in equation [15] represents the potential development of desired and achievable states **y**(t) subjected to the action or control **u**(t) of the government contingent on behavioral and technical aspects of an agricultural system. It summarizes the impact analysis of government public goods provision on private constraints via control and state variables. The dynamic constraint can be combined with the previous objective function to establish the central problem of intertemporal optimization.

## 5 OPTIMIZATION

### 5.1 MODELING ASPECTS

The objective function (Section 3, equation [7]) together with the intertemporal constraint (Section 4, equation [15]) specify the dynamic control problem that implies to use directly a Hamilton function presentation. The Hamilton function includes, as parts, the time dependent objective function. Part 1 is the constrained welfare function, and part 2 is the intertemporal technical and behavioral functions. Part 2 is moreover the description of underlying intertemporal conditions for potential exploitation of intensification benefits (part 1). Mathematically, the Hamilton function (Tu, 1992) depends on state variables y(t), control variables u(t) and the constraint assessment λ(t) that can be interpreted as intertemporal Lagrange factor function:

[16] $$H(t)=\exp\{\omega * t\}[0.5y(t)\Pi_1 y(t)+y(t)\Pi_2 u(t)+u(t)\Pi_3 u(t)+z(t)\Pi_4 y(t)+z(t)\Pi_5 u(t) + \pi_0' y(t) + p^{W'}\Pi_6\, y(t) - w' u(t)] + \lambda'(t)[\Psi\, y(t) + N\, u(t) + \Gamma\, e(t)]$$

Such a system explicitly satisfies the conditions of an optimal control problem (Tu, 1992). If the curvature of the objective function is proper, the equation [16] can be directly used for policy evaluation. However, the property of the curvature maybe theoretically satisfied by the general economic assumptions on profit functions, i.e. a social welfare functions normally exhibit normally semi-definit conditions, and should be mathematically guaranteed. For solution (Tu, 1992), a control theory problem has to fulfill three conditions to obtain a maximum:

[17] $$H_y = -\dot{\lambda}\,; \quad (8.2)\ H_u = 0\,; \quad (8.3)\ H_\lambda = \dot{y}$$

These conditions are applied to the stated Hamilton function. By that we get:

[18.a] $$\Pi_1\, y(t) + \Pi_2\, u(t) + \Pi_4\, z(t) + \Pi_6\, p^{W'} + \pi_0 + \Psi' \lambda(t) = -\dot{\lambda}(t) - \omega(t)\, \lambda(t)$$

[18.b] $$\Pi_2'\, y(t) + \Pi_3\, u(t) + \Pi_5\, z(t) + N' \lambda(t) + w = 0$$

[18.c] $\Psi\,(t) + N\,u(t) + \Gamma\,e(t) = \dot{y}(t)$

(for comparison see Ehui, S., Hertel, T. (1989, p.703-701). The conditions can be rearranged to a solvable differential equation system if equation [18.c] is changed to

[18.c'] $u(t) = N^{-1}\dot{y}(t) - N^{-1}\Psi\,y(t) - N^{-1}G\,e(t)$

Inserting in [18.a] results in

[19.a] $[\Pi_1 - \Pi_2 N^{-1}\Psi]y(t) + [\Psi' - \omega]\,\lambda(t) + \Pi_4\,z(t) - \Pi_2 N^{-1}\Gamma\,e(t) + \Pi_6 p^{w\prime} + \pi_0 = -\,\dot{\lambda}(t) - \Pi_2 N^{-1}\dot{y}(t)$

[19.b] $[\Pi_2' - \Pi_3 N^{-1}\Psi]\,y(t) + N'\,\lambda(t) + \Pi_5\,z(t) - \Pi_3 N^{-1}\Gamma\,e(t) + w = \quad -\Pi_3 N^{-1}\dot{y}(t)$

Equation [19] is a set of interdependent differential equations and offers the empirical solution.

For confirmation, applying techniques of multiple differential equations, the time path of *all* endogenous variables in the vectors y(t) and λ(t) can be mathematically derived. Technically speaking equations in the system [18] are used to come up with a combined sets of time dependent variables λ(t), y(t) and correspondingly u(t) (using equation [18.c′]. Generally speaking, the solution for u(t) prescribes the optimal choice on investments in the system, i.e. choices in the intensification strategy: The path of i(t), the infrastructure investments, n(t), the knowledge creation, f(t), the fertilizer and s(t), the seed provision. The specific path of l(t) describes the periodical reallocation of land to intensifying farms; i.e. u(t)= [l(t), i(t), n(t), f(t), s(t)]′ can be determined. Recursively, the state variable vector y(t), that reflect government's desires to improve the availability of infra structure I(t) and the knowledge situation N(t) can be projected.

## 5.2 IMPLEMENTATION ASPECTS

From the point of view of implementation of an intensification and modernization strategy and its potential, the variables Z(t) and L(t) are structural characteristics of the state of agriculture, since they reflect internal movements of soil quality (less soil degradation) and the land allocation L(t). Apparently, these variables and the price movements $p^{c,l}(t)$ and $p^{f,l}(t)$ in vector y(t), are not subject to the direct government responsibility. Rather, they are indirectly determined within the system and reflect the response of the socio-economic system (demand, supply, knowledge, technology and infrastructure). To be clear, the model *does not* provide a blueprint to be executed by government. Choices are complementary to emerging private structure in the modernization process of agriculture. The agricultural sector will be gradually put in to the position to develop its markets at minimal economic intervention and costs for the society.

Note, that the suggested provision of infrastructure and extension investments, as well as input market subsidization, go beyond traditional (non-liberal) control policies, such as price policy etc. Moreover, it is highly noticeable that the presented approach limits the role of governments to the provision of public goods such as investment in infrastructure i(t), extension n(t), and "assistance" in procurement of fertilizer f(t) and seeds s(t) on the one side. From the point of view of practical application of the modeling results and for modeling of real policy, notice, that the "assistance" to farmers in procurement of fertilizer and seeds can be subsequently translated into subsidy rates. However, that would need further description. Technically speaking a recalculation of the shadow prices for fertilizer and seeds, which are derived from the behavioral equations and which are the derivatives of the constrained profit function in turn, allows to determine subsidies on fertilizer and seed prices. Again, the specified model limits government interference to infrastructure investment, extension services, fertilizer and seed subsidies; no further price policies are involved.

## 6 DATA REQUIREMENTS

The estimations of necessary coefficients in the model corresponds to supply and demand estimations in analyses of a constrained farm sectors (Jayne et al., 1994). Examples for the application of the theoretical work, however, can also draw on elasticities derived from similar studies. The empirical setting could be based on a typical situation of a medium sized African country, in order to provide further hints for research needs. Moreover micro-census information on soil degradation can be integrated. For example, as Gretton & Salma (1997) have shown that soil degradation measurement can be made practical. But unless this information is not linked to more structural explanatory work, such as the one of Coxhead (1995), systematic investigations of equation [14] become a bottleneck for the empirical application of the suggested model.

The paper primarily shows the potential for empirical application and demonstrates which parameters are needed. Future work, intended for an improved application of the model, may operate with data derived from country studies that have been collected according to the theoretical thinking. Moreover, data from the marketing sector are in short supply and several studies are under progress that follow the outlined scheme, especially with regard to price transmission subject to prevalent infrastructure in these countries. Empirical research will deliver the necessary coefficients of the objective function and constraint functions due to constraints. In particular, vice versa, the paper shows the necessity to develop empirical analyses on constrained marketing in developing countries and outlines the corresponding framework. Potentially, different types of empirical analyses such as price transmission, knowledge accumulation, and constrained supply and demand analysis have to be applied.

**REFERENCES**

ARROW, K.J. (1992) 'The Economic implications of learning by doing'. In Review of Economic Studies, vol. 29, pp.155-173.

ARROW, K.J., M. Kurz (1970) Public investment, the rate of return, and optimal fiscal policy. Baltimore.

BADIANE, O., G.E. SHIVELY (1996) 'The importance of market integration for the response of local prices to policy changes'. In Bauer, S., Herrmann, R., Kuhlmann, F. (eds.) *Maerkte der Agrar- und Ernährungswirtschaft*, 37th. German conference of agricultural economist. Muenster-Hiltrup, pp.415-460,

BARBIER, E.B. (1990) 'The farm level economics of soil conservation: The uplands of Java'. In *Land Economics*, vol. 66, no. 2, pp.198-211.

BARBIER, B. (1996) 'Impact of market and population pressure on production, income and natural resources in the dryland savannas of West Africa. Bioeconomic modeling at the village level'. In *EPTD Discussion paper*, no.2, International Food Policy Research Institute, November.

BREMAN, H. (1997) 'Building soil fertility in Africa: Constraints and perspectives'. In *Paper presented at the "International workshop on Development of national strategies for soil fertility recapitalization in Sub-Saharan Africa.* Lome (Togo), 22-25 April

BURT, O.R. (1981) 'Farm level economics of soil conservation in the Palousa area of the North West'. In *American Journal of Agricultural Economics*, vol. 63, no.1, pp.83-92.

CHAMBERS, R (1988) *Production Economics*. New York.

COXHEAD, I. (1995) 'Economic Modeling of Land degradation in developing countries'. In *University of Wisconsin-Madison, Agricultural Economics Staff paper series*, no. 385, July.

COXHEAD, I., S.K. JAYASURIYA (1994) 'Technical change in agriculture and land degradation in developing countries: a general equilibrium analysis'. In *Land Economics*, vol. 70, no. 1, February, pp.20-37.

DIEWERT, W.E. (1986) *The measurement of the economic benefits of infrastructure services*. Berlin.

EHUI, S., T. HERTEL (1989) 'Deforestation and Agricultural Productivity in the Cote d'Ivoire'. In *American Journal of Agricultural Economics*, vol.71, no.4, pp.703-701

GOLAN, A, G. JUDGE, D. MILLER (1996) *Maximum Entropy Econometrics: Robust Estimation with Limited Data.* Chichester.

GOETZ, R. (1997) 'Diversification in Agricultural Production: A Dynamic Model of Optimal Cropping to Manage Soil Erosion'. In *American Journal of Agricultural Economics,* vol. 79, No. 2, pp.341-356.

GRETTON, P., U. SALMA (1997) Land degradation: links to agricultural output and profitability' In *Australian Journal of Agricultural and Resource Economics,* vol. 41, no.2, pp.209-225

HARTMANN, A. (1992) 'Structural Adjustment programs. An Evaluation of results'. In Schmitt, G., Tangermann, S (eds.) *Internationale Agrarpolitik und Entwicklung der Weltagrarwirtschaft. Schriften der Gesellschaft für Wirtschafts und Sozialwissenschaften,* Münster-Hiltrup, pp.229-240.

HOWITT, R.E. (1995) 'Positive Mathematical Programming'. In *American Journal of Agricultural Economics*, vol. 77, no.2, pp.329-342

JAEGER, W.K. (1992) 'The effects of Economic Policies on African Agriculture'. In *World Bank Discussion Papers, African Technical Department Series.* Washington.

JAYNE, T.S., Y. KHARTRI, C. THIRTLE, T. REARDON (1994) 'Determinants of Productivity Change Using a Profit Function: Smallholder Agriculture in Zimbabwe'. In: *American Journal of Agricultural Economics,* vol.76, no.4, pp.613-618.

LALL, S. (1996) 'Paradigms of Development: The East Asian Debate': In *Oxford Development Studies*, vol. 24, no.2, pp.111-131

NUPPENAU, E-.A. (1995) 'Land Reform and Government Expenditures for Emerging Small Scale Farmers: A Mathematical Analysis of Optimal Timing'. In *Discussion papers in agricultural economics. University of Giessen*, No. 28, Giessen.

PLATTEAU, J. P., (1996) 'Physical Infrastructure as a constraint on Agricultural Growth. The case of Sub-Saharan Africa': In *Oxford Development Studies,* vol. 24, no.3, October, pp.189-201

PUTTERMAN, L. (1996) 'Economic Reform and smallholder agriculture in Tanzania: A Discussion of Recent Market Liberalization, Road Rehabilitation, and Technology Dissemination Efforts'. In *World Development*, vol. 23, no.2, pp.311-326.

REARDON, T., et. al. (1996) 'Sustainable Capital-led Intensification in Sahel Agriculture: Addressing Structural Constraints after Macroeconomic Policy Reform'. *Paper presented at an International Symposium on "Food Security and Innovations: Successess and Lessons Learned"*. University of Hohenheim, Stuttgart/Germany. March 11-13 1996.

REARDON ,T., V. KELLY, E. CRAWFORD, B. DIAGENA, J. DIONE, K. SAVADOGO, and D. BOUGHTON (1997) 'Promoting Sustainable Intensification and productivity growth in Sahel agriculture after macroeconomic reform. In *Food Policy*, Vol. 22, No.4, p.317-327

SAVADOGO, K., T. REARDON, K. PIETOLA (1994) 'Farm productivity in Burkina Faso: Effects of animal traction and nonfarm income'. In *American Journal of Agricultural Economics*, vol.76, no. 3. pp.608-612

SCHERR, S., J., G. BERGERON, J. PENDLER, B. BARBIER (1996) 'Policies for sustainable development in fragile lands: methodological overview'. In *Report on "Policies for Sustainable Development in the Cental American Hillsides. IFPRI Environment and Production Technology Division (EPTD "*, Washington.

SCHIFF, M., C. E. MONTENEGRO (1997) 'Aggregate Agricultural Supply Response in Developing Countries: A Survey of Selected Issues'. In *Economic Development and Cultural Change*, vol. 45, no.2, p.393-410

THIRTLE, C.J., J. ATKINS, G. BOTTOMLY, N. GONESE, J. GOVEREH, Y. KHARRI (1994) 'Agricultural Productivity in Zimbabwe, 1970-89'. In *Economic Journal,* vol. 103, pp.474-480.

TU, P.N.V. (1992) *Introductory Optimization Dynamics*. Berlin.

VARIAN, H.L. (1994) *Micro Economics*, Amsterdam.

**APPENDIX:**

As an example for empirical evaluation a corresponding quadratic objective function, $\Pi^{i}_{t}(...)$, applied to farmers intensifying their practice, can be written in matrix form such as:

$$\Pi^{i}_{t}(L_t, h^{i}(I_t, N_t, Z^{i}_{t}, f_t, s_t), P^{c,l}_{t}, P^{f,l}_{t}, v^{i}_{t}) = 0.5 \begin{bmatrix} L(t) \\ I(t) \\ N(t) \\ Z(t) \\ P^{c,l}(t) \\ P^{f,l}(t) \end{bmatrix}' \begin{bmatrix} \pi^{i,1}_{1,1} & \pi^{i,1}_{1,2} & \pi^{i,1}_{1,3} & \pi^{i,1}_{1,4} & \pi^{i,1}_{1,5} \\ \pi^{i,1}_{2,1} & \pi^{i,1}_{2,2} & \pi^{i,1}_{2,3} & \pi^{i,1}_{2,4} & \pi^{i,1}_{2,5} \\ \pi^{i,1}_{3,1} & \pi^{i,1}_{3,2} & \pi^{i,1}_{3,3} & \pi^{i,1}_{3,4} & \pi^{i,1}_{3,5} \\ \pi^{i,1}_{4,1} & \pi^{i,1}_{4,2} & \pi^{i,1}_{4,3} & \pi^{i,1}_{4,4} & \pi^{i,1}_{4,5} \\ \pi^{i,1}_{5,1} & \pi^{i,1}_{5,2} & \pi^{i,1}_{5,3} & \pi^{i,1}_{5,41} & \pi^{i,1}_{5,5} \end{bmatrix} \begin{bmatrix} L(t) \\ I(t) \\ N(t) \\ Z(t) \\ P^{c,l}(t) \\ P^{f,l}(t) \end{bmatrix}$$

$$+ \begin{bmatrix} l(t) \\ i(t) \\ n(t) \\ f(t) \\ s(t) \end{bmatrix}' \begin{bmatrix} \pi^{i,2}_{1,1} & \pi^{i,2}_{1,2} & \pi^{i,2}_{1,3} & \pi^{i,2}_{1,4} \\ 0 & 0 & 0 & 0 \\ \pi^{i,2}_{3,1} & \pi^{i,2}_{3,2} & \pi^{i,2}_{3,3} & \pi^{i,2}_{3,4} \\ \pi^{i,2}_{4,1} & \pi^{i,2}_{4,2} & \pi^{i,2}_{4,2} & \pi^{i,2}_{4,4} \\ \pi^{i,2}_{5,4} & \pi^{i,2}_{5,4} & \pi^{i,2}_{5,4} & \pi^{i,2}_{5,5} \end{bmatrix} \begin{bmatrix} L(t) \\ I(t) \\ N(t) \\ Z(t) \\ P^{c,l}(t) \\ P^{f,l}(t) \end{bmatrix} + 0.5 \begin{bmatrix} l(t) \\ i(t) \\ n(t) \\ f(t) \\ s(t) \end{bmatrix}' \begin{bmatrix} \pi^{i,3}_{1,1} & 0 & \pi^{i,3}_{1,3} & \pi^{i,3}_{1,4} \\ 0 & 0 & 0 & 0 \\ \pi^{i,3}_{3,1} & 0 & \pi^{i,3}_{3,3} & \pi^{i,3}_{3,4} \\ \pi^{i,3}_{4,1} & 0 & \pi^{i,3}_{4,3} & \pi^{i,3}_{4,4} \end{bmatrix} \begin{bmatrix} l(t) \\ i(t) \\ n(t) \\ f(t) \\ s(t) \end{bmatrix} +$$

$$\begin{bmatrix} L(t) \\ I(t) \\ N(t) \\ Z(t) \\ P^{c,l}(t) \\ P^{f,l}(t) \end{bmatrix}' \begin{bmatrix} \pi^{i,4}_{1,1} \\ \pi^{i,4}_{1,2} \\ \pi^{i,4}_{1,3} \\ \pi^{i,4}_{1,4} \\ \pi^{i,4}_{1,5} \\ \pi^{i,4}_{1,6\ldots} \end{bmatrix} y_1 + \begin{bmatrix} l(t) \\ i(t) \\ n(t) \\ f(t) \\ s(t) \end{bmatrix}' \begin{bmatrix} \pi^{i,5}_{1,1} & \ldots \\ 0 & \ldots \\ 0 & \ldots \end{bmatrix} y_1$$

where $p^{i,1}_{i,i}$ coefficients, expressed in matrix notation, are coefficients derived from a multiple econometric supply analysis. In dual theory, the first derivative of the profit function provides linear functions and coefficients are derived from regressing product supply to prices and constraints. The coefficients can also be interpreted as elasticities of supply response. In the same vein the other profit and surplus functions of the intertemporal cost benefit analysis can established.

# ABOUT THE CONTRIBUTORS

**Lilibeth A. Acosta** (Faculty of Agriculture, University of Kyoto, Japan) is an agricultural economist with a PhD from Bonn University (1998). She worked as economic development specialist at the National Economic Development Authority in the Philippines for almost five years. Her tasks included evaluation of national agricultural development projects, preparation of policy-related research papers and the assessment of national agricultural policies. Her past research activities have dealt with price and marketing policies of the Philippine grain sector as well as agricultural trade. She has also worked on research related to agricultural trade policies in ASEAN countries. Her current research interests are quantitative analysis of international agricultural trade and economic integration policies, specifically AFTA, APEC and WTO agreements.

**Ousmane Badiane** (World Bank and International Food Policy Research Institute IFPRI, Washington DC, USA) is an agricultural economist with a PhD from Christian-Albrecht University in Kiel, Germany (1988). As a Senior Research Fellow at IFPRI, he developed and led the institute's research programme on Output Market Development and Reforms until 1998, when he joined the Africa Region of the World Bank as a Senior Agricultural Economist. He is also adjunct professor at the School of Advanced International Studies of Johns Hopkins University. He has worked on the adjustment process of local markets to pricing and marketing policy reforms and has developed a methodology to estimate the time path of local prices following changes in sectoral and economy-wide policies. He has published widely on the implications of pricing and trading policies for agricultural and economy-wide growth. His articles have been published in professional journals including the Journal of African Economies, the Journal of Development Economics, and the Quarterly Journal of International Agriculture.

**Houinsou Dedehouanou** (National University of Benin/Amsterdam Research Institute for Global Issues and Development Studies, the Netherlands) graduated in 1982 from the National University of Benin. He obtained a Master's degree in Agricultural Economics from the University of Reading (UK) in 1990. He is now a lecturer in 'Regional Analysis' and 'Project Management' at the Faculty of Agriculture in Benin. He also holds a Master's degree in Regional Planning from the Institute of Social Studies (The Hague, Netherlands). Currently he is completing a PhD thesis at the University of Amsterdam. His research interests are in the field of the New Institutional Economics and Organisational Studies in developing countries.

**Katell Le Goulven** (National School for Agronomy and National Institute for Research in Agronomy, Rural Sociology and Economics Department, Montpellier, France) is a PhD candidate in Agricultural Economics. She has carried out her research in partnership with the GRET (Research and Technological Exchange Group), a French non-governmental organisation working at the interface of agricultural research and development in Vietnam. The objective of her research is to link empirical tests of co-integration of hog prices in spatial rural markets to the markets' underlying institutions.

**Barbara Harriss-White** is professor of development studies in the University of Oxford (UK) and fellow of Wolfson College. An agricultural economist by background, since 1971, intermittently and by way of primary field research in India, Sri Lanka and Bangladesh and secondary research in Francophone West Africa, she has been studying the structure, evolution, behaviour and politics of domestic grain markets. She is also interested in regional development and in social welfare and has published widely on all these subjects. She has been a research consultant to seven UN agencies and to the UK Department for International Development.

**Nico Heerink** (Wageningen University, the Netherlands) is assistant professor in macroeconomics and economic policy analysis for LDCs. He obtained his PhD degree in Economics at the European University Institute in Florence (Italy) in 1991. The topic of his PhD research was the relationship between population growth and income distribution. His current research activities concentrate on the impact of economic policy reforms on food security and sustainable land use. He is the project leader of two international research projects on the impact of economic policy reforms on soil degradation in South-East China and East Africa (Kenya, Ethiopia), financed by the Netherlands Ministry of Development Co-operation and the European Union, respectively. He also contributes to research activities in West Africa on food security and land degradation.

**J.G.M. (Hans) Hoogeveen** graduated in economics from the University of Amsterdam, the Netherlands with a Master's thesis based on field work in Benin. In 1992 he joined the Vrije Universiteit (Free University) in Amsterdam, first as a researcher at the Economic and Social Institute, and later at the Institute for Environmental Studies where he focussed on the collection and analysis of survey data in Zimbabwe, India and the Netherlands. He has been affiliated with the Department of General and Development Economics of the Vrije Universiteit as a PhD student and is preparing a thesis on risk and insurance by smallholder farmers in Zimbabwe. He is also assisting in setting up a survey-based monitoring system of Zimbabwe's land reform programme for the World Bank.

**Xianjing Huang** (College of Land Management of Nanjing Agricultural University, P.R. China) is an associate professor in land economics and land policy analysis for China. He obtained his PhD degree in Management Science at Nanjing Agricultural University (NAU) in P.R.China in 1995. The topic of his PhD research was 'Prices of Farmland'. He is the project leader at NAU of two international research projects on the impacts of economic policy reforms on soil degradation in South-east China, financed by the Netherlands Ministry of Development Co-operation and the European Union, respectively. He also contributes to research activities in China on sustainable land use, sustainable agricultural development and urban economics.

**Michael Kopsidis** (Institute of Agricultural Development in Central and Eastern Europe, Halle/Saale, Germany) is an economist with a PhD from the University of Münster (1994). He has worked extensively on the impact of market integration on agricultural development in Western Europe during the 18th and 19th centuries. Since 1995 his main research interest has been the functioning of domestic agricultural markets in the Former Soviet Union. Besides the effects of poorly developed market structures on production, his research deals with the relationship between development economics theory and the emergence of agricultural markets in transition economies of the Former Soviet Union. At present, he is analysing regional differences in the degree of integration into food markets within Russia using an econometric approach. His articles have been published in Jahrbuch für Wirtschaftsgeschichte, Communist Economies & Economic Transformation and several proceedings.

**W. Erno Kuiper** (Wageningen University, the Netherlands) is an agricultural economist with a PhD from Erasmus University Rotterdam (1994). His research interests include time-series analysis for forecasting purposes and testing of long-run equilibrium relationships among variables in marketing models. He has focused in particular on market arbitrage among geographically dispersed food markets and vertical coordination in marketing channels. His articles have been published in Revista de Estudios Agro-Sociales, European Review of Agricultural Economics, European Financial Management and the Journal of Regional Science.

**Arie Kuyvenhoven** is professor of development economics at Wageningen University, the Netherlands, and director at the Netherlands Economic Institute, Rotterdam. Previously he taught at Erasmus University where he obtained his PhD in economics. He has served as a visiting professor at the University of Lagos and Cornell University. He has worked in more than 15 developing countries, including Nigeria, Pakistan, Indonesia and China, and has been a consultant to many international organisations. He is the author of 75 articles and has written or edited 10 books. His current research areas include development and trade, sustainable agriculture, and food security.

**Clemens Lutz** (University of Groningen, the Netherlands) is an economist with a PhD from the University of Amsterdam. He has worked extensively on the functioning of food markets in African countries including Benin, Burkina Faso, Togo, Ghana and Rwanda. He has published articles on this topic in the Journal of African Economies, the European Review of Agricultural Economics and the Journal of Regional Science. At present he is a lecturer in marketing and specialises on organisational issues in marketing channels.

**Matthew Meulenberg** is professor emeritus of marketing and consumer behaviour at Wageningen University, the Netherlands. His research interests are in the field of food distribution and consumer behaviour. In recent years he has published widely in journals including Agribusiness, European Review of Agricultural Economics, Journal of Agricultural Economics, Journal of Futures Markets, European Journal of Marketing and Trends in Food Science and Technology.

**Henk A.J. Moll** is associate professor in the Development Economics Group of Wageningen University, the Netherlands. After graduating from Wageningen University he worked for the Food and Agricultural Organisation and the Netherlands Ministry of Development Cooperation in Tanzania, Sri Lanka and Indonesia. Since 1982 he has worked at Wageningen University. Rural institutions and the economics of crop and livestock production are his main fields of research. His PhD research focused on rural financial institutions.

**Ernst-August Nuppenau** (University of Giessen, Germany) is an agricultural economist with a PhD from the University of Kiel, Germany (1986). He has worked extensively on international trade theory and policy, stabilisation of agricultural markets, price formation on agricultural markets and development and trade. His other interests include regional modelling and development economics, food marketing and parallel markets, rural development, cost benefit analysis and analysis of socio-economic policies in emerging market of developing countries. Recently, he has become interested in agro-environmental policy formation, institutions for sustainable development, compensatory payment design, political economy modelling in agro-environment policy, resource economics, flexibilisation of instruments in environmental economics and corresponding policy application. He has published on quantitative methods, application of control theory and design of appropriate models for policy advice. His articles have been published in World Development, Food Policy, Oxford Development Studies and various conference proceedings.

**Futian Qu** is professor of resource economics and sustainable development at Nanjing Agricultural University (NAU) in P.R. China. He obtained his PhD degree from the College of Economics and Trade at NAU in 1990. The topic of his PhD research was rural land institution reform in China. Since then, he has published extensively on land use policy in China and on economic reform and sustainable agricultural development in China. He was a visiting scholar at Wageningen University, the Netherlands for 13 months from 1993 - 1994. He has been Dean of the College of Land Management at Nanjing Agricultural University since 1996.

**Ruerd Ruben** is a senior lecturer in development economics at Wageningen University, the Netherlands and a visiting research fellow at the International Institute for Food Policy Research (IFPRI), Washington DC. He has worked extensively on the analysis of the economics of farm household behaviour in developing countries, mainly in Central America and West Africa. At Wageningen University he is in charge of the interdisciplinary research programme 'Sustainable land use and food security', focusing on the development of integrated bio-economic models for the appraisal of the farmers' supply response to agrarian policy instruments. He has published widely on rural institutional issues related to agrarian cooperatives, social capital, rural financial markets and rural employment.

**Max Spoor** (PhD, University of Amsterdam, Faculty of Economics and Econometrics, 1991), is associate professor at the Institute of Social Studies (ISS), the Netherlands, and co-ordinator of the ISS Centre for the Study of Transition and Development (CESTRAD). He is member of the staff group 'Rural Development, Environment and Population Studies' where he teaches rural policy analysis and data analysis. He has published widely on transitional economies and on agrarian transformation and environment in World Development, the Journal of Peasant Studies, Food Policy and the Journal of Development Studies. His most recent books are 'From State Intervention to Neo-Liberalism in Nicaragua' (1995, MacMillan/St.Martin's Press), and the edited volume 'The 'Market Panacea': Agrarian Transformation in LDCs and Former Socialist Economies' (IT Publications, 1997). He has also worked on consultancy missions and institutional capacity building projects (in Nicaragua, Vietnam, China and Kazakhstan). He is currently head of the Office of Research, Projects and Advisory Services, at the ISS.

**Erik Thorbecke** is the H.E. Babcock professor of economics and food economics and director of the Program on Comparative Economic Development at Cornell University, USA. He was awarded an honorary doctorate degree by the University of Ghent in 1981. He has made contributions in the areas of economic and agricultural development, the measurement and analysis of poverty and malnutrition, the social accounting matrix and general equilibrium modelling, and international economic policy. He is the author and co-author of more than 20 books and 150 articles. He has been an economic adviser to numerous U.S. and international agencies, and foreign governments, including USAID, the Food and Agricultural Organisation (Rome), the International Labour Organisation (Geneva), the World Bank and the OECD.

**Aad van Tilburg** graduated in econometrics from Erasmus University Rotterdam, the Netherlands (1971). He joined Wageningen University in 1974 and defended his PhD thesis on 'Consumer choice of cut flowers and pot plants' in 1984. His research covers several topics in agricultural marketing, particularly related to the performance of food marketing systems in developing countries. Recently, he co-edited 'Agricultural marketing and consumer behavior in a changing world' (1997, Kluwer Academic Publishers) and 'Agricultural marketing in tropical Africa' (1999, Ashgate). He has published chapters in books as well as articles in journals including the European Review of Agricultural Economics and the Journal of Regional Science.

**Paul Quarles van Ufford** (Amsterdam Research Institute for Global Issues and Development Studies, the Netherlands) graduated in Human Geography from the University of Amsterdam (1994). He defended his PhD thesis on trade and traders in the cattle market in the Republic of Benin, West Africa at the University of Amsterdam in December 1999. His main research interests include livestock trade and pastoral livestock marketing in Africa.

# AUTHOR INDEX

**J**



**K**



**L**



**M**



**N**

MIX
Papier aus verantwortungsvollen Quellen
Paper from responsible sources
FSC® C105338

If you have any concerns about our products,
you can contact us on
**ProductSafety@springernature.com**

In case Publisher is established outside the EU,
the EU authorized representative is:
**Springer Nature Customer Service Center GmbH**
**Europaplatz 3, 69115 Heidelberg, Germany**

Printed by Libri Plureos GmbH
in Hamburg, Germany